Advances in Wood Composites II

Advances in Wood Composites II

Editor

Antonios N. Papadopoulos

MDPI • Basel • Beijing • Wuhan • Barcelona • Belgrade • Manchester • Tokyo • Cluj • Tianjin

Editor
Antonios N. Papadopoulos
International Hellenic University
Greece

Editorial Office
MDPI
St. Alban-Anlage 66
4052 Basel, Switzerland

This is a reprint of articles from the Special Issue published online in the open access journal *Polymers* (ISSN 2073-4360) (available at: https://www.mdpi.com/journal/polymers/special_issues/wood_Compos_II).

For citation purposes, cite each article independently as indicated on the article page online and as indicated below:

LastName, A.A.; LastName, B.B.; LastName, C.C. Article Title. *Journal Name* **Year**, *Article Number*, Page Range.

ISBN 978-3-03943-521-0 (Hbk)
ISBN 978-3-03943-522-7 (PDF)

© 2020 by the authors. Articles in this book are Open Access and distributed under the Creative Commons Attribution (CC BY) license, which allows users to download, copy and build upon published articles, as long as the author and publisher are properly credited, which ensures maximum dissemination and a wider impact of our publications.

The book as a whole is distributed by MDPI under the terms and conditions of the Creative Commons license CC BY-NC-ND.

Contents

About the Editor vii

Antonios N. Papadopoulos
Advances in Wood Composites II
Reprinted from: *Polymers* **2020**, *12*, 1552, doi:10.3390/polym12071552 1

Pavlo Bekhta, Ján Sedliačik and Nataliya Bekhta
Effects of Selected Parameters on the Bonding Quality and Temperature Evolution Inside Plywood During Pressing
Reprinted from: *Polymers* **2020**, *12*, 1035, doi:10.3390/polym12051035 5

Eugenia Mariana Tudor, Anna Dettendorfer, Günther Kain, Marius Catalin Barbu, Roman Réh and Ľuboš Krišťák
Sound-Absorption Coefficient of Bark-Based Insulation Panels
Reprinted from: *Polymers* **2020**, *12*, 1012, doi:10.3390/polym12051012 21

Hamid R. Taghiyari, Roya Majidi, Ayoub Esmailpour, Younes Sarvari Samadi, Asghar Jahangiri and Antonios N. Papadopoulos
Engineering Composites Made from Wood and Chicken Feather Bonded with UF Resin Fortified with Wollastonite: A Novel Approach
Reprinted from: *Polymers* **2020**, *12*, 857, doi:10.3390/polym12040857 33

Liuyang Han, Juan Guo, Kun Wang, Philippe Grʹonquist, Ren Li, Xingling Tian and Yafang Yin
Hygroscopicity of Waterlogged Archaeological Wood from Xiaobaijiao No.1 Shipwreck Related to Its Deterioration State
Reprinted from: *Polymers* **2020**, *12*, 834, doi:10.3390/polym12040834 49

Yan Wu, Jichun Zhou, Qiongtao Huang, Feng Yang, Yajing Wang and Jing Wang
Study on the Properties of Partially Transparent Wood under Different Delignification Processes
Reprinted from: *Polymers* **2020**, *12*, 661, doi:10.3390/polym12030661 65

Pavlo Bekhta, Ján Sedliačik and Nataliya Bekhta
Effect of Veneer-Drying Temperature on Selected Properties and Formaldehyde Emission of Birch Plywood
Reprinted from: *Polymers* **2020**, *12*, 593, doi:10.3390/polym12030593 79

Mariana Domnica Stanciu, Daniela Sova, Adriana Savin, Nicolae Ilias and Galina A. Gorbacheva
Physical and Mechanical Properties of Ammonia-Treated Black Locust Wood
Reprinted from: *Polymers* **2020**, *12*, 377, doi:10.3390/polym12020377 93

Ayoub Esmailpour, Roya Majidi, Hamid R. Taghiyari, Mehdi Ganjkhani, Seyed Majid Mohseni Armaki and Antonios N. Papadopoulos
Improving Fire Retardancy of Beech Wood by Graphene
Reprinted from: *Polymers* **2020**, *12*, 303, doi:10.3390/polym12020303 111

Murtada Abass A. Alrubaie, Roberto A. Lopez-Anido and Douglas J. Gardner
Flexural Creep Behavior of High-Density Polyethylene Lumber and Wood Plastic Composite Lumber Made from Thermally Modified Wood
Reprinted from: *Polymers* **2020**, *12*, 262, doi:10.3390/polym12020262 125

Anuj Kumar, Tuula Jyske and Veikko Möttönen
Properties of Injection Molded Biocomposites Reinforced with Wood Particles of Short-Rotation Aspen and Willow
Reprinted from: *Polymers* **2020**, *12*, 257, doi:10.3390/polym12020257 139

Yan Xia, Chengye Ma, Hanmin Wang, Shaoni Sun, Jialong Wen and Runcang Sun
Multiple Analysis and Characterization of Novel and Environmentally Friendly Feather Protein-Based Wood Preservatives
Reprinted from: *Polymers* **2020**, *12*, 237, doi:10.3390/polym12010237 153

Gabriela Craciun, Elena Manaila, Daniel Ighigeanu and Maria Daniela Stelescu
A Method to Improve the Characteristics of EPDM Rubber Based Eco-Composites with Electron Beam
Reprinted from: *Polymers* **2020**, *12*, 215, doi:10.3390/polym12010215 167

Antonio Pizzi, Antonios N. Papadopoulos and Franco Policardi
Wood Composites and Their Polymer Binders
Reprinted from: *Polymers* **2020**, *12*, 1115, doi:10.3390/polym12051115 185

About the Editor

Antonios N. Papadopoulos is a specialist in Wood Science, Chemistry, and Technology. He is Associate Professor and Head of the Department of Forestry and Natural Environment at the International Hellenic University. He is an M.Sc. and Ph.D. holder, both from the University of North Wales, Bangor. His M.Sc. thesis focused on wood composites, and his Ph.D. thesis focused on chemical and thermal modification of wood. His main areas of research interest include chemical and thermal modification of wood, nanotechnology and wood science, composites, and wood-based panels and adhesives. He has published more than 200 peer-reviewed papers, books, and book chapters. He is on various Editorial Boards of journals in the field of wood science and technology.

Editorial

Advances in Wood Composites II

Antonios N. Papadopoulos

Laboratory of Wood Chemistry and Technology, Department of Forestry and Natural Environment, International Hellenic University, GR-661 00 Drama, Greece; antpap@for.ihu.gr

Received: 22 June 2020; Accepted: 5 July 2020; Published: 13 July 2020

The main advantage of wood composites is that they can be designed for specific performance requirements or specific qualities, since they are man-made. Therefore, they can be used in a very wide array of applications, from small-home to industrial-scale projects. This, in turn, enables many more options in design without sacrificing structural requirements. At the same time, their ability to be tailored to a number of uses makes them a very viable option for reducing the use of solid wood. This Special Issue, Advances in Wood Composites II, is a continuation of the "Advances in Wood Composites I" [1] and presents recent progress in the enhancement and refinement of the performance and properties of wood composites. This Special Issue, with a collection of 13 original contributions, provides selected examples of recent Advances in Wood Composites.

An excellent and updated review on wood composites and the polymer binders used for their manufacture is published in this Special Issue [2]. The review focuses on the most important aspects to look out for in manufacturing excellent wood composites and on binders that currently dominate in industry. The enormous progress made in this field during the last few years is highlighted, and an exciting and fascinating research future is expected.

It is known that the two major drawbacks of wood composites are their dimensional instability and their susceptibility to biotic and abiotic factors [3,4]. These can be as addressed by the chemical or thermal modification of the raw material [5–17]. An alternative means, probably more attractive, is so-called nanotechnology. Nanomaterials have unique characteristics, penetrate effectively and deeply into the wood substrate and improve fundamental properties [18–31].

Wollastonite, a silicate mineral, was applied in order to fortify urea formaldehyde resin for the manufacture of two types of composites (medium density fiber board and particleboard) made from chicken feather and wood [32]. It was reported that wollastonite behaved as a reinforcing filler, and therefore, most of the physicomechanical properties of the boards were improved.

In another study, nano-wollastonite and graphene were mixed in a water-based paint in order to investigate the fire properties of beech wood [33]. The results indicated that graphene presents a high potential to be applied as a fire retardant in the protection of wood and wood composites.

An interesting study investigated the performance of wood treated with ammonium hydroxide [34]. It is known that the dark color of wood in ammonia fuming is due to reactions among wood compounds and ammonia gas. These chemical reactions affect the physicomechanical properties of the treated wood. The study demonstrated that ammonia can be successfully used to change the uneven color of black locust wood.

Xia et al. [35] used feather as a source of protein in combination with copper and boron salts, and they made wood preservatives using nano-hydroxyapatite or nano-graphene oxide as nano-carriers. The results revealed that the penetration into the wood structure was successful, and it was concluded, based on the decay experiments, that the protein-based preservatives had the potential to be considered as a low-cost and environmental friendly alternative for wood preservation.

Bekhta et al. [36] studied the effects of selected process variables on temperature evolution and bond quality during the hot pressing of plywood. Their goal was to present an optimization of the manufacturing process in order to reduce both energy and adhesive consumption. The main finding of the study was that the use of densified veneers can reduce the press cycle (time and pressure) and resin consumption. In the same context, they also studied the possible effects of the veneer drying temperature on formaldehyde emission and on key board properties [37]. The results indicated that when elevated temperatures are applied in the drying of veneers, the formaldehyde emission from the boards is reduced significantly. It was concluded that a 185 °C in-steam dryer could be considered as optimum for industrial applications.

The sound absorption properties of wood composites have received limited attention by researchers to date. Noise is a fundamental issue today, especially in big cities. The paper by Tudor et al. [38] tackled this issue. They manufactured insulation panels from bark and studied their sound absorption properties. Bark from spruce and larch was used. They concluded that bark can successfully be used for the manufacture of insulation panels and may substitute materials currently used in sound applications.

Wood plastic composites receive great attention in this Special Issue. This interesting field is a combination of processing techniques for filler and fiber preparation and polymer science. In this context, the application of wood plastic composites in marine applications sounds very interesting. Alrubaie et al. [39] presented an innovative wood plastic composite lumber made from thermally modified wood. Its overall performance was compared to the performance of high-density polyethylene lumber (HDPE). The main conclusion of this study was that this innovative type of composite could successfully replace HDPE in structural applications.

In the area of composites, Cracium et al. [40] presented a natural-fiber-reinforced eco-composite. This was based on wood wastes (obtained by electron beam irradiation) and on a rubber monomer (ethylene-propylene-diene monomer). Sawdust was incorporated as a filler in order to improve the mechanical and physical characteristics of the composite. It was reported that 300 kGy irradiation seemed to be sufficient for obtaining the effect of reinforcement in the composite.

Kumar et al. [41] prepared injection-molded biocomposites in a biopolymeric matrix using raw material from short-rotation species, such as aspen and willow. This study highlighted the technical feasibility of preparing such a type of biocomposite and concluded that short-rotation species have the potential to be an alternative for biocomposite manufacture.

Another interesting topic addressed in this issue is the transparency of wood. Wu et al. [42] used the orthogonal test method to find the best way of partly delignifying wood. They concluded that this type of transparent wood retained most of the wood texture and color and had a certain degree of light, which means that it may used as a functional decorative material.

Last but not least, the issue of waterlogged archeological wood is addressed. Such studies are limited in the literature. Han et al. [43] addressed this issue, in a case study in China, by studying the hygroscopic properties of less decayed and moderately decayed waterlogged archeological wood, collected from marine shipwrecks. It was found that moderately decayed wood possessed higher hygroscopicity than that of less decayed wood. Based on the results of this study, which measures are necessary for shipwreck restoration can be decided.

It can surely and definitely be said that the field of wood composites is a fascinating one and has a bright future. What is presented here and in the previous editorial of the first series of "Advances in Wood Composites" is only a very short overview of what will happen in the near future [1]. Progress and recent developments in this field have been accelerating, and new approaches and ideas are continuously increasing, implying an exciting and interesting research future. Therefore, after the successful Special Issues "Advances in Wood Composites I" and "Advances in Wood Composites II", which both collected innovative papers from well-known scientists worldwide, a third part of this series is now available and open for submission.

Funding: This research received no external funding.

Conflicts of Interest: The author declares no conflict of interest.

References

1. Papadopoulos, A.N. Advances in Wood Composites. *Polymers* **2020**, *12*, 48. [CrossRef] [PubMed]
2. Pizzi, A.; Papadopoulos, A.N.; Policardi, F. Wood composites and their polymer binders. *Polymers* **2020**, *12*, 1115. [CrossRef] [PubMed]
3. Papadopoulos, A.N. Chemical modification of solid wood and wood raw materials for composites production with linear chain carboxylic acid anhydrides: A brief review. *BioResources* **2010**, *5*, 499–506.
4. Papadopoulos, A.N.; Bikiaris, D.N.; Mitropoulos, A.C.; Kyzas, G.Z. Nanomaterials and chemical modification technologies for enhanced wood properties: A review. *Nanomaterials* **2019**, *9*, 607. [CrossRef] [PubMed]
5. Papadopoulos, A.N.; Militz, H.; Pfeffer, A. The biological behaviours of pine wood modified with linear chain carboxylic acid anhydrides against soft rot decay. *Int. Biodeterior. Biodegrad.* **2010**, *64*, 409–412. [CrossRef]
6. Papadopoulos, A.N.; Duquesnoy, P.; Cragg, S.M.; Pitman, A.J. The resistance of wood modified with linear chain carboxylic acid anhydrides to attack by the marine wood borer Limnoria quadripunctata Hothius. *Int. Biodeterior. Biodegrad.* **2008**, *61*, 199–202. [CrossRef]
7. Papadopoulos, A.N.; Avtzis, D.; Avtzis, N. The biological effectiveness of wood modified with linear chain carboxylic acid anhydrides against the subterranean termites Reticulitermes flavipes. *Holz Roh Werkst.* **2003**, *66*, 249–252. [CrossRef]
8. Papadopoulos, A.N.; Avramidis, S.; Elustondo, D. The sorption of water vapour by chemically modified softwood: Analysis using various sorption models. *Wood Sci. Technol.* **2005**, *39*, 99–112. [CrossRef]
9. Papadopoulos, A.N. Decay resistance of acetylated OSB in ground stake test. *Holz Roh Werkst.* **2006**, *64*, 245–246. [CrossRef]
10. Papadopoulos, A.N. Moisture adsorption isotherms of two esterified Greek hardwoods. *Holz Roh Werkst.* **2004**, *63*, 123–128. [CrossRef]
11. Papadopoulos, A.N. Natural durability in ground stake test of propionylated particleboards. *Holz Roh Werkst.* **2007**, *65*, 171–172. [CrossRef]
12. Papadopoulos, A.N. The effect of acetylation on bending strength of finger jointed beech wood (*Fagus sylvatica* L.). *Holz Roh Werkst.* **2008**, *66*, 309–310. [CrossRef]
13. Papadopoulos, A.N. Natural durability and performance of hornbeam cement bonded particleboard. *Maderas Cienc Tecnol.* **2008**, *10*, 93–98. [CrossRef]
14. Papadopoulos, A.N. An investigation of the suitability of some Greek wood species in wood-cement composites manufacture. *Holz Roh Werkst.* **2007**, *65*, 245–246. [CrossRef]
15. Papadopoulos, A.N.; Tountziarakis, P. The effect of acetylation on the Janka hardeness of pine wood. *Eur. J. Wood Prod.* **2011**, *69*, 499–500. [CrossRef]
16. Papadopoulos, A.N. Sorption of acetylated pine wood decayed by brown rot, soft rot and white rot: Different fungi—Different behaviours. *Wood Sci. Technol.* **2012**, *46*, 919–926. [CrossRef]
17. Hung, K.-C.; Wu, T.-L.; Wu, J.-H. Long-term creep behavior prediction of sol-gel derived SiO_2- and TiO_2-wood composites using the stepped isostress method. *Polymers* **2019**, *11*, 1215. [CrossRef]
18. Mantanis, G.; Papadopoulos, A.N. The sorption of water vapour of wood treated with a nanotechnology compound. *Wood Sci. Technol.* **2010**, *44*, 515–522. [CrossRef]
19. Mantanis, G.; Papadopoulos, A.N. Reducing the thickness swelling of wood based panels by applying a nanotechnology compound. *Eur. J. Wood Wood Prod.* **2010**, *68*, 237–239. [CrossRef]
20. Papadopoulos, A.N.; Kyzas, G.Z.; Mitropoulos, A.C. Lignocellulosic composites from acetylated sunflower stalks. *Appl. Sci.* **2019**, *9*, 646. [CrossRef]
21. Taghiyari, H.R.; Soltani, A.; Esmailpour, A.; Hassani, V.; Gholipour, H.; Papadopoulos, A.N. Improving Thermal Conductivity Coefficient in Oriented Strand Lumber (OSL) Using Sepiolite. *Nanomaterials* **2020**, *10*, 599. [CrossRef] [PubMed]
22. Papadopoulos, A.N.; Taghiyari, H.R. Innovative wood surface treatments based on nanotechnology. *Coatings* **2019**, *9*, 866. [CrossRef]
23. Taghiyari, H.; Esmailpour, A.; Papadopoulos, A. Paint Pull-Off Strength and Permeability in Nanosilver-Impregnated and Heat-Treated Beech Wood. *Coatings* **2019**, *9*, 723. [CrossRef]
24. Bayani, S.; Taghiyari, H.R.; Papadopoulos, A.N. Physical and mechanical properties of thermally-modified beech wood impregnated with silver nano-suspension and their relationship with the crystallinity of cellulose. *Polymers* **2019**, *11*, 1538. [CrossRef] [PubMed]

25. Hassani, V.; Papadopoulos, A.N.; Schmidt, O.; Maleki, S.; Papadopoulos, A.N. Mechanical and Physical Properties of Oriented Strand Lumber (OSL): The Effect of Fortification Level of Nanowollastonite on UF Resin. *Polymers* **2019**, *11*, 1884. [CrossRef] [PubMed]
26. Esmailpour, A.; Taghiyari, H.R.; Najafabadi, R.M.; Kalantari, A.; Papadopoulos, A.N. Fluid Flow in Cotton Textile: Effects of Wollastonite Nanosuspension and *Aspergillus Niger* Fungus. *Processes* **2019**, *7*, 901. [CrossRef]
27. Taghiyari, H.R.; Hosseini, G.; Tarmian, A.; Papadopoulos, A.N. Fluid Flow in Nanosilver-Impregnated Heat-Treated Beech Wood in Different Mediums. *Appl. Sci.* **2020**, *10*, 1919. [CrossRef]
28. Taghiyari, H.R.; Bayani, S.; Militz, H.; Papadopoulos, A.N. Heat Treatment of Pine Wood: Possible Effect of Impregnation with Silver Nanosuspension. *Forests* **2020**, *11*, 466. [CrossRef]
29. Taghiyari, H.R.; Esmailpour, A.; Majidi, R.; Morrell, J.J.; Mallaki, M.; Militz, H.; Papadopoulos, A.N. Potential Use of Wollastonite as a Filler in UF Resin Based Medium-Density Fiberboard (MDF). *Polymers* **2020**, *12*, 1435. [CrossRef]
30. Taghiyari, H.R.; Nouri, P. Effects of nano-wollastonite on physical and mechanical properties of medium-density fiberboard. *Maderas Cienc. Tecnol.* **2015**, *17*, 833–842. [CrossRef]
31. Esmailpour, A.; Taghiyari, H.R.; Majidi, R.; Morrell, J.J.; Mohammad-Panah, B. Nano-wollastonite to improve fire retardancy in medium-density fiberboard (MDF) made from wood fibers and camel-thorn. *Wood Mater Sci. Eng.* **2019**. [CrossRef]
32. Taghiyari, H.R.; Majidi, R.; Esmailpour, A.; Samadi, Y.S.; Jahangiri, A.; Papadopoulos, A.N. Engineering Composites Made from Wood and Chicken Feather Bonded with UF Resin Fortified with Wollastonite: A Novel Approach. *Polymers* **2020**, *12*, 857. [CrossRef]
33. Esmailpour, A.; Majidi, R.; Papadopoulos, A.N.; Ganjkhani, M.; Armaki, S.M.; Papadopoulos, A.N. Improving Fire Retardancy of Beech Wood by Graphene. *Polymers* **2020**, *12*, 303. [CrossRef] [PubMed]
34. Stanciu, M.D.; Sova, D.; Savin, A.; Ilias, N.; Gorbacheva, G.A. Physical and Mechanical Properties of Ammonia-Treated Black Locust Wood. *Polymers* **2020**, *12*, 377. [CrossRef] [PubMed]
35. Xia, Y.; Ma, C.; Wang, H.; Sun, S.; Wen, J.; Sun, R. Multiple Analysis and Characterization of Novel and Environmentally Friendly Feather Protein-Based Wood Preservatives. *Polymers* **2020**, *12*, 237. [CrossRef] [PubMed]
36. Bekhta, P.; Sedliačik, J.; Bekhta, N. Effects of Selected Parameters on the Bonding Quality and Temperature Evolution inside Plywood during Pressing. *Polymers* **2020**, *12*, 1035. [CrossRef]
37. Bekhta, P.; Sedliačik, J.; Bekhta, N. Effect of Veneer-Drying Temperature on Selected Properties and Formaldehyde Emission of Birch Plywood. *Polymers* **2020**, *12*, 593. [CrossRef]
38. Tudor, E.M.; Dettendorfer, A.; Kain, G.; Barbu, M.C.; Réh, R.; Krišťák, Ľ. Sound-Absorption Coefficient of Bark-Based Insulation Panels. *Polymers* **2020**, *12*, 1012. [CrossRef]
39. Alrubaie, M.A.A.; Lopez-Anido, R.A.; Gardner, D.J. Flexural Creep Behavior of High-Density Polyethylene Lumber and Wood Plastic Composite Lumber Made from Thermally Modified Wood. *Polymers* **2020**, *12*, 262. [CrossRef]
40. Craciun, G.; Manaila, E.; Ighigeanu, D.; Stelescu, M.D. A Method to Improve the Characteristics of EPDM Rubber Based Eco-Composites with Electron Beam. *Polymers* **2020**, *12*, 215. [CrossRef]
41. Kumar, A.; Jyske, T.; Möttönen, V. Properties of Injection Molded Biocomposites Reinforced with Wood Particles of Short-Rotation Aspen and Willow. *Polymers* **2020**, *12*, 257. [CrossRef] [PubMed]
42. Wu, Y.; Zhou, J.; Huang, Q.; Yang, F.; Wang, Y.; Wang, J. Study on the Properties of Partially Transparent Wood under Different Delignification Processes. *Polymers* **2020**, *12*, 661. [CrossRef] [PubMed]
43. Han, L.; Guo, J.; Wang, K.; Grönquist, P.; Li, R.; Tian, X.; Yin, Y. Hygroscopicity of Waterlogged Archaeological Wood from Xiaobaijiao No.1 Shipwreck Related to Its Deterioration State. *Polymers* **2020**, *12*, 834. [CrossRef] [PubMed]

 © 2020 by the author. Licensee MDPI, Basel, Switzerland. This article is an open access article distributed under the terms and conditions of the Creative Commons Attribution (CC BY) license (http://creativecommons.org/licenses/by/4.0/).

Article

Effects of Selected Parameters on the Bonding Quality and Temperature Evolution Inside Plywood During Pressing

Pavlo Bekhta [1],*, Ján Sedliačik [2] and Nataliya Bekhta [1]

[1] Institute of Woodworking and Computer Technologies and Design, Ukrainian National Forestry University, 79057 Lviv, Ukraine; n.bekhta@nltu.edu.ua
[2] Faculty of Wood Sciences and Technology, Technical University in Zvolen, 960 01 Zvolen, Slovakia; jan.sedliacik@tuzvo.sk
* Correspondence: bekhta@nltu.edu.ua; Tel.: +38-032-2384499

Received: 25 March 2020; Accepted: 30 April 2020; Published: 2 May 2020

Abstract: This research optimizes the process of plywood production to determine its effectiveness in reducing energy and adhesive consumption for more efficient production with the required quality. The influence of selected parameters including veneer treatment (non-densified and densified), plywood structure, temperature, time and pressure of pressing, on the bonding quality and temperature evolution within the veneer stacks during hot pressing was investigated. Rotary-cut, non-densified and densified birch veneers and phenol formaldehyde (PF) adhesive were used to manufacture plywood samples. The effect of pressure and time of pressing on bonding quality of the plywood was determined. Bonding quality was evaluated by determining the shear strength of the plywood samples. The temperature evolution inside the veneer stacks was measured for birch veneers for different pressing temperatures and pressures for different numbers of veneer layers. The heating rate of the veneer stacks increased as the pressing temperature increased and decreased markedly with an increasing number of veneer layers. At a high pressing pressure, the heating rate of the densified veneer stacks was faster than that of non-densified veneers at the same pressure. The use of densified veneers for the production of plywood can lead to a shorter pressing time (17–50% reduction), lower glue consumption (33.3% reduction) and a lower pressing pressure (22.2% reduction) without negatively impacting the bonding strength of the plywood.

Keywords: plywood; densification; core layer temperature; bonding quality; hot pressing; veneer stack heating

1. Introduction

The structure and properties of plywood are formed in the hot-pressing process. Hot pressing is one of the most important operations in the production of plywood, which has an impact on the properties of plywood. This operation is also important from an economic point of view. The hot press step determines the performance of the pressing line and defines the capacity of the factory. Hot pressing is one of the most energy-consuming processes in plywood manufacturing after the veneer-drying process. Hot-pressing parameters such as pressing time, pressure and temperature are key factors that directly affect the properties of plywood panels [1–3]. Despite that plywood was the first created wood-based composite, few individual studies have been conducted to investigate the effect of process variables on hot-pressing of plywood [1,3–8]. The temperature evolution within the panel during hot-pressing is important for the chemical and physical processes that contribute to the properties of the panel. The temperature evolution within the panel depends on wood species, their density and moisture content, compression of the veneers, glue spreading rate, and the pressing temperature, pressure and time. The interactions among these parameters are however, complicated and unclear [9].

Pressing pressure depends on wood species, physical properties of the wood, characteristics of the veneer surface and on the type of adhesive, its viscosity, the pressing temperature, etc. The application of pressure helps the adhesive to wet and penetrate the wood surface by forcing it into the void spaces of the wood [10,11]. However, too high of pressure should be avoided as the adhesive largely squeezes out [12]. Bonding quality is influenced by the amount of adhesive penetration into the wood substrate during the manufacture of wood composites, i.e., plywood [10]. Optimal adhesive penetration is required to repair damaged wood surfaces, and it provides better contact with the inner surface for chemical bonding or blocking and transfers stresses between the laminates [10,12], promoting a more efficient use of adhesive and providing a reliable thickness of the adhesive layer [10].

Pressing pressure and adhesive spread rate are the main factors for determining the thickness of the adhesive layer. The thickness of the adhesive layer should be controlled because it directly affects the strength of the wood composites. Insufficient pressure in the production process results in a thick adhesive layer [13]. As a rule, the thick adhesive layer of many common adhesives is characterized by insufficient strength [11,13]. Pressure should be applied evenly and adequately because wood adhesives based on synthetic resin, such as phenol formaldehyde (PF), are not capable of forming strong bonds in the thick and variable thickness of the adhesive layers due to their low viscosity [12].

In low-density wood species, high pressure causes the adhesive to penetrate so deep into the wood that there is insufficient adhesive to fill the bonding line; this excessive penetration can lead to lower bond strength [11]. By contrast, low pressures can cause a decrease in the shear strength and do not provide close contact between the surfaces, resulting in a poor adhesive layer [13,14].

Thermosetting PF adhesive is typically used in the manufacturing of plywood destined for use in exterior structural applications. To harden the adhesive and form a strong adhesive bond, the temperature inside the veneer stack (in the core layer of the stack) must be greater than 100 °C [14]. The heat from the hot plates of the press should be transferred to the core layer of the stack as quickly as possible. If the heat transfer process to reach a central layer temperature greater than 100 °C is prolonged, then the adhesive layers closer to the plates will be exposed to higher temperatures longer than required, and this can lead to premature hardening of the adhesive in these layers and even destruction of the adhesive. Therefore, for plywood, the total pressing time is determined by the time it takes to reach a sufficient temperature in the inner glue layer to cure the resin (i.e., the temperature at which the resin hardens) [15]. On one hand, to maximize process efficiency, the pressing time should be as short as possible. This is enabled by increasing the pressing temperature. Even a slight increase in the optimum pressing temperature can adversely affect the surface quality and strength of the plywood panels. On the other hand, the pressing should be long enough to allow the glue to harden. Furthermore, shortening the time of hot pressing can effectively reduce energy consumption as well as time necessary for the production of wood composite materials. Therefore, choosing the optimum pressing parameters, i.e., the temperature, pressure and pressing time, is very important, both technologically and economically.

The possibility of reducing hot pressing time has been studied in previous works on some types of wood composite materials, such as laminated veneer lumber (LVL) [15,16], particleboard [17], oriented strand board (OSB) [18] and medium-density fiberboard [19]. However, there are only a few studies in the existing literature regarding the effect of pressing time on the properties of plywood [16,20–23]. Shortening the duration of plywood pressing can be realized by steam injection [20] or by veneer incising [16,21]. However, the rapid increase in core temperature, typical of steam-injected particleboard, does not occur in steam-injected plywood [20]. Mirski et al. [22] demonstrated that the application of PF resin modified with ethyl malonate enables the production of plywood with good mechanical properties and bond quality in a pressing time shortened by 38% and can reduce the pressing temperature by 20 °C. Li et al. [23] investigated the effects of hot-pressing parameters (temperature, pressure, time and veneer layers) on the shear strength of multi-plywood using modified soy protein adhesives.

The authors found that the heating rate of the plywood core layer increased with the increase of hot-pressing temperature and decreased noticeably with an increased number of veneer layers.

Another approach for shortening the heating time and the overall pressing time is the preliminary thermal compression of veneers. In recent years, a number of studies have shown that introducing a veneer preparation process via thermo-mechanical compression prior to applying the adhesive reduces the pressure and time of pressing and also significantly reduces the consumption of adhesive without negatively impacting the bonding quality of the plywood [24–26]. Kurowska et al. [27] concluded that veneer densification shortened the total pressing time by 12–25% in comparison to control samples. Bekhta and Salca [8] found that the multilayered plywood made of densified birch veneers with an adhesive spread rate of 150 g/m² was heated faster compared to plywood made of non-densified veneers.

Thus, we hypothesized that thermal compression of the veneers would lead to reaching the curing temperature of the glue in the core layer more rapidly, even at a lower adhesive spread rate, compared with panels made from non-densified veneers, and this will subsequently shorten the pressing time. The purpose of this study was to obtain a better understanding of the heating process of veneer stacks and of the temperature evolution within the plywood panels during hot pressing when using different types of veneer (non-densified and densified), different pressures, time of pressing and pressing temperatures and different numbers of veneer layers. We also investigated the optimal pressing parameters to reduce glue consumption and studied how this will affect the quality of the plywood bonding and increased production efficiency.

2. Materials and Methods

2.1. Materials

In this study, we used rotary-cut birch (*Betula verrucosa* Ehrh.) veneers (LLC «ODEK» Ukraine) with dimensions of 300 mm × 300 mm and thickness of 1.6 mm and density of 625 kg/m³. The average moisture content of the non-densified veneers was 5.7%.

Half of the veneer sheets were densified by the application of heat and pressure between the smooth and carefully cleaned heated plates of an open-system laboratory press at a temperature of 150 °C and a pressure of 2 MPa for 1 min. After densification, the samples were removed from the hot press and allowed to cool to room temperature. The average thickness and moisture content of the densified birch veneers were 1.5 mm and 1.4%, respectively.

The commercial PF adhesive Fenokol 43 EX (Chemko, a. s. Slovakia), with a solid content of 47% (at 105 °C), a viscosity of 278 mPa·s, a gel time of 24 s (at 150 °C), a free phenol content of 0.013%, a free formaldehyde content of 0.032% and a hydrogen ion concentration (pH11), was used to bond the veneers. The PF resin was used for plywood panel manufacturing without any filler or additive.

2.2. Experimental Procedure

In this study, two series of experiments were performed.

During the first series of experiments, the temperature evolution of the adhesive-free plywood samples during hot pressing was measured:

- inside a stack of common non-densified birch veneers at different pressing temperatures (100, 120, 130, 140 and 150 °C);
- inside a stack of either non-densified or densified birch veneers at different pressing pressures (1.0, 1.4 and 1.8 MPa);
- inside a stack of either non-densified or densified birch veneers at different pressing temperatures (100, 130 and 150 °C) and containing different numbers of veneer layers (3, 5 and 7 layers).

In this series of experiments, the measurements of the core layer temperature inside the veneer stacks were carried out without the presence of glue. This was done to determine the real impact of each studied factor on the temperature evolution inside the sample. It was difficult to measure the temperature when adhesive was used because the adhesive in a liquid or solid state would significantly affect the result.

During the second series of experiments, the influence of the pressing parameters, in particular the pressing time (120, 180, 240, 300 and 360 s) and the pressing pressure (1.0, 1.4 and 1.8 MPa), on the properties of PF adhesive-bonded three-layer plywood made from either non-densified or densified birch veneers with reduced adhesive consumption (100 g/m^2) was studied.

2.3. Core Layer Temperature Testing

A thermocouple was placed inside the plywood sample to measure the glue line inner temperature. A 5 mm × 150 mm (width × length) groove was opened in the middle of the core veneer to install the thermocouple. The temperatures were measured with the thermocouple at the center of the central veneer sheet (layer) of the stack and were monitored every 5 s to record the temperature evolution in the panel during pressing. Data collection was initiated when the surface veneer began to receive pressure. When the temperature inside the stack reached the pressing temperature, the test was completed and the data were saved to a computer.

2.4. Preparation of Plywood Samples

The three-layer plywood samples from densified veneer were made in an electrically heated hydraulic laboratory press under the following conditions: 150 °C pressing temperature; and different specified values of pressing pressure (1.0, 1.4 and 1.8 MPa) and pressing time (120, 180, 240, 300 and 360 s) and glue spreading (100 and 150 g/m^2). For comparison, plywood control samples from non-densified veneer were made at the same pressing conditions and pressing pressure of 1.0 and 1.8 MPa. During the last 30 s of the press cycle, the pressure was continuously reduced to 0 MPa. The adhesive was applied onto one side of every uneven ply. The plies were assembled perpendicular to each other (veneer sheets were laid tight/loose) to form plywood of three/five/seven plies. Adhesive was applied onto the veneer surface with a hand roller spreader.

2.5. Shear Strength Test

During the experiment, all plywood samples were conditioned prior to testing for 2 weeks at 20 ± 2 °C and 65% ± 5% relative humidity. The panels were cut to extract test samples according to the standard requirements. The shear strength was determined according to methods EN 314-1 [28] and EN 314-2 [29] after pretreatment for their intended use in exterior conditions. For the shear strength test, PF plywood test pieces were immersed for 4 h in boiling water, followed by drying in a ventilated oven for 16 h at 60 ± 3 °C, immersion in boiling water for 4 h, and finally, immersion in cool water at 20 ± 3 °C for at least 1 h. Ten samples were used for each variant shear strength mechanical test.

3. Results

During hot pressing, heat is first, transferred from the hot press plates to the outer veneer layers by conduction and then continues to migrate to the middle. The effective porosity in veneer panels was only 0.05–0.5% compared to the total panel voids, which ranged from 50% to 70%. The rate of convection is negligible; thus, heat conduction is dominant [30]. At the same time, upon contact of the surface veneer layers with the hot press plates, the moisture present in these layers turns into steam. The steam migrates to the middle of the panel. However, during hot pressing of veneer panels, it is veneer compression that results in layered and uniform barriers to moisture movement [30]. The interior vapor pressure increases as the steam continues to migrate from the hot surfaces toward the colder middle. In the middle zone of the panel (from which moisture is virtually free to evaporate and, under certain conditions, is contained there in the form of a superheated steam–water mixture), the temperature is continually increasing, approaching the temperature of the press plates [9]. However, the rate of temperature rise in the outer and inner layers of the middle zone of the panel is different [4,9].

3.1. Influence of Pressing Temperature on the Heating Rate of the Veneer Stacks

Figure 1 shows how the core layer temperature inside the three-layer birch veneer stack depends on the plywood pressing temperature and Table 1 presents the time required to reach a temperature of 100 °C inside the three-layer panel when applying the different pressing temperatures.

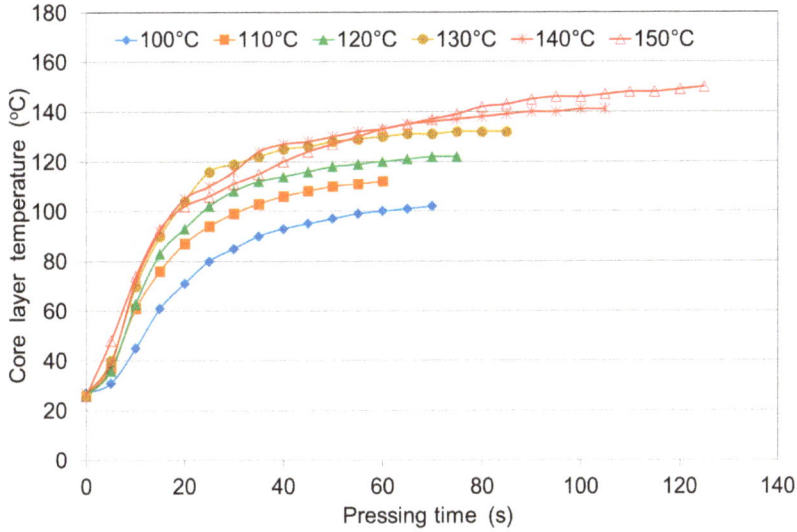

Figure 1. Core layer temperature curves at different pressing temperatures and 1.8 MPa of three-layer plywood made from non-densified birch veneers.

Table 1. Time needed for the core layer of three-layer plywood bonded with PF adhesive to reach 100 °C and the pressing temperature.

Pressing Temperature (°C)	Time to Reach 100 °C (s)	Time to Reach the Pressing Temperature (s)
100	60	60
110	30	50
120	25	60
130	19	60
140	19	90
150	19	125

The core layer temperature rose from an initial temperature of 25–30 °C to approximately 100 °C in approximately 20 s when the press was closed, but it took almost 125 s to reach a pressing temperature of 150 °C.

During the gradual heating of the veneer stacks, the wood underwent a temperature change and the water contained within the wood was also altered. At 100 °C, some of the moisture was converted to steam, filling all the spaces within the wood and between the veneer sheets. The rapid heating of the middle of the panel to 100 °C can be explained by the fact that at this temperature liquid water is transformed into steam, which moves from the outer layers to the inner layers, quickly heating the panel to 100 °C.

It is natural that the interior temperature reaches 100 °C most rapidly at the higher pressing temperatures of 130–150 °C. The veneer stack heated the slowest at a pressing temperature of 100 °C. In this case, it took 60 s for the panel interior to reach 100 °C, which is equal to the temperature of the press plates (pressing temperature). By contrast, at 150 °C, the veneer stack was heated three times faster. These results are in good agreement with a previous study [4] that found that a considerable amount of thermal energy is needed to cure PF adhesives, i.e., the application of high temperatures (135–150 °C) or long pressing times (45–60 s/mm).

At low pressing temperatures of 100–120 °C, we observed gradual and slow heating of the panels. At the higher pressing temperatures of 130–150 °C, during the first 20 s, the panels heated very quickly before the moisture evaporated after which the heating rate slowed. Table 1 shows that the time required for the core layer of the three-layer panel to reach 100 °C was significantly reduced when the temperature was increased from 100 °C (60 s) to 120 °C (25 s) or 150 °C (19 s). After the interior temperature reached the temperature of water evaporation, the temperature inside the panel increased slowly until the core layer temperature was close to the pressing temperature.

Throughout the period of heat transfer between the sample and the hot press plate, the system remained at the veneer-heating stage. The heat was transmitted from the outside to inside, and the plywood gradually reached the pressing temperature, increasing from the surface layer to the core layer. At high pressing temperatures, such as 150 °C, for a short time (approximately 20 s), the core layer rapidly heated to 100 °C due to steam moving from the outer layers to the middle. However, during this time, the panel was densified, and its density increased while the porosity decreased, making it more difficult for the steam concentrated inside the panel to escape. If the steam cannot escape from the center of the panel, which is compressed between the press plates, then the steam will continue to increase the temperature of the center zone to the temperature of the press plates. A similar pattern of heating was previously described [4]. If the temperature increases from 20 to 100 °C, the conductivity slightly increases up to 14% and 24% in the longitudinal and transverse directions, respectively [31].

3.2. The Effect of Pressing Pressure on the Heating Rate of Non-Densified and Densified Veneer Stacks

Based on the results of the previous series of experiments, birch veneer and a plywood pressing temperature of 150 °C were used in subsequent studies.

Figure 2 shows the dependence of the core layer temperature on the pressing pressure of the three-layer birch non-densified and densified veneer stacks. Table 2 shows the time required to reach a core layer temperature of 100 °C or a pressing temperature of 150 °C inside the three-layer panels when applying different pressures. There was practically no difference between densified and non-densified veneer stacks in the time required to reach an interior temperature of 100 °C when applying the different pressing pressures.

At pressures of 1.0, 1.4 and 1.8 MPa, a core layer temperature of 150 °C was reached after 110 s, 85 s and 125 s for panels made from non-densified veneers and after 115 s, 90 s and 75 s, for panels made from densified veneers, respectively. The data show that at the lower pressing pressures of 1.0 and 1.4 MPa, the panels heated at the same rate regardless of whether they were made of non-densified or densified veneer. The reason may be that panels have more porosity in the stack and make it relatively easy for steam to penetrate/migrate to the core while panels pressed at 1.8 MPa may mainly rely on heat conduction to raise the core temperature.

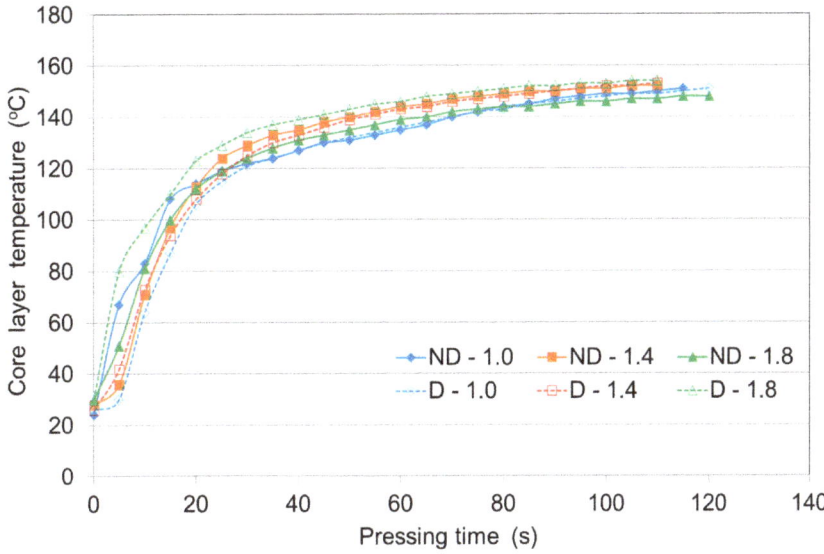

Figure 2. Core layer temperature curves of three-layer plywood made from non-densified (ND) or densified (D) birch veneers at different pressing pressures and 150 °C.

Table 2. Time needed for the core layer of three-layer plywood bonded with PF adhesive to reach 100 °C or the temperature of pressing.

Type of Veneer	Pressure of Pressing (MPa)	Time to Reach 100 °C (s)	Time to Reach 150 °C (s)
ND	1.0	14	110
ND	1.4	16	85
ND	1.8	19	125
D	1.0	18	115
D	1.4	17	90
D	1.8	11	75

It is believed that the moisture content of the veneer stack affects the heat transfer from the outer layers to the panel core. Usually, higher moisture contents increase the thermal conductivity, which will accelerate the heat transfer. This may also be valid for low-density panels. The non-densified veneer had a higher moisture content (5.7%) than the densified veneer (1.4%). Arruda and Del Menezzi [32] also stated that thermomechanical treatment provided lower equilibrium moisture content of veneers. Therefore, for a panel made of non-densified veneer, we expected that its moisture content would have a significant effect on the rate of temperature rise. Data from Table 2 shows that densified veneer under high-pressure pressing was heated faster than the non-densified veneer at the same pressure. For the panel made of densified veneer, core temperatures of 100 °C and 150 °C were reached after 11 and 75 s, respectively, compared with 19 and 125 s for the non-densified veneer stack. Kurowska et al. [27] found that veneer densification shortened total pressing time by 12–25% in comparison to control samples. This phenomenon is caused by lower total moisture content in the veneer stack. Furthermore, faster stack internal temperature gain is caused by more dense wood substance. Wood compression mainly causes a reduction in empty spaces between cells and cell lumen [32]. Cai et al. [33] also showed that the panel moisture content does not have a significant effect on the time to reach the maximum core temperature.

The effective thermal conductivity of compressed wood was found to be lower than that of uncompressed wood [34]. On the other hand, Hrazsky and Kral [4] showed that the rate of heat passage increased with the increasing working pressure (within a certain pressing temperature range). However, in our study, this was only true for densified veneer. For non-densified veneers, it took longer to reach the pressing temperature inside the panel when applying high pressure. One of the reasons for this finding may be the higher moisture content of non-densified veneers compared with densified veneers. It is known that the rate of heat passage decreases with an increasing moisture content [4].

The most rapid heat transfer occurs when steam can pass through cracks and voids in the veneer. High pressures restrict these passages [20]. Wang et al. [30] similarly stated that during hot pressing the small deformations of the veneer ply effectively act as barriers to gas and moisture movement rather than the curing glue line acting in this regard. These barriers caused a sealing effect that led to a low rate of convection within the panel [30].

3.3. The Effect of the Number of Layers on the Heating Rate of Non-Densified and Densified Veneer Stacks

Figure 3 shows the effect of the number of veneer layers (3, 5 and 7) on the heating rate of the non-densified and densified veneer stacks at different pressing temperatures (100 °C, 130 °C and 150 °C). The heating rate decreased significantly with an increased number of veneer layers, not only at the stage of rapid heating, but also at the stage of slow heating (Figure 3). With an increasing number of veneer layers, the time required for the core layer to reach 100 °C increased, as did the time required for the core layer to reach the pressing temperature (Figure 3).

With an increasing number of veneer layers, the temperature at which moisture evaporation occurred decreased with an increasing number of layers and the time period over which moisture evaporation occurred was prolonged, probably because the moisture evaporation required more heat due to the increased number of layers. The heat transfer rate decreased with the increase of veneer layers, so more time was required to reach 100 °C or the pressing temperature in the core layer, leading to an increasing time difference between reaching 100 °C and the pressing temperature (Figure 3). When there were too many veneer layers during hot pressing, the core layer temperature could not reach the pressing temperature even with an unlimited increase in pressing time, although the core layer temperature could approach the pressing temperature.

Figure 3a shows that the three-, five- and seven-layer panels of non-densified veneer heated faster at a pressing temperature of 100 °C than those of densified veneers. At the pressing temperature of 130 °C, the 3, 5 and 7-layer panels reached a core temperature of 100 °C in nearly the same time, regardless of densification. These results were also apparent at the pressing temperature of 150 °C. At pressing temperatures of 130–150 °C, for a further increase in core temperature above 100 °C, the densified veneer stacks heated to the pressing temperature faster than the non-densified veneer stacks and the difference between the pressing temperatures for the non-densified and densified veneers was already quite large. For example, the pressing temperature of 130 °C was reached in the core after 140, 225 and 325 s for the 3-, 5- and 7-layer non-densified veneers and after 70, 135 and 325 s for the densified veneers, respectively.

The pressing temperature of 150 °C was reached in the panel core after 125, 300 and 500 s for the 3-, 5- and 7-layer non-densified veneer stacks and after 85, 210 and 400 s for the 3-, 5- and 7-layer densified veneer stacks, respectively. Thus, the 3-, 5- and 7-layer panels of both the non-densified and densified veneers at pressing temperatures of 130 and 150 °C reached a core temperature of 100 °C in nearly the same time, but upon further heating, the densified panels heated much faster. The heating rate of the 3-, 5- and 7-layer panels to a temperature of 130 °C at a pressing temperature of 130 °C was faster by 50%, 40% and 0%, respectively, for the densified veneers compared with the non-densified veneers. The rate of heating to 150 °C at a pressing temperature of 150 °C for the 3-, 5- and 7-layer panels was faster by 32%, 30% and 20%, respectively, for the densified veneers compared with the non-densified veneers.

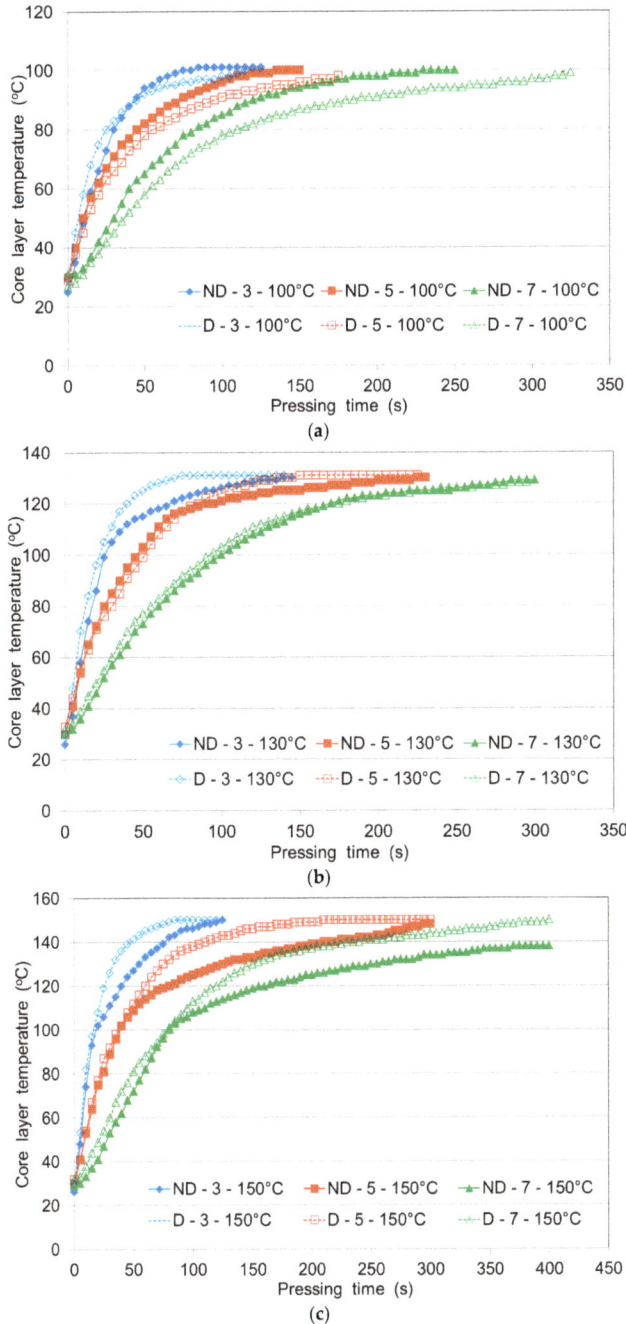

Figure 3. Core layer temperature curves of 3-, 5- and 7-layer plywood panels made from non-densified (ND) or densified (D) birch veneers at different pressing temperatures: (**a**) 100 °C; (**b**) 130 °C; (**c**) 150 °C.

The heating rate decreased markedly with an increasing number of veneer layers. The effect of increasing the number of layers was mainly observed in the continuation of the first stage of constant temperature and the increase in time required for the core temperature to reach the pressing temperature. The heating rate of the core layer increased with increasing pressing temperature. In any case, a pressing temperature of 100 °C cannot be recommended for PF-bonded plywood panels for both technological and economic reasons. These findings are in good agreement with previous studies [4] who showed that temperatures of 135–150 °C or longer pressing times are required to cure PF adhesives.

Considering the curves (Figures 1–3), we can conclude that the core temperature evolution can be divided into three stages as follows: in the first stage the core temperature remains constant for approximately 25–30 s after the platen reaches the veneer face surface; the second stage refers to the rapid increase in the core temperature due to convective heat flow; in the third stage the temperature remains nearly constant during the moisture vaporization. A similar core temperature distribution in the middle layers of multi-ply veneer assemblies of either non-densified or densified veneer stacks, as well as for when adhesives are not used, was previously observed [8]. The results of the temperature evolution inside the plywood samples for both densified and non-densified veneers were found to be quite similar to the heat transfer during hot pressing of particleboards and fiberboards [35,36].

3.4. The Influence of Pressing Pressure and Time on the Bonding Strength of Plywood Samples

The average shear strength values of the samples, along with the Duncan's test results, are depicted in Figure 4. The shear strengths of the plywood samples composed of either densified or non-densified veneers were higher than 1.2 MPa and met the requirements of the EN 314-2 standard [29]. The highest shear strength was obtained by plywood samples composed of densified veneers made at a pressure of 1.4 MPa and a press times of 6 min. The lowest shear capacity was observed in samples composed of densified veneers made at a pressure 1.8 MPa using a 2 min press time. For densified veneers, the shear strength increased with increasing press time.

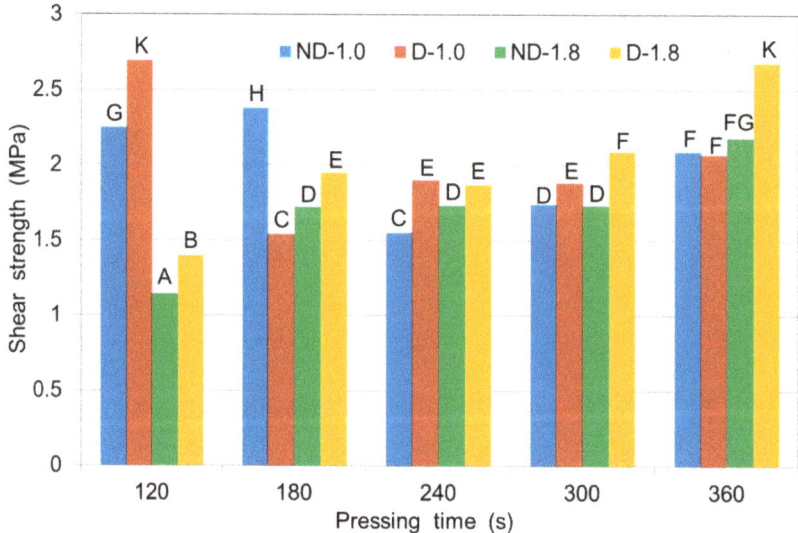

Figure 4. Shear strengths of 3-layer plywood samples made from birch veneers at various pressing pressures (1.0 and 1.8 MPa) for various durations at 150 °C and 100 g/m^2 of adhesive spread: ND—non-densified veneer; D—densified veneer. Latin letters A–K indicate Duncan group.

The shear strength values of the plywood samples made at a pressure of 1.8 MPa, but with different pressing times and at a lower adhesive spread rate (100 g/m^2) were higher for the densified veneers than for the non-densified veneers (Figure 4). Increasing the pressing time of the samples from 120 to 360 s increased the shear strength and all shear strength values met the requirements of EN 314-2 [29]. The images of densified wood shows a significant improvement in the glue line, which became thinner and more continuous [32].

At a pressing pressure of 1.0 MPa and a lower adhesive spread rate (100 g/m^2), the shear strengths of the plywood samples composed of densified veneers and made with pressing times of 120 s and 240–360 s were higher than shear strengths of the samples composed of non-densified veneers. In contrast, at a pressing time of 180 s, the shear strength values of the samples made from non-densified veneers were higher than of the samples from densified veneers (Figure 4).

For samples composed of densified veneers, the shear strength values for samples pressed at 1.0 MPa were lower than those made at 1.8 MPa. Even at 1.0 MPa, the shear strength values were high (>1.5 MPa) and met the requirements of EN 314-2 [29].

For plywood samples made from densified veneers at a lower glue consumption rate (100 g/m^2) and pressing times of 180–360 sec, higher pressures of 1.4 and 1.8 MPa led to higher shear strengths than when applying a pressure of 1.0 MPa (Figure 5). In contrast, for a pressing time of 120 s, the shear strength values of the samples at a pressure of 1.0 MPa were higher than the shear strengths of the samples made at 1.4 and 1.8 MPa.

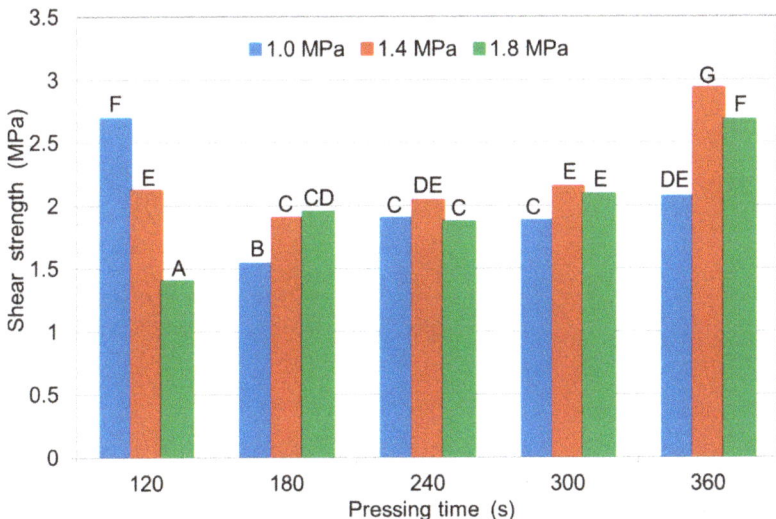

Figure 5. Shear strengths of 3-layer plywood samples made from densified birch veneers at various pressing pressures for various durations at 150 °C and 100 g/m^2 of adhesive spread. Latin letters A–G indicate Duncan group.

Long pressing times at high pressing pressure can increase the veneer compression ratio and reduce the productivity of plywood, which is unacceptable in the plywood manufacturing industry. In terms of economics and technology, it is possible to choose a pressure of 1.4 MPa and a pressing time of 180–300 s. This allows the pressing time to be reduced by 17–50% and the pressing pressure by 22.2% without negatively impacting the bonding strength of the plywood. In similar study it was also found that densified veneers, except 25% pressing time shortening, allow 25% glue load reduction without affecting glue bonds strength properties [27].

Figure 6 compares the bonding strengths of plywood samples made from either non-densified or densified veneers at reduced (100 g/m^2) and currently accepted (150 g/m^2) adhesive spread rates and at different pressing pressures. The highest bonding strength was observed for the densified veneers at a pressure of 1.4 MPa and the lower glue consumption of 100 g/m^2. A lower bonding strength was found for samples made from densified veneer at a pressure of 1.8 MPa and an adhesive spread rate of 150 g/m^2 than at a lower adhesive spread rate. Typically, the densification process smooths the surface of the veneer and decreases its roughness [37,38]; therefore, less adhesive is required for bonding. A lower adhesive consumption results in a reduced thickness of the adhesive layer and an increased bonding strength. An adhesive spread rate of 150 g/m^2 for densified veneer is too large; the adhesive is squeezed out of the panel, the thickness of the adhesive layer increases and as a consequence, the adhesive strength decreases. It is known [39] that with increasing glue line thickness, the bonding strength decreases; with a thicker glue line, higher internal stress is generated during glue shrinkage, which can lead to a lower shear strength. Moreover, at the high glue spread level, gas pressure increased significantly due to the high MC in the glue line, which generally led to blisters or blows, which could largely deteriorate plywood bond quality [40].

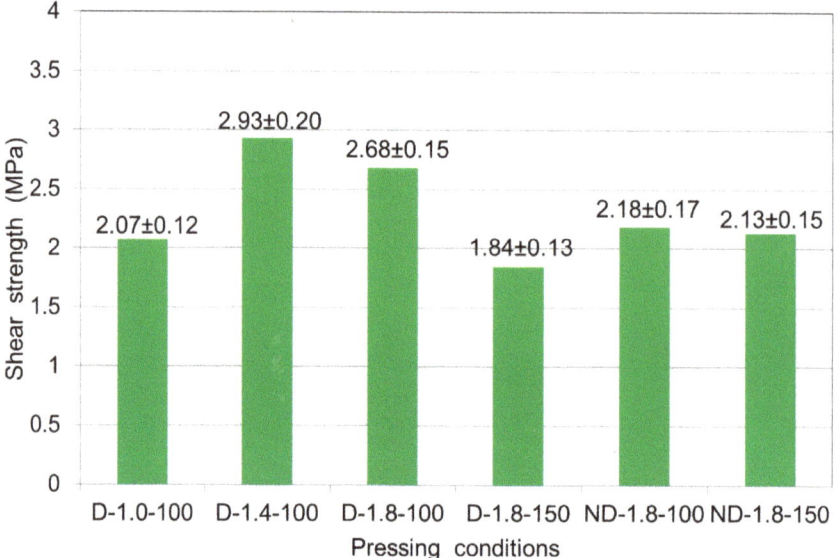

Figure 6. Shear strengths of 3-layer plywood samples made from birch veneers at various pressing pressures (1.0, 1.4 and 1.8 MPa) and a pressing time of 6 min for 150 and 100 g/m^2 of adhesive spread: ND—non-densified veneer; D—densified veneer.

At a pressing pressure of 1.4 MPa, the bonding strength was 2.93 MPa and was 8.5% higher than at a pressure of 1.8 MPa (2.68 MPa). At a pressing pressure of 1.8 MPa, but with an adhesive spread rate of 150 g/m^2, the bonding strength (1.84 MPa) was 31.3% lower than with the lower adhesive spread rate (100 g/m^2) at the same pressure. Wang et al. [40] also found that shear strength decreased as the pressing pressure increased. The high pressing pressure leads to reduced gas permeability and high internal gas pressure causing blisters or blows, which could completely destroy shear strengths of samples.

When comparing plywood samples made from densified and non-densified veneer made at a pressure of 1.8 MPa and an adhesive spread rate of 100 g/m^2, the bonding strength of the densified samples was 18.7% higher (2.68 MPa) than of the non-densified samples (2.18 MPa). The opposite pattern was observed for the adhesive spread rate of 150 g/m^2. The bonding strength of the non-densified samples was 13.6% higher (2.13 MPa) than for the densified samples (1.84 MPa). A previous study found that surface roughness will affect gluing and bonding between two layers of panels. It was observed that the adhesive was not distributed evenly on panels made from uncompressed veneer due to the effects of its rough surface in comparison with panels made from compressed veneer [41]. Non-densified veneer is rough and requires more glue, while densified veneer is smooth and has less roughness. The adhesive spread rate of 150 g/m² was too large for the smooth veneer, which led to thickening of the adhesive layer and consequently reducing the bonding strength. Practice shows that the thicker the adhesive layer, the more noticeable the influence of internal stresses and, as a rule, the lower the bonding strength. For high-density hardwood veneers, a smooth surface is a necessity; where there is no surface contact, there can be no adhesion [14,42].

The veneer roughness plays an important role in the depth of penetration and the uniform distribution of the adhesive and influences the bonding quality of veneers. Arruda and Del Menezzi [37] also found that increasing the temperature or time led to a significant reduction in roughness. According to these authors, the veneer roughness decreased by 43.4%, which contributed to reducing the stress points between the veneer surface and the adhesive layer. Several studies [38,43–46] also determined that improved surface roughness of veneers increased the shear strength of the plywood made from them. Images of thermo-mechanically treated wood show a significant improvement in the glue line, which became thinner and more continuous [32]. Moreover, the permeability of the PF glue line decreased during glue curing and the permeability of cured glue lines (films) decreased with increasing glue spread [30]. The thermal conductivity of plywood increases with an increasing glue spreading rate by using phenol formaldehyde resin adhesive [47].

4. Conclusions

As a current contribution of the performed research, we can draw the following conclusions.

The heating rate of the veneer stacks increased as the pressing temperature increased. The panels were the slowest to heat at a pressing temperature of 100 °C, whereas at a pressing temperature of 150 °C, the panels heated to a core temperature of 100 °C three times faster. The heating rate of the core layer increased with increasing pressing temperature.

Practically, there was no difference in the time required to heat the core to a temperature of 100 °C for panels made of non-densified vs. densified veneers at the different pressing pressures. This pattern changed when heating the core to the pressing temperature. In this case, the densified veneer stacks heated faster than the non-densified panels at high pressing pressures. The heating rate of both the non-densified and densified veneer stacks decreased markedly with an increasing number of veneer layers. The 3-, 5- and 7-layer panels, for both the non-densified and densified veneers, reached a core temperature of 100 °C in nearly the same time at pressing temperatures of 130 and 150 °C. However, upon further heating, the densified veneer stacks heated much faster. The rate of heating to 150 °C for the 3-, 5- and 7-layer panels at a pressing temperature of 150 °C was faster by 32%, 30% and 20%, respectively, for the densified veneer than for the non-densified veneer.

When using densified veneers for the production of plywood, a shorter pressing time (17–50% reduction), lower glue consumption (33.3% reduction) and a lower pressing pressure (22.2% reduction) can be used without negatively impacting the bonding strength of the plywood samples.

The findings of this study provide useful information necessary for optimizing the plywood manufacturing process by balancing product qualities and productivity.

Author Contributions: Conceptualization, P.B.; methodology, P.B., J.S. and N.B.; investigation, P.B., J.S. and N.B.; writing—original draft preparation, P.B.; writing—review and editing, P.B., J.S. and N.B. All authors have read and agreed to the published version of the manuscript.

Funding: This research was funded by the Slovak Academic Information Agency and by the Slovak Research and Development Agency under the contracts No. APVV-16-0177, APVV-17-0456, APVV-17-0583 and APVV-18-0378; and ITMS project code: 313011T720 "LignoPro".

Acknowledgments: The work was supported by the Slovak Academic Information Agency and by the Slovak Research and Development Agency under the contracts No. APVV-16-0177, APVV-17-0456, APVV-17-0583 and APVV-18-0378; and ITMS project code: 313011T720 "LignoPro".

Conflicts of Interest: The authors declare no conflict of interest.

References

1. Zavala, D.; Humphrey, P.E. Hot pressing veneer-based products: The interaction of physical processes. *For. Prod. J.* **1996**, *46*, 69–77.
2. Zombori, G.B.; Kamke, F.A.; Wastson, L.T. Simulation of the internal conditions during the hot-pressing process. *Wood Fiber Sci.* **2003**, *35*, 2–23.
3. Wang, B.J. Experimentation and Modeling of Hot Pressing Behaviour of Veneer-Based Composites. Ph.D. Thesis, The University of British Columbia, Vancouver, BC, Canada, 2007.
4. Hrazsky, J.; Kral, P. Determination of pressing parameters of spruce water-resistance plywood. *J. For. Sci.* **2007**, *53*, 231–242. [CrossRef]
5. Demirkir, C.; Özsahin, Ş.; Aydin, I.; Colakoglu, G. Optimization of some panel manufacturing parameters for the best bonding strength of plywood. *Int. J. Adhes. Adhes.* **2013**, *46*, 14–20. [CrossRef]
6. Antov, P.; Savov, V.; Neykov, N. Sustainable bio-based adhesives for eco-friendly wood composites. A review. *Wood Res.* **2020**, *65*, 51–62. [CrossRef]
7. Wei, P.; Rao, X.; Yang, J.; Guo, Y.; Chen, H.; Zhang, Y.; Chen, S.; Deng, X.; Wang, Z. Hot Pressing of Wood-Based Composites: A Review. *For. Prod. J.* **2016**, *66*, 419–427. [CrossRef]
8. Bekhta, P.; Salca, E.-A. Influence of veneer densification on the shear strength and temperature behavior inside the plywood during hot press. *Constr. Build. Mater.* **2018**, *162*, 20–26. [CrossRef]
9. Kamke, F.A. Physics of hot pressing. In *Proc. Fundamentals of Composite Processing*; Winandy, J., Kamke, F.A., Eds.; General Technical Report FPL–GTR–149; USDA Forest Service, Forest Products Laboratory: Madison, WI, USA, 2004; pp. 3–26.
10. Kamke, F.A.; Lee, J.N. Adhesive penetration in wood—A review. *Wood Fiber Sci.* **2007**, *39*, 205–220.
11. Frihart, C.R.; Hunt, C.G. Adhesives with wood materials, bond formation and performance. In *Wood Handbook—Wood As an Engineering Material*; General Technical Report FPL-GTR-190; U.S. Department of Agriculture, Forest Service, Forest Products Laboratory: Madison, WI, USA, 2010; Chapter 10; pp. 10–1–10–24.
12. Vick, C.B. Adhesive bonding of wood materials. In *Wood Handbook—Wood as an Engineering Material*; General Technical Report FPL–GTR–113; U.S. Department of Agriculture, Forest Service, Forest Products Laboratory: Madison, WI, USA, 1999; Chapter 9; pp. 9–1–9–24.
13. Kurt, R.; Cil, M. Effects of press pressures on glue line thickness and properties of laminated veneer lumber glued with phenol formaldehyde adhesive. *BioResources* **2012**, *7*, 5346–5354. [CrossRef]
14. Pizzi, A. Phenolic Resin Wood Adhesives. In *Wood Adhesives: Chemistry and Technology*; Pizzi, A., Ed.; Marcel Dekker, Inc.: New York, NY, USA; Basel, Switzerland, 1983; Chapter 3; pp. 105–176.
15. Pangh, H.; Doostoseini, K. Optimization of press time and properties of laminated veneer lumber panels by means of a punching technique. *BioResources* **2017**, *12*, 2254–2268. [CrossRef]
16. Wang, B.; Dai, C.; Troughton, G. Development of a new incising technology for plywood/LVL production. Part 2. Effect of incising on LVL strength properties. *For. Prod. J.* **2003**, *53*, 99–102.
17. Ashori, A.; Nourbakhsh, A. Effect of press cycle time and resin content on physical and mechanical properties of particleboard panels made from the underutilized low-quality raw materials. *Ind. Crop. Prod.* **2008**, *28*, 225–230. [CrossRef]
18. Yapici, F.; Esen, R.; Yörür, H. The effects of press time and press pressure on the modulus of rupture and modulus of elasticity properties of oriented strand board (OSB) manufactured from scots pine. *PRO Ligno* **2013**, *9*, 532–535.
19. Zhong, Z.; Sun, S.; Fang, X.; Ratto, J.A. Adhesive strength of guanidine hydrochloride-modified soy protein for fiberboard application. *Int. J. Adhes. Adhes.* **2002**, *22*, 267–272. [CrossRef]
20. Jokerst, R.W.; Geimer, R.L. 1994. Steam-assisted hot-pressing of construction plywood. *For. Prod. J.* **1994**, *44*, 34–36.
21. Kurowska, A.; Borysiuk, P.; Mamiński, M.Ł. Simultaneous veneers incising and lower pressing temperatures—the effect on the plywood pressing time. *Eur. J. Wood Prod.* **2011**, *69*, 495–497. [CrossRef]
22. Mirski, R.; Dziurka, D.; Łęcka, J. Potential of shortening pressing time or reducing pressing temperature for plywood resinated with PF resin modified using alcohols and esters. *Eur. J. Wood Prod.* **2011**, *69*, 317–323. [CrossRef]
23. Li, H.; Li, C.; Chen, H.; Zhang, D.; Zhang, S.; Li, J. Effects of hot-pressing parameters on shear strength of plywood bonded with modified soy protein adhesives. *BioResources* **2014**, *9*, 5858–5870. [CrossRef]

24. Bekhta, P.A.; Marutzky, R. Reduction of glue consumption in the plywood production by using previously compressed veneer. *Holz Roh Werkst.* **2007**, *65*, 87–88. [CrossRef]
25. Bekhta, P.; Niemz, P.; Sedliačik, J. Effect of pre-pressing of veneer on the glueability and properties of veneer-based products. *Eur. J. Wood Prod.* **2012**, *70*, 99–106. [CrossRef]
26. Bekhta, P.; Sedliačik, J.; Jones, D. Effect of short-term thermomechanical densification of wood veneers on the properties of birch plywood. *Eur. J. Wood Prod.* **2018**, *76*, 549–562. [CrossRef]
27. Kurowska, A.; Borysiuk, P.; Maminski, M.; Zbiec, M. Veneer densification as a tool for shortening of plywood pressing time. *Drvna Ind.* **2010**, *61*, 193–196.
28. EN 314-1. *Plywood. Bonding Quality. Part. 1: Test. Methods*; European Committee for Standardization: Brussels, Belgium, 2004.
29. EN 314-2. *Plywood. Bonding Quality. Part. 2: Requirements*; European Committee for Standardization: Brussels, Belgium, 1993.
30. Wang, B.J.; Zhou, X.; Dai, C.; Ellis, S. Air permeability of aspen veneer and glueline: Experimentation and implications. *Holzforschung* **2006**, *60*, 304–312. [CrossRef]
31. Suleiman, B.; Larfeldt, J.; Leckner, B.; Gustavsson, M. Thermal conductivity and diffusivity of wood. *Wood Sci. Technol.* **1999**, *33*, 465–473. [CrossRef]
32. Arruda, L.M.; Del Menezzi, C.D.S. Properties of a Laminated Wood Composite Produced with Thermomechanically Treated Veneers. *Adv. Mater. Sci. Eng.* **2016**, *2016*, 8458065. [CrossRef]
33. Cai, Z.; Muehl, J.H.; Winandy, J.E. Effects of panel density and mat moisture content on processing medium density fiberboard. *For. Prod. J.* **2006**, *56*, 20–25.
34. Asako, Y.H.; Kamikoga, H.; Nishimura, H.; Yamaguchi, Y. Effective thermal conductivity of compressed woods. *Int. J. Heat Mass Transfer.* **2002**, *45*, 2243–2253. [CrossRef]
35. Liu, Z.T.; Wang, J.Y.; Yu, H. Study on factors influencing the heat-transfer process in hot-pressing of wood particleboard. *J. Beijing For. Univ.* **1995**, *17*, 267–272.
36. Garcia, R.; Cloutier, A. Characterization of heat and mass transfer in the mat during the hot pressing of MDF panels. *Wood Fiber Sci.* **2005**, *37*, 23–41.
37. Arruda, L.; Del Menezzi, C.H.S. Effect of thermomechanical treatment on physical properties of wood veneers. *Int. Wood Prod. J.* **2013**, *4*, 217–224. [CrossRef]
38. Bekhta, P.; Proszyk, S.; Krystofiak, T.; Mamonova, M.; Pinkowski, G.; Lis, B. Effect of thermomechanical densification on surface roughness of wood veneers. *Wood Mater. Sci. Eng.* **2014**, *9*, 233–245. [CrossRef]
39. Pizzi, A.; Mittal, K.L. *Handbook of Adhesive Technology*, 2nd ed.; Dekker: New York, NY, USA, 2003.
40. Wang, B.J.; Dai, C. Hot-pressing stress graded aspen veneer for laminated veneer lumber (LVL). *Holzforschung* **2005**, *59*, 10–17. [CrossRef]
41. Nordin, N.A.; Sulaiman, O.; Hashim, R.; Salim, N.; Sato, M.; Hiziroglu, S. Properties of laminated panels made from compressed oil palm trunk. *Compos. Part B Eng.* **2013**, *52*, 100–105. [CrossRef]
42. Kúdela, J.; Lagaňa, R.; Andor, T.; Csiha, C. Variations in beech wood surface performance associated with prolonged heat treatment at 200 °C. *Acta Facultatis Xylologiae Zvolen* **2020**, *62*, 5–17. [CrossRef]
43. Faust, T.D.; Rice, J.T. Effect of veneer surface roughness on glue bond in Southern Pine plywood. *For. Prod. J.* **1986**, *36*, 57–62.
44. Aydin, I. Activation of wood surfaces for glue bonds by mechanical pre-treatment and its effects on some properties of veneer surfaces and plywood panels. *Appl. Surf. Sci.* **2004**, *233*, 268–274. [CrossRef]
45. Biadała, T.; Czarnecki, R.; Dukarska, D. Water resistant plywood of increased elasticity produced from European wood species. *Wood Res.* **2020**, *65*, 111–124. [CrossRef]
46. Candan, Z.; Hiziroglu, S.; Mcdonald, A.G. Surface quality of thermally compressed Douglas fir veneer. *Mater. Des.* **2010**, *31*, 3574–3577. [CrossRef]
47. Ferrtikasari, N.; Tanaka, T.; Yamada, M. Relationship between thermal conductivity and adhesive distribution of phenol-formaldehyde visualized with x-ray computed tomography on sugi (*Cryptomeria japonica* D.Don) heartwood plywood. *IOP Conf. Ser. Mater. Sci. Eng.* **2019**, *593*, 012003. [CrossRef]

© 2020 by the authors. Licensee MDPI, Basel, Switzerland. This article is an open access article distributed under the terms and conditions of the Creative Commons Attribution (CC BY) license (http://creativecommons.org/licenses/by/4.0/).

Article

Sound-Absorption Coefficient of Bark-Based Insulation Panels

Eugenia Mariana Tudor [1,2], Anna Dettendorfer [3], Günther Kain [1], Marius Catalin Barbu [1,2], Roman Réh [4] and Ľuboš Krišťák [4,*]

1. Forest Products Technology and Timber Construction Department, Salzburg University of Applied Sciences, Markt 136a, 5431 Kuchl, Austria; eugenia.tudor@fh-salzburg.ac.at (E.M.T.); guenther.kain@aon.at (G.K.); marius.barbu@fh-salzburg.ac.at (M.C.B.)
2. Faculty of Wood Engineering, Transilvania University of Brasov, Bld. Eroilor nr.29, 500036 Brasov, Romania
3. Steelcase, Brienner Str. 42, 80333 München, Germany; anna.dettendorfer@gmx.de
4. Faculty of Wood Sciences and Technology, Technical University in Zvolen, T. G. Masaryka 24, SK-960 01 Zvolen, Slovakia; reh@tuzvo.sk
* Correspondence: kristak@tuzvo.sk

Received: 22 March 2020; Accepted: 24 April 2020; Published: 29 April 2020

Abstract: The objective of this study was to investigate the sound absorption coefficient of bark-based insulation panels made of softwood barks Spruce (*Picea abies* (L.) H. Karst.) and Larch (*Larix decidua* Mill.) by means of impedance tube, with a frequency range between 125 and 4000 Hz. The highest efficiency of sound absorption was recorded for spruce bark-based insulation boards bonded with urea-formaldehyde resin, at a level of 1000 and 2000 Hz. The potential of noise reduction of larch bark-based panels glued with tannin-based adhesive covers the same frequency interval. The experimental results show that softwood bark, an underrated material, can substitute expensive materials that involve more grey energy in sound insulation applications. Compared with wood-based composites, the engineered spruce bark (with coarse-grained and fine-grained particles) can absorb the sound even better than MDF, particleboard or OSB. Therefore, the sound absorption coefficient values strengthen the application of insulation panels based on tree bark as structural elements for the noise reduction in residential buildings, and concurrently they open the new ways for a deeper research in this field.

Keywords: spruce and larch bark; sound absorption coefficient; impedance tube; biomass; up-cycling

1. Introduction

Noise control is an important issue in modern life. A lot of factors contribute to its increase, e.g., population growth, expansion of the urban centers, densification of the housing sector, correlated to the number of vehicles, the development of automatic machines in industrial companies and devices [1–4].

Noise pollution is the second most important environmental factor in Europe, North America and South-East Asia, contributing to different diseases after air pollution. People of all age groups are becoming more vulnerable to mental stress, heart diseases, sleep disturbance, tinnitus, learning disabilities etc. [5–7]. The range of frequencies for the human voice and musical sound is mostly from 125 to 3000 Hz [8,9]. The human audible frequency range extends up to 15 kHz for most persons, and can reach 20 kHz for children and young people [10]. The sound absorption coefficient gives information about the acoustical effectiveness of a material and is defined as the fraction of the energy of incident sound waves absorbed by the material [1,4]. The values of the sound absorption coefficient are between 0 (no absorption) and 1 (complete absorption, e.g., acoustical walls in recording studios) [11–13]. Sound insulation (expressed as the transmission loss factor) and absorption are two different properties. Materials that are effective as sound insulators are mostly not useful as sound absorbers and vice versa. Parameters that influence values of these sound insulation and absorption include density, porosity and material thickness. Thicker, denser, and heavier material have higher transmission loss factor values, and porous materials are more effective at sound absorption [14–17].

Noise control in buildings is ensured through insulation from external sound sources and absorption of sound generated within a space by blocking the transmission of sound from a room to another [18].

Non-woven materials have been analyzed as sound absorption materials by various researchers who studied cotton [19], cellulose [20] and needle [21]. Rwawiire et al. [22] investigated the sound insulation characteristics of non-woven fabric from the inner bark of three Ficus species. The measurements revealed that the sound absorption of the bark cloths have higher sound absorption properties at higher frequencies, with an improved absorption coefficient when increasing the bark cloth fiber layers [22]. The planks of birch bark were used for a hundred years as sound absorbers under turf roofs in Sweden, serving also as waterproof membrane [23]. Natural materials such as bark, jute, flax, kenaf, hemp, coir fiber, wood, wool, coconut, straw, cane and corn husk can be designed as thermal and sound absorbers with the advantage of availability [24–27], sustainability [28–35] and biodegradability [36–42].

This paper presents some aspects about sound absorption properties of tree bark insulation panels made of larch (*Larix decidua* Mill.) and spruce (*Picea abies* (L.) H.Karst.), with different particle sizes. The sound absorption coefficient of these panels was compared with wood-based composites from a previous research conducted by Smardzewski et al. [9]. This is an example of the upcycling application of woody biomass as resource for sound and thermal insulation boards that can diminish the noise level in a building. In the context of the scarcity of raw materials, the cascading use of wood and forest residues plays an important role, and should be weighed as a basic concept within the circular economy [43,44]. Bio-economy can be considered environmentally beneficial only if the bio-based resources are managed sustainably [45].

2. Materials and Methods

The spruce and larch bark were collected in a local sawmill in Salzburg County, Austria. The bark planks were ground by means of a 4-shaft shredder RS40 at Untha Co. (Kuchl, Austria), with a mesh of 30 mm. Subsequently, the bark particles were dried at 60 °C and 200 to 250 mbar in a vacuum kiln dryer Brunner-Hildebrand High VAC-S, HV-S1 from 65% to 9.0% moisture content.

The material was repeatedly screened according to EN 15149-1:2011 [46] with a sieve shaker Retsch AS 200, to obtain particles in a size spectrum of 8–13 mm and 10–30 mm.

Four types of bark-based insulation boards were manufactured (Table 1). The spruce bark was bonded with 10% urea formaldehyde type Prefere 10F102 (MetaDynea Austria, Krems, Austria). The larck bark from Graggaber sawmill, Unternberg, Austria was glued with 10% tannin-based adhesive. The formulation included Mimosa tannin extract powder (*Acacia mearnsii*) from Phenotan, Tanac, Brasil, hexa-methylenetetramine (hexamine) from Merck Schuchardt, Hohenbrunn, Germany (C99 %) and sodium hydroxide solution (C32 %) from Carl Roth, Karlsruhe, Germany. A total of 50% tannin extract powder and 50% water were stirred with a mechanical mixer at 700 and 1500 rpm. A total of 10% of hexamine was added and sodium hydroxide was used to adjust the pH value to 9. The boards were pressed at 180 °C for five minutes with a press factor of 24 s/mm [47,48].

Table 1. Particle size, dimensions, density levels and adhesive type of the bark insulation boards.

Board Type	Bark Particle Size (mm)	Board Thickness (mm)	Boards Dimension (mm)	Board Density (kg/m^3)	Adhesive
Spruce fine	8–13	21	500 × 500	500	10% UF
Spruce coarse	10–30	20	500 × 500	414	10% UF
Larch coarse, thin	10–30	11	500 × 500	690	10% tannin
Larch coarse, thick	10–30	19	500 × 500	571	10% tannin

100 mm diameter samples were cut from bark-based insulation boards manufactured in the laboratories of Salzburg University of Applied Sciences. The thermal conductivity, mechanical properties, microstructures and volatile organic compounds (VOC) emissions of these panels were analyzed in publications by Kain et al. [48–52].

The samples were prepared from different areas of the boards and cut according to EN 326-1:1994 [53] and ISO16999 [54]. The average moisture content of the samples was 8–9%, according to EN 322:2005 [55].

Acoustical Measurements in Impedance Tube

The acoustical properties of bark-based insulation panels (Figure 1) were measured with an impedance tube system at Krämer & Stegmaier, Berlin, Germany, according to EN ISO 10534-2:2001 (transfer function method) [56]. The system consists of two different sized pipes, which are optimized for high and low frequencies. This covers a frequency range from 50 to 5000 Hz. In contrast to measurements in the reverberation chamber, only small material samples are required for the impedance tube (30 or 150 mm edge length) [57].

Figure 1. Larch bark particles (8–13 mm) (left) and bark based composite specimens (right) for acoustic measurements.

The test specimen was introduced in a stationary acoustic field generated by a speaker under normal incidence. The absorption coefficient and the impedance were determined using the transfer function between two microphones. By evaluating the incident and reflected sound energy, the sound absorption capacity of the material was determined [58]. The precision of results depends on the design specifications of the impedance tube. These are the diameter of the tube, the distance of the microphones from the samples and the distance between the microphones [59].

The share of sound absorbed by the bark-based insulation samples was calculated using Equations (1) and (2):

$$\alpha = \frac{I_i}{I_r} \equiv \frac{|P_i|^2 - |p_r|^2}{|p_i|^2} = 1 - \left[\frac{n-1}{n+1}\right]^2 = \frac{4n}{(1+n)^2} \quad (1)$$

$$n = \frac{p_{max}}{p_{min}} \quad (2)$$

where

α is he sound absorption coefficient

I_i and I_r are intensities of incident and reflected waves

p_i and p_r are the pressures of incident and reflected waves

n is the standing wave ratio (the ratio of the maximum p_{max} to minimum p_{min} pressure of the sound wave) [60] cited by [61].

3. Results and Discussion

The measurements were carried out with and without wall clearance. The assessments with wall clearance are relevant for products such as multi-layered acoustic panels with cavities.

The acoustical properties of materials can be determined by means of impedance tube. In order to calculate the sound absorption capacity of the larch bark samples, a sound wave was emitted in the direction of the test specimen; and then the reflected sound energy was determined.

Figure 2 shows how much sound was absorbed by the individual samples at a frequency ranging from 125 to 4000 Hz, without wall clearance.

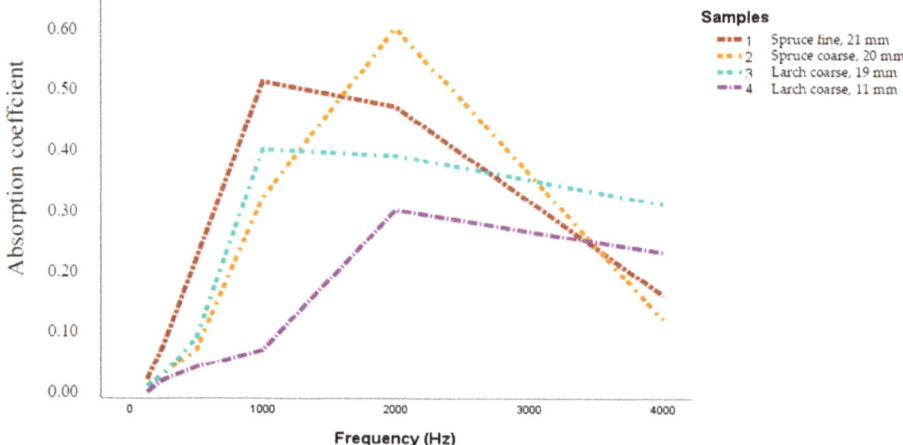

Figure 2. Sound absorption coefficient for spruce and larch bark insulation boards, measured without wall clearance.

The thickness of the board plays an important role for the sound absorption coefficient (α), considering here the samples with 19, 20 and 21 mm, compared with the one with 11 mm. The contribution of bark insulation boards to a better sound absorption can be observed in the frequency range of 1000–2000 Hz [9,42,62] when α increases significantly and has a peak at 2000 Hz, at $\alpha = 0.61$ for the 21 mm sample of spruce board at a density level of 414 kg/m^3, compared to $\alpha = 0.31$ for the 11 mm larch bark sample with a density of 693 kg/m^3. Two other peaks were recorded at 1000 Hz for the spruce bark board ($\alpha = 0.52$) made with fine particles (500 kg/m^3) and for the larch bark thick board ($\alpha = 0.41$) with coarse particles (571 kg/m^3). After 2000 Hz α decreases for all testing specimens, meaning that the boards are able to absorb noise at a level of absorption coefficient smaller than 0.62. This experiment shows that the sound absorbing properties of bark based insulating materials can be enhanced by reducing density and increasing thickness. These results are in compliance with Arenas and Crocker [16] and McMullan [17]. Based on these studies, materials that possess a high value of sound absorption are usually porous materials. Sound absorption behavior observed for low-density particleboard showed that this board had higher porosity compared to medium-density particleboard.

In case of using wall clearance (Figure 3), the sound absorption effect is high at low frequencies, ranging from 125 Hz (very close values, from 0.65 to 0.7). The peak is reached by the 20 mm thick sample made of spruce, with a level of α = 0.79 at 250 Hz. After this frequency, the absorption coefficient decreases abrupt until 1000 Hz, with a short increment up to 0.44 for spruce bark sample with fine-grained particle size (8–13 mm). After that α ranges from 0.36 to 0.06 at 4000 Hz. The lowest value means almost no sound absorption.

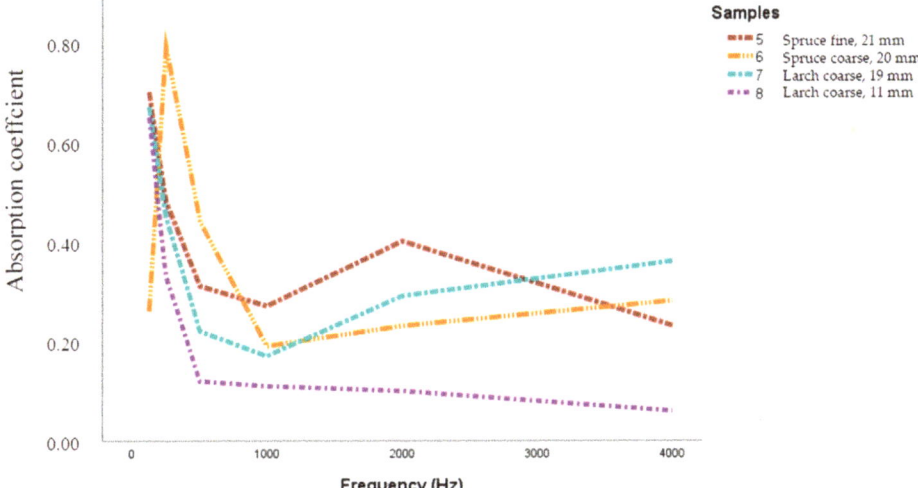

Figure 3. Sound absorption coefficient for spruce and larch bark insulation boards, measured with wall clearance.

Smardzewski et al. [9] analyzed the sound absorption of 17 different wood-based materials. For a thickness of 7.9 < x < 11 mm, the highest absorption coefficients were recorded between 1000 Hz and 2000 Hz for honeycomb (T07, for a 10 mm thickness: α = 0.10 (1000 Hz), α = 0.25 (2000 Hz)), honeycomb + veneer (T05, for a 9.8 mm thickness: α = 0.09 (1000 Hz), α = 0.25 (2000 Hz)), honeycomb + oak + texture (T06, for a 10.5 thickness: α = 0.10 (1000 Hz), α = 0.24 (2000 Hz)) and honeycomb + Lloyd loom mat (T03, for a 10.2 mm thickness: α = 0.11 (1000 Hz), α = 0.23 (2000 Hz)), compared to α = 0.08 and α = 0.31 at 1000, respectively 2000 Hz for the 11 mm larch bark board, that performed better for this interval of frequencies (Figure 4).

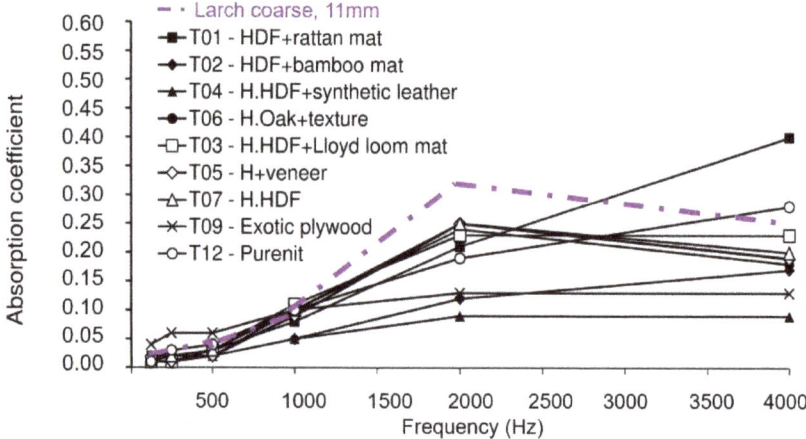

Figure 4. Dependence of the sound absorption coefficient on frequency for wood-based composites with a thickness 7.9 < x < 11 mm (after Smardzewski et al. [9]).

In the case of medium thicknesses (16 < x < 18.4mm), the wood-based materials with densities from 481 kg/m^3 (T10, particleboard), 515 kg/m^3 (T08, poplar plywood), 558 kg/m^3 (T11, medium density fibreboard, MDF) and 613 kg/m^3 (T13, oriented strand board) did not reach sound absorption coefficients higher than 0.15 (Figure 5). The spruce bark samples with similar density levels (414 kg/m^3 for spruce coarse and 500 kg/m^3 for spruce fine) can absorb sound better at frequencies between 250 and 4000 Hz, with two peaks at 0.52 (spuce fine, at 1000 Hz) and 0.61 (spruce coarse, at 2000 Hz).

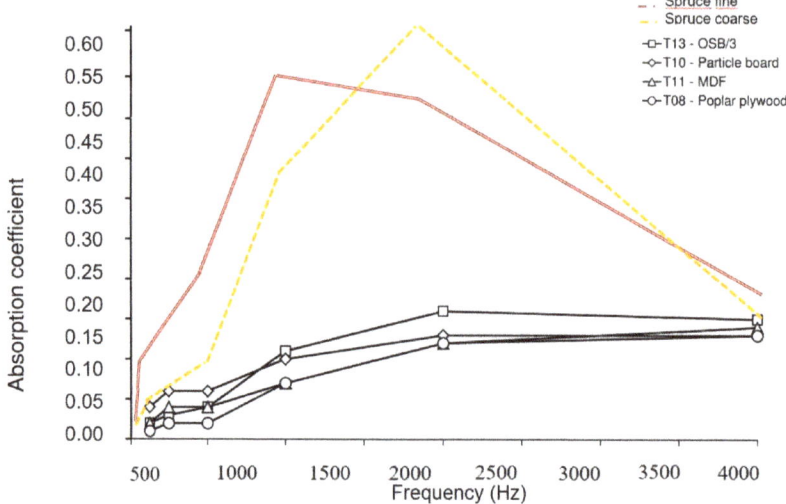

Figure 5. Dependence of the sound absorption coefficient on frequency for wood-based composites with a thickness 16 < x < 18.4 mm and a density 500 < y < 600 kg/m^3 (after Smardzewski et al. [9]).

For thickness interval 22.3 < x < 28 mm, the wood-based materials with densities that raged from 220 kg/m^3 (T17, tubular particle board), 270 kg/m^3 (T14, DendroLight—core), 459 kg/m^3 (T15, DendroLight—planked with HDF) and 493 kg/m^3 (T16, DendroLight—planked with birch plywood) performed better to the larch bark board with a thickness of 19.3 mm (Figure 5). The best values of α

coefficient (0.06, 0.24, 0.56, 0.64, 0.5 and 0.36) were recorded for the T17 panels, with the best acoustic absorbability for all frequencies for the interval between 125 and 4000 Hz. At 1 kHz the 19 mm larch bark panel recorded an α coefficient with 35% smaller than the T17 tubular particleboard, and at 2 KHz was 20% reduced, but these results are consistent with the data shown in Figure 6. These outputs correspond with experimental data of Karlinasari et al. [63], Yang et al. [64], Zulkifli et al. [65,66] and Smardzewski et al. [67], where the maximum sound absorption coefficient was achieved by wood and wood composites at middle frequencies (about 1500–3000 Hz). These results are also in compliance to authors [9,14,68–70], in that the best sound absorption capabilities of wood based composites were achieved by the samples with low density of surface layers and high porosity.

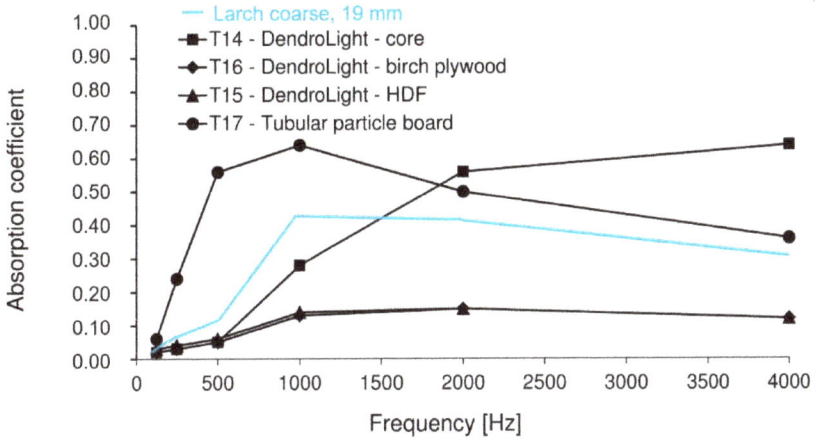

Figure 6. Dependence of the sound absorption coefficient on frequency for wood-based composites with a thickness 22.3 < x < 28 mm and a density 220 < y < 500 kg/m^3 (after Smardzewski et al. [9]).

The human ear is most sensitive to noise at central frequencies [71], therefore bark-based insulation panels can be recommended as sound insulation applications.

4. Conclusions

This paper reveals that the sound insulation properties of the tested larch and spruce bark panels open a new subject related to the advantages of the use of wood biomass. The thickness of the board plays an important role for the sound absorption coefficient.

The sound absorption coefficient for the panel with a thickness of 19 mm were compared with wood-based composites with similar thicknesses and it was found that the spruce bark panels (with coarse and fine particles) can absorb sound even better than OSB, particleboard, MDF and poplar plywood, therefore the values of the sound absorption coefficient confine their application as structural elements for reducing the noise effects in residential buildings, and open new ways for a deeper research in this field.

The spruce composite panels with densities lower than 500 kg/m^3 are able to attenuate up to 60% more sound, compared to densified larch agglomerated boards (600, respectively 700 kg/m^3).

In case of using a clearance wall (Figure 3), the sound absorption effect is high (0.7 for the panels with spruce with fine-grained particles and with larch bark) at low frequencies, starting from 125 Hz. These findings encourage further investigations about sound insulation properties of the bark-based insulation panels.

This research brings an expanding knowledge in the area of composites based on softwood bark and emphasizes on the advantages of an underrated material, namely tree bark, that can substitute wood and other materials that involve more grey energy that affect their life cycle assessment.

Author Contributions: Funding acquisition and data analysis, A.D. and E.M.T.; experiment design, G.K. and A.D.; investigation, A.D.; resources: A.D., G.K., M.C.B.; supervision, R.R.; writing and proofreading, E.M.T., M.C.B., and Ľ.K. All authors have read and agreed to the published version of the manuscript.

Funding: This research was supported by the Slovak Research and Development Agency under contracts no. APVV-17-0583 and APVV-18-0378, and VEGA 1/0717/19.

Acknowledgments: We would like to express our special thanks to Jerzy Smardzewski for his support with the diagrams of experimental data from reference [9]. Many thanks to Graggaber Co. in Unternberg for providing the bark, Untha Co. in Kuchl for size reduction of bark and to MetaDynea Co., Krems, Austria for providing the adhesives. Many thanks to Krämer & Co., Berlin, Germany, for the measurements of sound absorption coefficient in impedance tube.

Conflicts of Interest: The authors declare no conflict of interest

References

1. Bohatkiewicz, J. Noise Control Plans in Cities–Selected Issues and Necessary Changes in Approach to Measures and Methods of Protectin. *Transp. Res. Procedia* **2016**, *14*, 2744–2753. [CrossRef]
2. Murphy, E.; King, E.A. Strategic environmental noise mapping: Methodological issues concerning the implementation of the EU Environmental Noise Directive and their policy implications. *Environ. Int.* **2010**, *36*, 290–298. [CrossRef] [PubMed]
3. Godshall, W.; Davis, J. *Acoustical Absorption Properties of Wood-Based Panel Materials*; USDA Forest Service: Madison, WI, USA, 1969.
4. Casas-Ledón, Y.; Salgado, K.D.; Cea, J.; Arteaga-Pérez, L.E.; Fuentealba, C. Life cycle assessment of innovative insulation panels based on eucalyptus bark fibers. *J. Clean. Prod.* **2020**, *249*, 119356. [CrossRef]
5. Moszynski, P. Who warns noise pollution is a gowing harard health in Europe. *BMJ* **2011**, *342*, d2114. [CrossRef]
6. Oudin, A. Short review: Air pollution, noise and lack of greenness as risk factors for Alzheimer's disease–epidemiologic and experimental evidence. *Neurochem. Int.* **2020**, *134*, 104646. [CrossRef]
7. Klompmaker, J.O.; Janssen, N.A.; Bloemsma, L.D.; Gehring, U.; Wijga, A.H.; Brink, C.V.D.; Lebret, E.; Brunekreef, B.; Hoek, G. Residential surrounding green, air pollution, traffic noise and self-perceived general health. *Environ. Res.* **2019**, *179*, 108751. [CrossRef]
8. Pépiot, E. Male and female speech: A study of mean f0, f0 range, phonation type and speech rate in Parisian French and American English speakers. *Speech Prosody* **2014**, *7*, 305–309. Available online: https://halshs.archives-ouvertes.fr/halshs-00999332 (accessed on 23 February 2020).
9. Smardzewski, J.; Kamisiński, T.; Dziurka, D.; Mirski, R.; Majewski, A.; Flach, A.; Pilch, A. Sound absorption of wood-based materials. *Holzforschung* **2015**, *69*, 431–439. [CrossRef]
10. Hornsby, B.; Ricketts, T. The effects of hearing loss on the contribution of high- and low-frequency speech information to speech understanding. II. Sloping hearing loss. *J. Acoust. Soc. Am.* **2006**, *119*, 1752–1763. [CrossRef]
11. Peng, L. Sound absorption and insulation functional composites. In *Advanced High Strength Natural Fibre Composites in Construction*; Woodhead Publishing: Cambridge, UK, 2017; pp. 333–373.
12. Tsalagkas, D.; Börcsök, Z.; Pásztory, Z. Thermal, physical and mechanical properties of surface overlaid bark-based insulation panels. *Eur. J. Wood Prod.* **2019**, *77*, 721–730. [CrossRef]
13. Pásztory, Z.; Börcsök, Z.; Tsalagkas, D. Density optimization for the manufacturing of bark-based thermal insulation panels. In *IOP Conference Series: Earth and Environmental Science, Proceedings of the 5th International Conference on Environment and Renewable Energy, Ho Chi Minh City, Vietnam, 25–28 February 2019*; IOP Publishing Ltd.: Bristol, UK, 2019; Volume 307.
14. Karlinasari, L.; Hermawan, D.; Maddu, A.; Martianto, B.; Lucky, I.K.; Nugroho, N.; Hadi, Y.S. Acoustical properties of particleboards made from betung bamboo as building construction material. *Bioresources* **2012**, *7*, 5700–5709. [CrossRef]
15. Bucur, V. *Acoustic of Wood*, 2nd ed.; Springer: Berling/Heidelberg, Germany, 2006; pp. 7–36.
16. Arenas, J.; Crocker, M.J. Recent trends in porous sound-absorbing materials. *Sound Vib.* **2010**, *44*, 12–18.
17. McMullan, R. *Environmental Science in Building*, 5th ed.; Palgrave: New York, NY, USA, 2002; pp. 200–251.
18. Mehta, M.; Johnson, J.; Rocafort, J. *Architectural Acoustics Principles and Design*; Pearson: London, UK, 1999; pp. 301–306.

19. Jiang, N.; Chen, J.Y.; Parikh, D.V. Acoustical evaluation of carbonized and activated cotton nonwovens. *Bioresour. Technol.* **2009**, *100*, 6533–6536. [CrossRef] [PubMed]
20. Krucińska, I.; Gliścińska, E.; Michalak, M.; Ciechańska, D.; Kazimierczak, J.; Bloda, A. Sound-absorbing green composites based on cellulose ultrashort/ultra-fine fibers. *Text Res. J.* **2014**, *85*, 646–657. [CrossRef]
21. Shahani, F.; Soltani, P.; Zarrebini, M. The analysis of acoustic characteristics and sound absorption coefficient of needle punched nonwoven fabrics. *J. Eng. Fibers Fabr.* **2014**, *9*, 84–92. [CrossRef]
22. Rwawiire, S.; Tomkova, B.; Gliscinska, E.; Krucinska, I.; Michalak, M.; Militky, J.; Jabbar, A. Investigation of sound absorption properties of bark cloth nonwoven fabric and composites. *Autex Res. J.* **2015**, *15*, 173–180. [CrossRef]
23. Berge, B. *The Ecology of Building Materials*, 2nd ed.; Elsevier: Oxford, UK, 2001; p. 427.
24. Cao, L.; Fua, Q.; Yang, S.; Dinga, B.; Yu, J. Porous materials for sound absorption. *Compos. Commun.* **2018**, *10*, 25–35. [CrossRef]
25. Barbu, M.C.; Reh, R.; Çavdar, A.D. *Non-Wood Lignocellulosic Composites. Materials Science and Engineering: Concepts, Methodologies, Tools, and Applications*; IGI Global: Hershey, PA, USA, 2017. [CrossRef]
26. Réh, R.; Igaz, R.; Krišťák, Ľ.; Ružiak, I.; Gajtanska, M.; Božíková, M.; Kučerka, M. Functionality of Beech Bark in Adhesive Mixtures Used in Plywood and Its Effect on the Stability Associated with Material Systems. *Materials* **2019**, *12*, 1298. [CrossRef]
27. Hassan, T.; Jamshaid, H.; Mishra, R.; Khan, M.Q.; Petru, M.; Novak, J.; Choteborsky, R.; Hromasova, M. Acoustic, Mechanical and Thermal Properties of Green Composites Reinforced with Natural Fibers Waste. *Polymers* **2020**, *12*, 654. [CrossRef]
28. Danihelová, A.; Němec, M.; Gergel, T.; Gejdoš, M.; Gordanová, J.; Sčensný, P. Usage of recycled technical textiles as thermal insulation and an acoustic absorber. *Sustainability* **2019**, *11*, 2968. [CrossRef]
29. Němec, M.; Igaz, R.; Gergel, T.; Danihelová, A.; Ondrejka, V.; Krišťák, L.; Gejdoš, M.; Kminiak, R. Acoustic and thermophysical properties of insulation materials based on wood wool. *Akustika* **2019**, *33*, 115–123.
30. Tudor, E.M.; Barbu, M.C.; Petutschnigg, A.; Réh, R.; Krišťák, Ľ. Analysis of Larch-Bark Capacity for Formaldehyde Removal in Wood Adhesives. *Int. J. Environ. Res. Public Health* **2020**, *17*, 764. [CrossRef] [PubMed]
31. Bekhta, P.; Sedliačik, J. Environmentally-Friendly High-Density Polyethylene-Bonded Plywood Panels. *Polymers* **2019**, *11*, 1166. [CrossRef] [PubMed]
32. Balali, A.; Hakimelahi, A.; Valipour, A. Identification and prioritization of passive energy consumption optimization measures in the building industry: An Iranian case study. *J. Build. Eng.* **2020**, *30*, 101239. [CrossRef]
33. Grazieschi, G.; Gori, P.; Lombardi, L.; Asdrubali, F. Life cycle energy minimization of autonomous buildings. *J. Build. Eng.* **2020**, *30*, 101229. [CrossRef]
34. Koezjakov, A.; Urge-Vorsatz, D.; Crijns-Graus, W.; van den Broek, M. The relationship between operational energy demand and embodied energy in Dutch residential buildings. *Energy Build.* **2018**, *165*, 233–245. [CrossRef]
35. Lorincová, S.; Hitka, M.; Štarchoň, P.; Stachová, K. Strategic Instrument for Sustainability of Human Resource Management in Small and Medium-Sized Enterprises Using Management Data. *Sustainability* **2018**, *10*, 3687. [CrossRef]
36. Jing, W. Life-Cycle Assessment of Envelope Structure of Typical Residential Buildings in Cities and Towns in Severe Cold Areas. Ph.D. Thesis, Tongji University, Shanghai, China, 2005.
37. Brás, A.; Gomes, V. LCA implementation in the selection of thermal enhanced mortars for energetic rehabilitation of school buildings. *Energy Build.* **2015**, *92*, 1–9. [CrossRef]
38. Yang, Q.; Kong, L.; Tong, H.; Wang, X. Evaluation Model of Environmental Impacts of Insulation Building Envelopes. *Sustainability* **2020**, *12*, 2258. [CrossRef]
39. Silvestre, J.D.; Pargana, N.; De Brito, J.; Pinheiro, M.D.; Durão, V. Insulation Cork Boards—Environmental Life Cycle Assessment of an Organic Construction Material. *Materials* **2016**, *9*, 394. [CrossRef]
40. Nowoświat, A.; Dulak, L. Impact of Cement Dust Pollution on the Surface of Sound-Absorbing Panels on Their Acoustic Properties. *Materials* **2020**, *13*, 1422. [CrossRef]
41. Hýsek, Š.; Neuberger, P.; Sikora, A.; Schönfelder, O.; Ditommaso, G. Waste Utilization: Insulation Panel from Recycled Polyurethane Particles and Wheat Husks. *Materials* **2019**, *12*, 3075. [CrossRef] [PubMed]

42. Stanciu, M.; Curtu, I.; Cosereanu, C.; Vasile, O.; Olarescu, C. Evaluation of Absorption Coefficient of Biodegradable Composite Materials with Textile Inserts. *Rom. J. Acoust. Vib.* **2011**, *2*, 99–102.
43. Tong, X.; Wang, T.; Chen, Y.; Wang, Y. Towards an inclusive circular economy: Quantifying the spatial flows of e-waste through the informal sector in China. *Resour. Conserv. Recycl.* **2018**, *135*, 163–171. [CrossRef]
44. Fan, Y.; Lee, C.; Lim, J.; Klemeš, J. Cross-disciplinary approaches towards smart, resilient and sustainable circular economy. *J. Clean. Prod.* **2019**, *232*, 1482–1491. [CrossRef]
45. Anastasiades, K.; Blom, J.; Buyle, M.; Audenaerta, A. Translating the circular economy to bridge construction: Lessons learnt from a critical literature review. *Renew. Sustain. Energy Rev.* **2020**, *117*, 1–11. [CrossRef]
46. *EN 15149-1:2011 Solid Biofuels-Determination of Particle size Distribution—Part 1: Oscillat-ing Screen Method Using Sieve Apertures of 1 mm and Above, -Test Method*; CEN, European Committee for Standardization: Brussels, Belgium, 2011.
47. Tudor, E.; Barbu, M.; Petutschnigg, A.; Réh, R. Added-value for wood bark as a coating layer for flooring tiles. *J. Clean. Prod.* **2017**, *180*, 1354–1360. [CrossRef]
48. Kain, G.; Lienbacher, B.; Barbu, M.; Senck, S.; Petutschnigg, A. Water vapour diffusion resistance of larch (Larix decidua) bark insulation panels and application considerations based on numeric modelling. *Constr. Build. Mater.* **2018**, *164*, 308–316. [CrossRef]
49. Kain, G.; Lienbacher, G.; Barbu, M.C.; Richter, K.; Petutschnigg, A. Larch (Larix decidua) bark insulation board: Interactions of particle orientation, physical-mechanical and thermal properties. *Eur. J. Wood Wood Prod.* **2018**, *76*, 489–498. [CrossRef]
50. Kain, G.; Barbu, M.C.; Richter, K.; Plank, B.; Tondi, G.; Petutschnigg, A. Use of tree bark as insulation material. *For. Prod. J.* **2015**, *65*, S16–S25.
51. Kain, G.; Stratev, D.; Tudor, E.; Lienbacher, B.; Weigl, M.; Barbu, M.C.; Petutschnigg, A. Qualitative investigation on VOC-emissions from spruce (Picea abies) and larch (Larix decidua) loose bark and bark panels. *Eur. J. Wood Wood Prod.* **2020**, *78*, 403–412. [CrossRef]
52. Kain, G.; Charwat-Pessler, J.; Barbu, M.C.; Plank, B.; Richter, K.; Petutschnigg, A. Analyzing wood bark insulation board structure using X-ray computed tomography and modeling its thermal conductivity by means of finite difference method. *J. Compos. Mater.* **2016**, *50*, 795–806. [CrossRef]
53. *EN 326-1:1994: Wood-Based Panels-Sampling, Cutting and Inspection—Part 1: Sampling and Cutting of Test Pieces and Expression of Test Results, -Test Method*; CEN, European Committee for Standardization: Brussels, Belgium, 1994.
54. *ISO16999, Wood-Based Panels—Sampling and Cutting of Test Pieces*; International Organization for Standardization: Geneva, Switzerland, 2003.
55. *EN 322:2005 Wood-Based Panels-Determination of Moisture Content, -Test Method*; CEN, European Committee for Standardization: Brussels, Belgium, 2005.
56. *ISO 10534-2:1998: Acoustics-Determination of Sound Absorption Coefficient and Impedance in Impedance Tubes—Part 2: Transfer-Function Method*; ISO/TC 43/SC2 Building Acoustics; CEN, European Committee for Standardization: Brussels, Belgium, 1998.
57. Stegmaier, K. Akustikbüro Krämer&Stegmaier, Ingenieurbüro für Schallschutz und Technische Akustik. 2018. Available online: http://www.akustik-berlin.de/ (accessed on 14 April 2018).
58. Dettendorfer, A. Hetta. Ein Ressourcenschonendes Möbel für IKEA. Bachelor's Thesis, Salzburg University of Applied Sciences, Kuchl, Austria, 2016.
59. Niresh, J.A.; Neelakrishnan, S.; Subharani, S.; Kannaian, T.; Prabhakaran, R. Review of acoustic characteristics of materials using impedance tube. *ARPN J. Eng. Appl. Sci.* **2015**, *10*, 3319–3326.
60. Kundt, A. Acoustic Experiments. *Philos. Mag. J. Sci.* **2009**, *35*, 41–48. [CrossRef]
61. Shahani, F.; Soltani, P.; Zarrebini, M. The analysis of acoustical characteristics and sound absorption coefficient of woven fabrics. *Text. Res. J.* **2012**, *82*, 875–882. [CrossRef]
62. Na, Y.; Agnhage, T.; Cho, G. Sound Absorption of Multiple Layers of Nanofiber Webs and the Comparison of Measuring Methods for Sound Absorption Coefficients. *Fibers Polym.* **2012**, *13*, 1348–1352. [CrossRef]
63. Karlinasari, L.; Hermawan, D.; Maddu, A.; Martiandi, B.; Hadi, Y.S. Development of particleboard for acoustical panel from tropical fast growing species. *J. Trop. For. Sci.* **2012**, *24*, 64–69.
64. Yang, H.S.; Kim, D.J.; Kim, H.J. Rice straw-wood particle composite for sound absorbing wooden construction materials. *Bioresour. Technol.* **2003**, *86*, 117–121. [CrossRef]

65. Zulkifli, R.; Mohd Nor, M.J.; Mat Tahir, M.F.; Ismail, A.R.; Nuawi, M.Z. Acoustic properties of multi-layer coir fibers sound absorption panel. *J. Appl. Sci.* **2008**, *8*, 3709–3714. [CrossRef]
66. Zulkifli, R.; Nor, M.J.M. Noise control using coconut coir fiber sound absorber with porous layer backing and perforated panel. *Am. J. Appl. Sci.* **2010**, *7*, 260–264. [CrossRef]
67. Smardzewski, J.; Batko, W.; Kamisinski, T.; Flach, A.; Pilch, A.; Dziurka, D.; Mirski, R.; Roszyk, E.; Majewski, A. Experimental study of wood acoustic absorption characteristics. *Holzforschung* **2013**, *68*, 160. [CrossRef]
68. Wassilieff, C. Sound absorption of wood-based materials. *Appl. Acoust.* **1996**, *48*, 339–356. [CrossRef]
69. Ghofrani, M.; Ashori, A.; Mehrabi, R. Mechanical and acoustical properties of particleboards made with date palm branches and vermiculite. *Polym. Test.* **2017**, *60*, 153–159. [CrossRef]
70. Gergel, T.; Danihelova, A.; Danihelova, Z. Acoustic comfort in wooden buildings made from cross laminated timber. *Akustika* **2016**, *25*, 29–37.
71. Ilgun, A.; Cogurcu, M.T.; Ozdemir, C.; Kalipci, E.; Sahinkaya, S. Determination of sound transfer coefficient of boron added waste cellulosic and paper mixture panels. *Sci. Res. Essay* **2010**, *5*, 1530–1535.

© 2020 by the authors. Licensee MDPI, Basel, Switzerland. This article is an open access article distributed under the terms and conditions of the Creative Commons Attribution (CC BY) license (http://creativecommons.org/licenses/by/4.0/).

Article

Engineering Composites Made from Wood and Chicken Feather Bonded with UF Resin Fortified with Wollastonite: A Novel Approach

Hamid R. Taghiyari [1,*], **Roya Majidi** [2], **Ayoub Esmailpour** [2], **Younes Sarvari Samadi** [3], **Asghar Jahangiri** [1] **and Antonios N. Papadopoulos** [4,*]

1. Wood Science and Technology Department, Faculty of Materials Engineering and New Technologies, Shahid Rajaee Teacher Training University, Tehran 1678815811, Iran; mohamad.mj8@yahoo.com
2. Department of Physics, Faculty of Sciences, Shahid Rajaee Teacher Training University, Tehran 1678815811, Iran; r.majidi@sru.ac.ir (R.M.); esmailpour@sru.ac.ir (A.E.)
3. Faculty of Wood Technology and Construction, Rosenheim University of Applied Sciences, Rosenheim 83024, Germany; younes.sarvari@gmail.com
4. Laboratory of Wood Chemistry and Technology, Department of Forestry and Natural Environment, International Hellenic University, GR-661 00 Drama, Greece
* Correspondence: htaghiyari@sru.ac.ir (H.R.T.); antpap@for.ihu.gr (A.N.P.)

Received: 19 March 2020; Accepted: 6 April 2020; Published: 7 April 2020

Abstract: Wood-composite panel factories are in shortage of raw materials; therefore, finding new sources of fibers is vital for sustainable production. The effects of chicken feathers, as a renewable source of natural fibers, on the physicomechanical properties of medium-density fiberboard (MDF) and particleboard panels were investigated here. Wollastonite was added to resin to compensate possible negative effects of chicken feathers. Only feathers of the bodies of chickens were added to composite matrix at 5% and 10% content, based on the dry weight of the raw material, particles or fibers. Results showed significant negative effects of 10%-feather content on physical and mechanical properties. However, feather content of 5% showed some promising results. Addition of wollastonite to resin resulted in the improvement of some physical and mechanical properties. Wollastonite acted as reinforcing filler in resin and improved some of the properties; therefore, future studies should be carried out on the reduction of resin content. Moreover, density functional theory (DFT) demonstrated the formation of new bonds between wollastonite and carbohydrate polymers in the wood cell wall. It was concluded that chicken feathers have potential in wood-composite panel production.

Keywords: engineering materials; composite panels; chicken feather; cell-wall polymers; thermal conductivity coefficient; wollastonite; wood; natural materials

---

1. Introduction

Fast-growing wood species are often used in the manufacture of composite panels, engineered and modified wood, and paper industries [1,2]. Therefore, their use is of advantage since they offer a homogeneous structure which is of great importance for many general and specific purposes [3–5]. Composite manufacturing factories have always been confronted with some ongoing issues, such as the emission of formaldehyde, heat transfer to the core of the mat, vulnerability to vapor, and biological susceptibility to fungi and insects [6–10]. Moreover, numerous studies have been focused on the limitation of formaldehyde emission and on the improvement of the resin bond [11]. The heat-transferring properties of metals and improving effects of different materials at micro- and nano-scales [12–17] were also found to decrease hot press time and to improve the physicomechanical properties in wood composites [6,18]. Under this frame, wollastonite (as a silicate mineral, $CaSiO_3$) was found to improve the biological and physicomechanical properties of both solid wood and wood based panels, as well as to improve the fire retardancy and to increase thermal conductivity coefficient in medium-density fiberboards (MDF) [19–26], therefore, the first aim of the present study was to find out possible effects that wollastonite may have on physical and mechanical properties of two engineering wood composites, namely medium-density fiberboards and particleboards. Based on potential positive results of the addition of wollastonite on properties of composite panels in the present study, future studies on decreasing urea-formaldehyde (UF) resin content, or even using an eco-friendly resin within a green framework, would be predictable and should be carried out, similar to what was previously achieved by the application of tannin in wood-composite panels [27–32].

At the same time, Iranian wood-composite manufacturing factories confront the problem of shortages in wood fiber or particle resources to maintain sustainable production, therefore, potential natural fibers should be considered in order to meet the constant need for raw materials. In this way, numerous chicken farms exist in Iran, and therefore a huge amount of chicken feathers are in stock. It is reported by the Ministry of Agriculture of Iran that the production of chicken feathers in 2012 was about 80,000 metric tons; this figure corresponds to a manufacture of approximately 20 million composite panels by incorporating 5% feather content in to the wood furnishes. It also has to be mentioned that two million metric tons of chicken feathers are produced annually in the United States [26–33] whereas a figure of 3.1 million tons of feather waste is reported for the European Union [34–36]. Nowadays, high amounts of chicken feathers are disposed of in landfills and only a very small portion are converted into low-nutritional-value animal food [34–36]. This solution does not utilize the potential that this neglected material possesses, and more importantly, the management of environmental and health concerns becomes more difficult as overall waste rises. As chicken feathers are considered a waste raw material, it may be a cheap and renewable source for wood-composite industry.

It is reported that chicken feathers were used as a reinforcement in manufacturing wood-cement composites, however no improvement in physicomechanical properties was found [37]. It is known that feathers consist of half quill and half fiber, by weight in approximate [33], which in turn consists of the hydrophobic protein keratin, which presents strength similar to that of nylon with a diameter smaller than that of the wood fiber. It is also worth mentioning that its covalent bonds stabilize the three-dimensional protein structure that is hard to break [38–40].

In the present study, chicken feathers were applied to the mat at a 5%- and 10%-dry-weight basis of wood fibers in the present research project. This approach would contribute to a more efficient use of natural resources and take advantage of this material that is produced in huge amounts and is currently underutilized by the poultry industry. It is intended that the use of materials from renewable resources contribute to sustainability and a reduction in the environmental impact associated with the incineration or disposing of poultry feathers into landfills. They are cheap, low density, abundantly available and renewable, delivering strong and stiff fibers, intrinsic characteristics of vital importance for the valorization of this waste for reinforcing material in composite materials.

The present study was, therefore, primarily carried out to find a new source of natural fibers to feed the MDF-manufacturing factories in Iran which are greatly suffering from a shortage of raw materials (natural wood fibers). For this purpose, urea-formaldehyde resin (UF) was used because melamine-urea-formaldehyde (MUF) and other resins are neither popular in Iran's market nor economical for the composite factories [6,20,21]. It should also be noted that the hydroxyl-groups of serine amino acids in feather fibers could possibly bind to wood fibers, contributing to the physical and mechanical properties of the MDF panels produced [24,33].

The separation of quill and feather-fiber was estimated to be costly and no wood composite manufacturing factories in Iran could afford such extra expenses. Therefore, in this study, the whole feather (quills and feather-fibers together) was used so that any potential positive results could directly be used at industrial scale. However, the quills of the wing feathers are not flexible and they created major problems in the preliminary tests. Therefore, only the body feather of chickens were used in the present study, considering their flexibility and small size of quills.

2. Materials and Methods

2.1. Specimen Procurement

Wood fibers were procured from Sanaye Choobe Khazar Company in Amol of Iran (MDF Caspian Khazar). The fibers consisted of a mixture of five species, namely beech (*Fagus orientalis*), alder (*Alnus glutinosa*), maple (*Acer hyrcanum*), hornbeam (*Carpinus betulus*) and poplar (mostly *Populus nigra*) species from local forests (Amol, Iran). The target board thickness was 16 mm and the target density was 0.67 g/cm^3. The temperature and the total nominal pressure of the plates were 175 °C and 160 bars respectively, whereas the press time was six minutes. Urea-formaldehyde resin (UF) was procured from Pars Chemical Industries Company, Tehran, Iran. UF content was 10% with 200–400 cP in viscosity, 47 s of gel time, and 1.277 g/cm^3 in density. Produced panels were conditioned (25 °C, and 40% ± 3% relative humidity) for three weeks before testing. The moisture content of the board specimens at the time of testing was 7.5%. Five replicate panels were produced for each treatment. The board manufacture parameters are summarized in Table 1.

Table 1. Board manufacture parameters.

Board Density	0.67 g/cm^3
Board Thickness	16 mm
Press Temperature	175 °C
Press Time	6 min
Pressure of Plates	160 bars
Resin Type and Content	10% urea formaldehyde (UF) resin
Resin Characteristics	200–400 cP in viscosity, 47 s of gel time, and 1.277 g/cm^3 in density.
Wollastonite Content	10% of UF resin (based on the dry weight of the resin)

Wood chips were procured from Shahid Dr. Bahonar Composite-board Company (Gorgan, Iran) to produce particleboards. The chips comprised the same species as were used for wood fibers mentioned above; only a 5%–7% pruning branches of the fruit gardens was added. Boards were 16 mm in thickness and 0.67 g/cm^3 in density; density was kept constant for all treatments. The same resin and production conditions were used in particleboard manufacturing program. Five boards were made for each treatment.

Feathers were procured from a commercial chicken farm located in Tehran, Iran. The preliminary evaluation of the costs revealed that separation of feather fibers from the quills would be costly and not encouraging for composite-manufacturing factories. It was decided that the whole feather would be used in this study so that any possible positive results could be directly used on a commercial scale. Therefore, only the feathers of the body, which are small and flexible enough for MDF production, were mixed with the wood fibers and chips in a drum-mixer to form the wood–chicken feather composite-mat. The length of the feathers ranged from one to three centimeters. The flow diagram of this experimental procedure is presented in Figure 1.

2.2. Wollastonite Application

Wollastonite gel was produced in close cooperation with Mehrabadi Manufacturing Company in Tehran, Iran. Chemical composition of wollastonite used in the present study is presented in Table 2. More than 90% of wollastonite particles ranged 1–4 μm in thickness and width, and 5–25 μm in length. A total of 10% of wollastonite gel was applied, based on the dry weight of resin. Wollastonite was mixed with the UF resin by a magnetic stirrer for 20 min. The mixture of UF + Wollastonite was sprayed on fibers in a rotary drum.

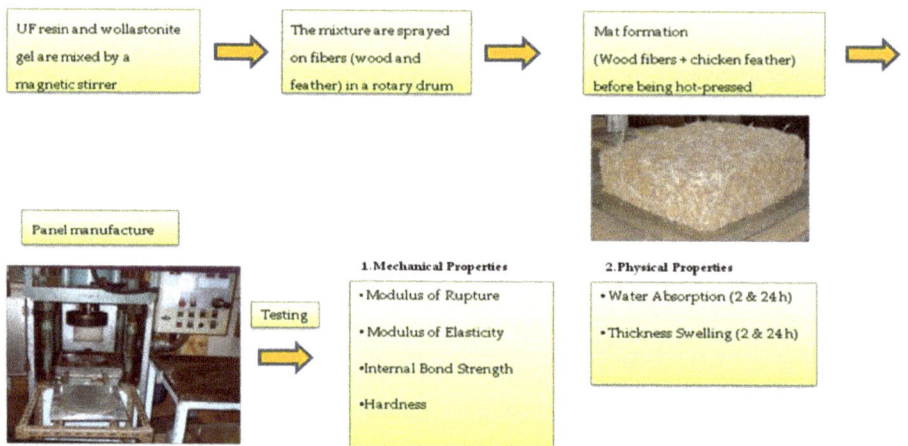

Figure 1. Flow diagram of the experimental procedure.

Table 2. Composition of the wollastonite used in the present study.

Component	Proportion (% w/w)
SiO_2	47.1
CaO	39.9
Al_2O_3	3.9
Fe_2O_3	2.8
TiO_2	0.2
K_2O	0.04
MgO	1.4
Na_2O	0.2
SO_3	0.05
Water	The rest

2.3. Temperature Measurement

A digital thermometer with a sensor probe was applied in order to measure the temperature, with 0.1 °C precision, at the core of the mat, at 5-s intervals (Figure S1, from the Supplementary Materilas). Its 4-mm diameter probe was directly inserted into the core of the mat (from the front edge boarder of the mat), in horizontal direction, for about 50 mm. Temperature measurement was started immediately after the two hot plates reached the stop-bars.

2.4. Physical and Mechanical Properties

Physical and mechanical properties were determined in accordance with the Iranian National Standard ISIRI 9044 PB Type P2 [41] (compatible with ASTM D1037-99) specifications. Mechanical properties were measured, using an INSTRON 4486 test machine. The physical properties included water absorption (WA) and thickness swelling (TS), after 2 and 24 h immersion in water. A digital scale with a 0.01 g precision was used for WA measurement. A digital caliper with a 0.01 mm precision was used for TS measurement. Five mechanical properties were also measured, including modulus of rupture, modulus of elasticity, brittleness, internal bond and hardness at 5.4 mm of penetration. Dimension of the specimens for physical properties (water absorption and thickness swelling), as well as internal bond test, were 50 mm × 50 mm. Dimension of modulus of rupture (MOR) and modulus of elasticity (MOE) specimens was 350 mm × 50 mm; the loading span was 320 mm. Specimens were loaded at a rate of three mm per minute. Brittleness was calculated based on Equation (1), in which the ratio (%) of the work absorbed in the elastic region divided by the total absorbed work is measured [42,43]. Once internal bond specimens were cut, the two faces in each of the test specimens were glued to an aluminum block, using hot-melt adhesive. The blocks were then pulled until failure. For hardness measurement, two specimens of 75 mm × 50 mm, each with a thickness of 16 mm, were bound together to prepare thickness of 32 mm according to the standard specifications. Hardness was then measured using a 11.28 mm diameter modified Janka ball, with a projected impact area of 100 mm^2.

$$Brittleness = \frac{Area1}{Area1 + Area2} \times 100 \ (\%) \quad (1)$$

2.5. Density Functional Theory

For a better understanding of how wollastonite reacted with carbohydrates in wood polymers, some simulations were performed. All simulations were carried out based on density functional theory (DFT) using the OpenMX3.8 package [25,26]. The exchange and correlation potential was described with generalized gradient approximation (GGA) of Perdew–Burke–Ernzerhof (PBE). The long-range Van der Waals interactions were included in the simulations by the DFT-D2 approach. The plane wave cutoff energy was uniformly set to 50 Ry in all three cell wall polymer calculations (cellulose, hemicellulose and lignin).

Adsorption energy, E_{ads}, was calculated by the Equation (2) in order to evaluate the interaction between wollastonite and hemicellulose or lignin,

$$E_{ads} = E_{\text{hemicellulose/lignin + Wollastonite}} - (E_{\text{hemicellulose/lignin}} + E_{\text{Wollastonite}}) \quad (2)$$

where $E_{\text{hemicellulose/lignin + Wollastonite}}$ is the total energy of hemicellulose or lignin with adsorbed W (wollastonite); $E_{\text{hemicellulose/lignin}}$ denotes the total energy of isolated hemicellulose or lignin; and $E_{\text{Wollastonite}}$ is the total energy of the isolated W. The negative adsorption energy represents the stable adsorption structure.

2.6. Modelling of Wollastonite

Wollastonite crystals contain silicate chains along with the chain axis, linked to a periodicity of three tetrahedral. The calcium is linked by irregular octahedral coordination to six of the oxygen [25,26].

2.7. Modelling of Hemicellulose

Hemicellulose is a branched polysaccharide consisting of shorter chains of around 200 sugar units. Twenty percent of the biomass contains hemicellulose molecules derived from different sugar monomers like glucose, xylose, mannose, galactose, rhamnose, and arabinose [44]. The model of hemicellulose introduced by Kaith et al. [44] was elaborated in the present project to evaluate the adsorption of wollastonite and water molecules.

2.8. Modelling of Lignin

Lignin is known as the second most abundant biopolymer on earth. It possesses a high content of aromatic groups. There are three monolignol building blocks in lignin, methoxylated to various degrees: p-coumaryl alcohol, coniferyl alcohol and sinapyl alcohol. These building blocks are incorporated into lignin in the form of the phenyl propanoids derivatives. In the present study, modeling was separately completed based on all three monolignols. This biopolymer contains small amounts of incomplete and modified monolignols, as well as other monomers. There is a wide range of different functional groups in lignin molecules, including aliphatic and aromatic hydroxyl groups, double bonds and phenyl groups [45].

2.9. Statistical Analysis

SAS software program was used to carry out statistical analysis in the present study (version 9.2; 2010, SAS Institute Inc., Cary, NC, USA). To discern significant difference among different treatments and produced panels, one-way analysis of variance was performed at 95% level of confidence. Then, Duncan's multiple range test (DMRT) was completed to group each property among treatments. In order to find degrees of similarities among different treatments based on all properties studied here, hierarchical cluster analysis from SPSS/18 (2010) software was used. For graphical statistics (fitted-line, contour and surface plots), Minitab software was utilized (version 16.2.2; 2010, Minitab Inc., State College, PA, USA).

3. Results and Discussion

3.1. Temperature of the Core of Composite Mats

Measurement of temperature at the core of composite mats revealed a significant difference between MDF panels without wollastonite and the three wollastonite-treated panels (Figure 2A). All treatments showed an almost identical increase up to 90 s; however, the three wollastonite-treated panels showed a clear higher temperature after the first 90 s. This clearly showed the effects of the higher thermal conductivity coefficient in wollastonite-treated panels on the heat transfer to the core section of the composite mat [20].

Measurement of the core section of the particleboard mats showed a significant lower temperature in comparison to the MDF mats (Figure 2B). This can be attributed to the higher contact surface among wood fibers (MDF matrix) in comparison to wood particles (particleboard matrix); that is, the surface-to-surface contact is higher between wood fibers in comparison to the contact between wood particles, so the heat of the hot-press plates could more rapidly be transferred to the core section in MDF mat.

3.2. Physical Properties

Water absorption (WA) was the same in the three MDF-treatments without wollastonite, both for 2 and 24 h immersion in water (Figure S2, from the Supplementary Materials). This showed that addition of feather to the MDF-matrix did not significantly affect the water absorption. Wollastonite-treated panels showed a significant decrease in water absorption for all the three treatments. It was previously reported that wollastonite-treated composite panels had lower gas and liquid permeability [20,26]. In this way, the reinforcement of UF resin by wollastonite caused higher integration of fiber in the composite-matrix, preventing water to easily pass through. Similar reinforcement in resin and paint was previously reported by the addition of wollastonite and graphene [15]. Moreover, the formation of bonds between wollastonite and wood polymers prevented wood hydroxyl groups to be actively involved in making bonds with water molecules [25,26], decreasing WA in all treatments.

In particleboards specimens, the procedure was somehow different; wollastonite decreased water absorption only after 24 h immersion in water (Figure S2, from the Supplementary Materials). The addition of feathers (both 5% and 10% contents) significantly increased water absorption after 2 h of immersion, and wollastonite could not compensate for it, probably because chicken feathers reached their maximum moisture content. However, wollastonite could control WA to some extent after 24-h immersion.

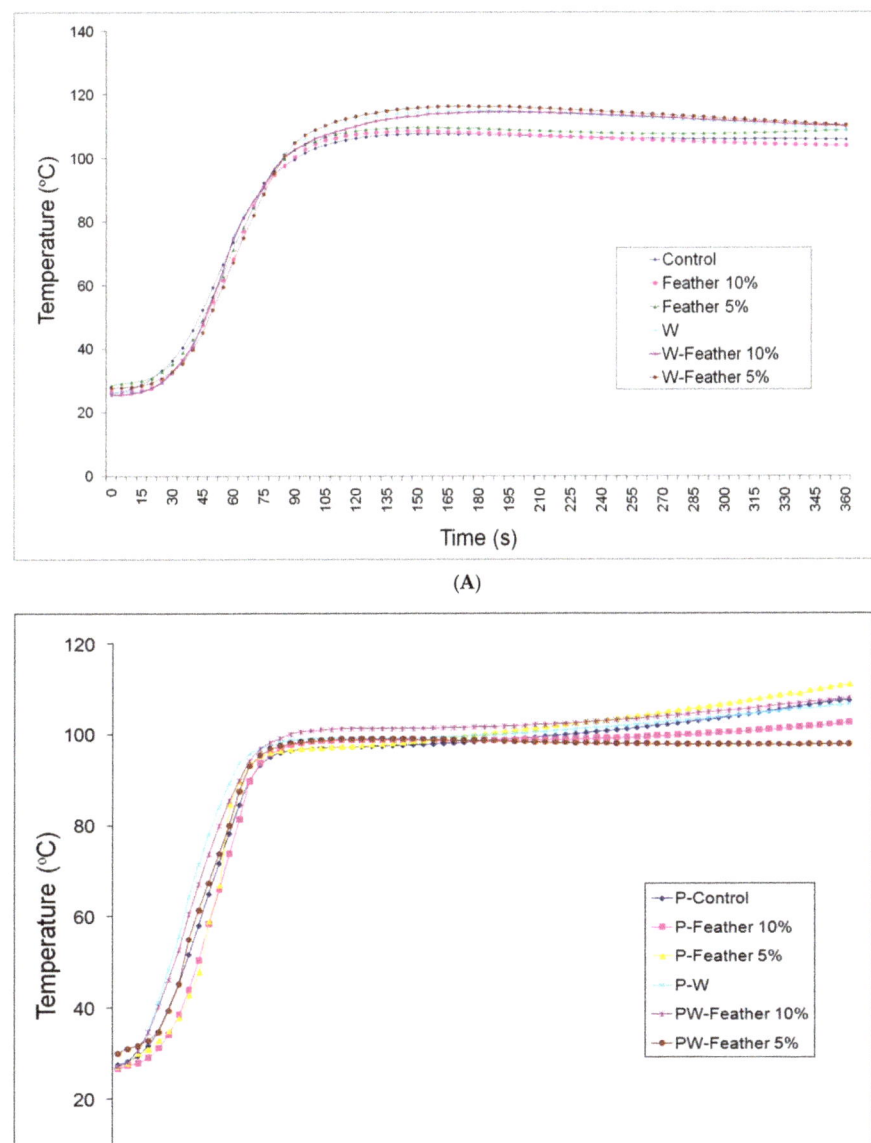

Figure 2. Temperature (Celsius) at the core section of the MDF (**A**), and the particleboard (**B**) at 5-s intervals (P = particleboard; MDF = medium-density fiberboard; W = wollastonite; S = time intervals).

The lowest thickness swelling occurred in wollastonite-treated 5%-feather content MDF panels. The addition of wollastonite or feathers at both 5% and 10% contents generally resulted in a significant decrease in thickness swelling in MDF panels after both 2 and 24 h immersion in water (Figure S3, from the Supplementary Materials). This was attributed to the reinforcing effect of wollastonite and formation of bonds between wollastonite and wood polymers; moreover, the hydrophobic properties of keratin in feathers contributed to this phenomenon.

In particleboard panels, the procedure was different again (Figure S3, from the Supplementary Materials). The highest thickness swelling occurred in 10%-feather content particleboard panels with no wollastonite content. In fact, lower surface-to-surface contact between wood-feather-matrix as well as the voids and spaces between the wood particles provided more opportunity for water to penetrate into the particleboard-matrix; however, by addition of wollastonite to panels, water penetration could be controlled significantly in 10%-feather content wollastonite-panels. In the meantime, the addition of 5% feathers could even improve thickness swelling at 2-h immersion. The lowest thickness swelling values were found in wollastonite-treated particleboard panels.

3.3. Mechanical Properties

The highest modulus of rupture was observed in wollastonite-treated MDF panels without feathers (15.5 MPa) (Figure S4, from the Supplementary Materials). Wollastonite improved modulus of rupture in all panels. This improvement was attributed to reinforcement of resin [15,46], as well as formation of new bonds between wollastonite and wood polymers [25,26]. The addition of feathers to the MDF panels significantly decreased modulus of rupture. In this connection, the level of decrease in 5%-feather content was small in comparison to the high significant decrease that occurred in the 10%-feather content panels. In fact, visible checks and cracks (internal blows) occurred in the core section of the mat in 10%-feather MDF panels (Figure 3). This clearly showed that this amount of feather content would not be suitable for MDP panels made with UF resin, as the UF resin is not compatible with keratin in chicken feathers. These cracks were reported to be the main reason for the higher mass loss values in specimens exposed to fungi attack [24]. It was concluded that 10%-feather content was too high, but 5%-feather could be considered suitable to satisfy the fiber shortage and keep the physical and MOR properties at satisfactory level. The addition of wollastonite decreased MOR in particleboards (Figure S4, from the Supplementary Materials). Only panels with 10% feather content showed an improvement by addition of wollastonite, maybe due to the higher hygroscopicity of chicken feather.

Figure 3. Cracks (blows) (↓) in the core layer of MDF-feather boards (F-10%).

The highest modulus of elasticity was found in wollastonite-treated MDF panels without feathers (1760 MPa) (Figure S5, from the Supplementary Materials). Panels with 5%-feather content showed significant increase in modulus of elasticity in comparison to panels with no feather, proving the elastic-increasing effect of feathers in the MDF-matrix. However, the 10%-feather content seemed to be too high and resulted in a significant decrease in MOE. The addition of wollastonite to panels manufactured with 10% feather content could improve MOE to as high as that of control panels. An almost identical trend was seen in particleboard panels; the highest MOE value was observed in wollastonite-treated panels without feathers, and the addition of feathers to the matrix significantly decreased modulus of elasticity (Figure S5, from the supplementary materials).

Brittleness was not significantly changed with either addition of wollastonite or feathers at 5% consumption level in MDF panels (Figure S6, from the Supplementary Materials). However, the addition of the 10%-feather caused a significant increase in brittleness in MDF panels. This again showed that the 10%-feather content was too high. Wollastonite clearly decreased brittleness in all three particleboard treatments (control, 5%- and 10%-feather contents) (Figure S6, from the Supplementary Materials). The addition of wollastonite to the wood fibers or wood particles significantly decreased internal bond in the MDF and particleboard panels without feather content (Figure S7, from the Supplementary Materials). This was attributed to the absorption or gathering of resin molecules by wollastonite particles, preventing them from being active in the process of sticking the strips together. Moreover, acting as a kind of filler, wollastonite improved modulus of rupture and hardness. However, the measurement of internal bond requires resins to be under pulling force. Wollastonite did not have improving effect on the pulling force of UF resin.

Hardness was measured at 3, 4, 5 and 5.4 mm penetration depths in order to gain a better understanding of the effects of the addition of wollastonite and feathers on the surface or inner parts of panels. Almost identical trends in increase and decrease were observed in all four depths of penetration of the steel ball in both MDF and particleboard panels, indicating that the effects were the same at different depths (Figure S8, from the Supplementary Materials). Feathers significantly decreased hardness in both MDF and particleboard panels, which were quite predictable due to the softness of feathers in comparison to wood fibers and particles. Furthermore, the particleboard panels showed significantly higher hardness values in comparison to MDF panels. The addition of wollastonite increased hardness in MDF-feather panels, although not significantly in 5%-feather content panels. However, in particleboard panels, no significant trend was observed.

3.4. Adsorption of Wollastonite on Wood Cell Wall Polymers

Different configurations of wollastonite on hemicellulose were constructed by floating and rotating wollastonite on the surface of hemicellulose and lignin molecules. First, wollastonite was placed far away from the cellulose or lignin surfaces. Then, the distance of wollastonite from the surface was gradually decreased so that the optimal adsorption distance was found based on minimum adsorption energy. The closest distance between wollastonite and the hemicellulose surface, and the adsorption energy of the most stable structure, were found to be 1.7 Å and −4.5 eV, respectively. This large adsorption energy revealed a strong adsorption of wollastonite on hemicellulose as a result of the formation of bonds between them, which proved to be effective in holding wollastonite on the hemicellulose surface.

The comparison between adsorption energies of wollastonite on cellulose and hemicellulose demonstrated a stronger energy bond between wollastonite and cellulose. The adsorption energies of wollastonite on cellulose and hemicellulose were −6.6 and −4.5 (eV), respectively. The stronger bond with cellulose was mainly attributed to the fact that cellulose is a straight chain polymer with no branches; consequently, it provided smoother surface for wollastonite to be easily adsorbed on it. However, hemicellulose is a branched chain polymer; the branches made it difficult for the wollastonite to make bonds with it. It should be noted that as far as water absorption is concerned, hemicellulose has a higher number of hydroxyl groups and is more hydrophilic in comparison to cellulose [47,48]. However, in regard to the adsorption energy of wollastonite on either cellulose or hemicellulose, a single bond between wollastonite and cellulose has more energy than that of hemicellulose.

Water molecules with various orientations were placed on different functional groups of hemicellulose to clarify adsorption of water on hemicellulose. The results showed formation of hydrogen bonds between water molecules and hemicellulose (OH$_{water}$ ... O$_{hemicellulose}$ and OH$_{water}$

... OH$_{hemicellulose}$). OH$_{water}$ indicated the hydroxyl group of adsorbed water molecule; O$_{hemicellulose}$ and OH$_{hemicellulose}$ represented hydroxyl groups of hemicellulose, respectively.

Though wollastonite demonstrated higher adsorption energy in comparison to water molecules, the increase of the number of adsorbed water molecules to twelve molecules gave a competition priority to water over W.

The calculated adsorption distance between wollastonite and lignin was 1.8 Å, and the most stable structures had an average adsorption energy of −2.6 eV. The large adsorption distance along with the small adsorption energy indicated that the adsorption of wollastonite on lignin was so weak that it can practically be ignored. This is quite consistent with the fact that lignin is considered a hydrophobic element [47,48].

Moreover, the adsorption of one water molecule was separately investigated on three different monolignols of lignin (namely, p-coumaryl alcohol, coniferyl alcohol and sinapyl alcohol). Adsorption energies of all three alcohols were positive. The positive energy is considered corroborating evidence of the hydrophobicity of lignin, implying than none of the three structures were energetically stable and, therefore, water molecules could not practically be adsorbed on them. This can be explained by the fact that hydrogen bonds between water and lignin cannot be formed because of the lack of hydroxyl groups in lignin. Ultimately, lignin is hydrophobic in nature. Still, it should be noted that the main reason for the hydrophobic nature of lignin could be the existence of more phenolic groups in its structure.

3.5. Relation between Physical and Mechanical Properties

A fitted-line plot between MOR versus MOE revealed a significant relation (R-square of 100%). This showed the direct effect of an increase or decrease in one property on the other. In addition, a high significant R-square was found between MOR versus brittleness and hardness, although not as high as that in MOE. A low R-square (63%) was found between MOR versus internal bond. With due consideration to the fact that the four properties of MOR, MOE, hardness and brittleness are mostly dependent on the surface layers of specimens rather than the core section, the high significant correlations are justified. The internal bond, however, is mainly dependent on the properties of the core of the composite panels, showing that the properties of the surface layers and core layer of composite panels may be quite independent to each other.

The cluster analysis of the MDF panels based on all physicomechanical properties studied (water absorption and thickness swelling after 2 and 24 h immersion in water, modulus of rupture, modulus of elasticity, brittleness, internal bond and hardness at 5.4 mm of penetration) showed a different clustering of control and wollastonite-treated panels. The cluster analysis identified the significant effects of wollastonite on the overall physical and mechanical properties of medium-density fiberboards (Figure 4A). Wollastonite–5%-feather treatment was closely clustered to wollastonite-treated panels; this clearly showed that, although there was an addition of feathers to the MDF-matrix, and a significant diminishment in properties was anticipated, wollastonite could compensate for the loss to a great extent. With due consideration to the mitigating effects of wollastonite on the overall properties, future studies on decreasing resin content are to be carried out, similar to what was previously achieved by the application of tannin in wood-composite panels [27–32]. Moreover, 5%-feather treatment was closely clustered to the control panels; this indicated that through addition of 5% of feathers to the MDF-matrix, the overall properties remained the same. Therefore, it can be concluded that chicken feathers can be used in MDF manufacturing programs. However, the addition of 10% of feathers to the MDF-matrix resulted in a significant difference in the overall panel properties; 10%-feather panels were remotely clustered to the rest of the treatments.

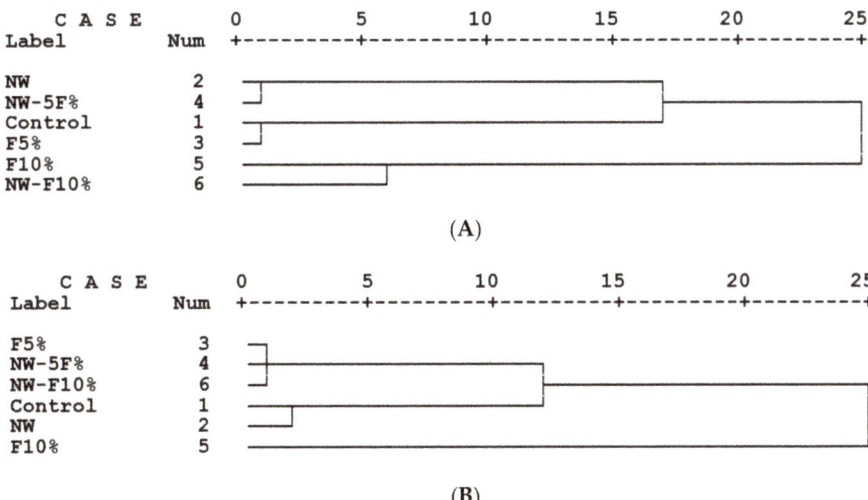

Figure 4. Cluster analysis in the medium-density fiberboard (**A**) and particleboard (**B**) panels based on all the physical and mechanical properties studied in this study (W = wollastonite; F = feather content).

In particleboard panels, control panels (without wollastonite or feather content) were closely clustered to the wollastonite-treated panels (Figure 4B). This showed that wollastonite did not have significant effects on the overall physicomechanical properties. Panels with 10% feather content were remotely clustered with all the other treatments, showing that this feather content was not suitable for the production of particleboards.

Contour plots showed an increasing relationship of hardness values versus internal bond and MOR values (Figure 5A). However, brittleness showed a completely inverse relationship with the mechanical properties of MOR and MOE (Figure 5B). Moreover, it was found that hardness had a straight relationship with internal bond values, but an inverse relationship with brittleness (Figure 5C). The contour plot of internal bond versus hardness at two depths (3 and 5.4 mm) demonstrated a direct relationship with both shallow and deeper penetrations of the hardness ball (Figure 5D); this implied that addition of wollastonite and feathers to mat had similar effects on different layers of composite panels.

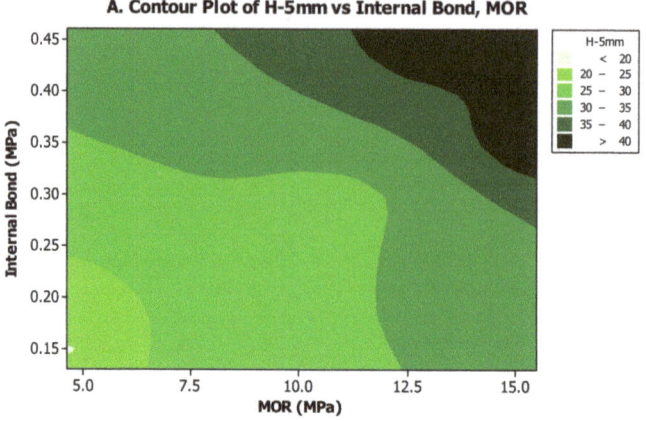
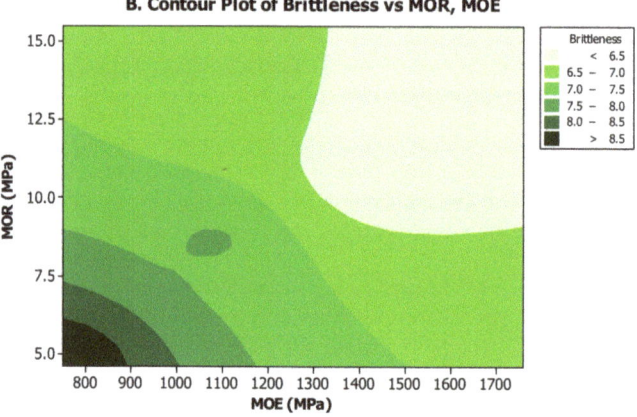
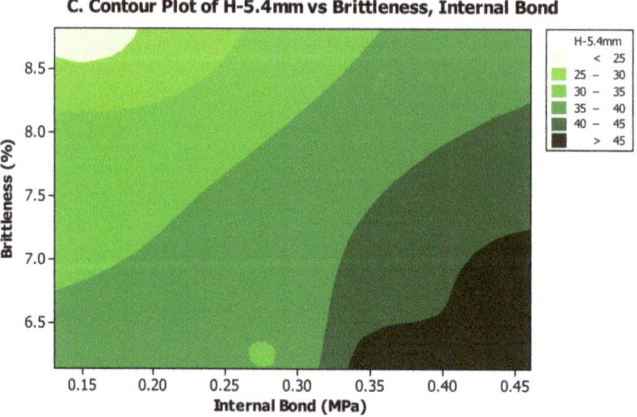

Figure 5. *Cont.*

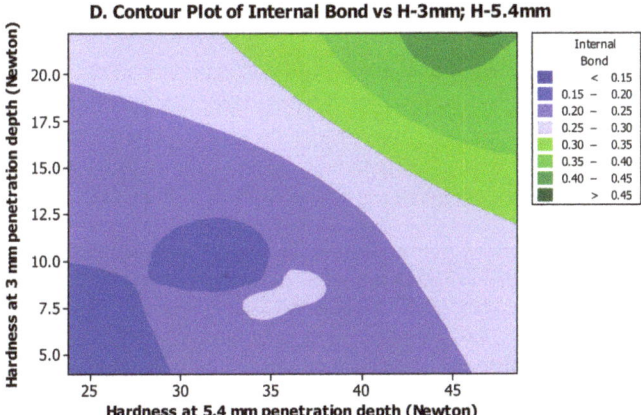

Figure 5. Contour plots among different properties of composite panels observed in this study. (**A**) among MOR and internal bond properties versus hardness at 5-mm penetration depth; (**B**) among MOR and MOE properties versus brittleness; (**C**) among brittleness and internal bond properties versus hardness at 5.4 mm penetration depth; (**D**) among hardness at 3 mm and 5.4 mm penetration depth versus internal bond. (MOR = modulus of rupture; MOE = modulus of elasticity).

4. Conclusions

Chicken feathers were mixed at 5% and 10% consumption levels with wood fibers and particles to produce medium-density fiberboard (MDF) and particleboard panels, in order to comply with the growing need for new sources of raw materials. Urea-formaldehyde (UF) resin was used as the binder. Wollastonite was mixed in UF resin to mitigate the potential negative effects of chicken feathers, and also to investigate if the addition of wollastonite has any potential in future studies to decrease resin content in composite panels in the same way that tannin was reported. The addition of 10%-feather resulted in significant negative effects on all physical and mechanical properties. A feather content of 5% showed some promising results. Wollastonite acted as reinforcing filler in the resin, improving most of the physical and mechanical properties. It was concluded that chicken feathers have potential in wood-composite production.

Supplementary Materials: The following are available online at http://www.mdpi.com/2073-4360/12/4/857/s1, Figure S1: Temperature measurement at the core of the mat with 5-second intervals with a digital thermometer using a thermocouple probe inserted into the center of the core of the MDF-mat; Figure S2: Water absorption (%) in medium-density fiberboard (A) and particleboard (B) panels after 2 and 24 hours immersion in distilled water (MDF=medium-density fiberboard; PB = particleboard panels; NW=nano-wollastonite; WA = water absorption); Figure S3: Thickness swelling (%) in medium-density fiberboard (A) and particleboard (B) panels after 2 and 24 hours immersion in distilled water (MDF=medium-density fiberboard; PB = particleboard panels; NW=nano-wollastonite; TS = thickness swelling); Figure S4: Modulus of rupture (MPa) in medium-density fiberboard (A) and particleboard (B) panels (MDF=medium-density fiberboard; PB = particleboard panels; NW=nano-wollastonite); Figure S5: Modulus of elasticity (MPa) in medium-density fiberboard (A) and particleboard (B) panels (MDF=medium-density fiberboard; (PB = particleboard panels; NW=nano-wollastonite); Figure S6. Brittleness (%) in medium-density fiberboard (A) and particleboard (B) panels (MDF = medium-density fiberboard; PB = particleboard panels; NW = nano-wollastonite); Figure S7. Internal bond (MPa) in medium-density fiberboard (A) and particleboard (B) panels (MDF = medium-density fiberboard; PB = particleboard panels; NW = nano-wollastonite; Figure S8. Hardness (N) in medium-density fiberboard (A) and particleboard (B) panels after 3, 4, 5, and 5.4 mm of penetration into the MDF-matrix (MDF = medium-density fiberboard; PB = particleboard panels; NW = nano-wollastonite).

Author Contributions: Methodology, H.R.T. and R.M.; Validation, H.R.T., R.M., and A.E.; Investigation, H.R.T., Y.S.S. and A.J.; Writing-Original Draft Preparation, H.R.T., R.M. and A.N.P.; Writing-Review and Editing, H.R.T., A.E. and A.N.P.; Visualization, H.R.T. and R.M.; Supervision, H.R.T and A.N.P. All authors have read and agreed to the published version of the manuscript.

Funding: This research received no external funding.

Acknowledgments: The first author appreciates constant scientific support of Jack Norton (Retired, Horticulture and Forestry Science, Queensland Department of Agriculture, Forestry and Fisheries, Australia).

Conflicts of Interest: The authors declare no conflict of interest.

References

1. Hubbe, M.A.; Smith, R.D.; Zou, X.; Katuscak, S.; Potthast, A.; Ahn, K. Deacidification of Acidic Books and Paper by Means of Non-aqueous Dispersions of Alkaline Particles: A Review Focusing on Completeness of the Reaction. *Bioresources* **2017**, *12*, 4410–4477. [CrossRef]
2. Papadopoulos, A.N. Chemical modification of solid wood and wood raw materials for composites production with linear chain carboxylic acid anhydrides: A brief Review. *BioResources* **2010**, *5*, 499–506.
3. Tajvidi, M.; Gardner, D.J.; Bousfield, D.W. Cellulose Nanomaterials as Binders: Laminate and Particulate Systems. *J. Renew. Mater.* **2016**, *4*, 365–376. [CrossRef]
4. Altuntas, E.; Narlioglu, N.; Alma, M.H. Investigation of the fire, thermal, and mechanical properties of zinc borate and synergic fire retardants on composites produced with PP-MDF wastes. *BioResources* **2017**, *12*, 6971–6983.
5. Hassani, V.; Papadopoulos, A.N.; Schmidt, O.; Maleki, S.; Papadopoulos, A.N. Mechanical and Physical Properties of Oriented Strand Lumber (OSL): The Effect of Fortification Level of Nanowollastonite on UF Resin. *Polymers* **2019**, *11*, 1884. [CrossRef]
6. Taghiyari, H.R.; Bibalan, O.F. Effect of copper nanoparticles on permeability, physical, and mechanical properties of particleboard. *Holz als Roh-und Werkst.* **2012**, *71*, 69–77. [CrossRef]
7. Bari, E.; Taghiyari, H.R.; Schmidt, O.; Ghorbani, A.; Aghababaei, H. Effects of nano-clay on biological resistance of wood-plastic composite against five wood-deteriorating fungi. *Maderas. Ciencia y tecnología* **2015**, *17*, 205–212. [CrossRef]
8. Bayatkashkoli, A.; Taghiyari, H.R.; Kameshki, B.; Ravan, S.; Shamsian, M. Effects of zinc and copper salicylate on biological resistance of particleboard against Anacanthotermes vgans termite. *Int. Biodeterior. Biodegrad.* **2016**, *115*, 26–30. [CrossRef]
9. Papadopoulos, A.N.; Taghiyari, H.R. Innovative wood surface treatments based on nanotechnology. *Coatings* **2019**, *9*, 866. [CrossRef]
10. Papadopoulos, A.N.; Bikiaris, D.N.; Mitropoulos, A.C.; Kyzas, G.Z. Nanomaterials and chemical modification technologies for enhanced wood properties: A review. *Nanomaterials* **2019**, *9*, 607. [CrossRef]
11. Mantanis, G.; Athanassiadou, E.T.; Barbu, M.C.; Wijnendaele, K. Adhesive systems used in the European particleboard, MDF and OSB industries. *Wood Mater. Sci. Eng.* **2017**, *13*, 104–116. [CrossRef]
12. Majidi, R. Electronic properties of graphyne nanotubes filled with small fullerenes: A density functional theory study. *J. Comput. Electron.* **2016**, *15*, 1263–1268. [CrossRef]
13. Harsini, I.; Matalkah, F.; Soroushian, P.; Balachandra, A.M.; Balach, A.M. Robust, Carbon Nanotube/Polymer Nanolayered Composites with Enhanced Ductility and Strength. *J. Nanomater. Mol. Nanotechnol.* **2017**, *6*, 6. [CrossRef]
14. Taghiyari, H.R.; Soltani, A.; Esmailpour, A.; Hassani, V.; Gholipour, H.; Papadopoulos, A.N. Improving Thermal Conductivity Coefficient in Oriented Strand Lumber (OSL) Using Sepiolite. *Nanomaterials* **2020**, *10*, 599. [CrossRef] [PubMed]
15. Esmailpour, A.; Majidi, R.; Papadopoulos, A.N.; Ganjkhani, M.; Armaki, S.M.; Papadopoulos, A.N. Improving Fire Retardancy of Beech Wood by Graphene. *Polymers* **2020**, *12*, 303. [CrossRef]
16. Bayani, S.; Taghiyari, H.R.; Papadopoulos, A.N. Physical and mechanical properties of thermally-modified beech wood impregnated with silver nano-suspension and their relationship with the crystallinity of cellulose. *Polymers* **2019**, *11*, 1535. [CrossRef]
17. Taghiyari, H.; Esmailpour, A.; Papadopoulos, A. Paint Pull-Off Strength and Permeability in Nanosilver-Impregnated and Heat-Treated Beech Wood. *Coatings* **2019**, *9*, 723.
18. Taghiyari, H.R.; Avramidis, S. Specific gas permeability of normal and nanosilver-impregnated solid wood species as influenced by heat-treatment. *Maderas Ciencia y tecnologia* **2019**, *21*, 89–96.
19. Karim, M.; Daryaei, M.G.; Torkaman, J.; Oladi, R.; Ghanbary, M.A.T.; Bari, E.; Yilgör, N. Natural decomposition of hornbeam wood decayed by the white rot fungus Trametes versicolor. *Anais da Academia Brasileira de Ciências* **2017**, *89*, 2647–2655. [CrossRef]

20. Taghiyari, H.R.; Mobini, K.; Samadi, Y.S.; Doosti, Z.; Nouri, P.; Reza, T.H. Effects of Nano-Wollastonite on Thermal Conductivity Coefficient of Medium-Density Fiberboard. *J. Nanomater. Mol. Nanotechnol.* **2013**, *2*, 1. [CrossRef]
21. Taghiyari, H.R.; Karimi, A.; Paridah, M.T. Nano-Wollastonite in Particleboard: Physical and Mechanical Properties. *Bioresources* **2013**, *8*, 5721–5732. [CrossRef]
22. Taghiyari, H.R.; Bari, E.; Sistani, A.; Najafian, M.; Ghanbary, M.A.T.; Ohno, K.M. Biological resistance of nanoclay-treated plastic composites with different bamboo contents to three types of fungi. *J. Thermoplast. Compos. Mater.* **2019**. [CrossRef]
23. Taghiyari, H.R.; Kalantari, A.; Kalantari, A.; Avramidis, S. Effect of wollastonite nanofibers and exposure to Aspergillus niger fungus on air flow rate in paper. *Measurement* **2019**, *136*, 307–313. [CrossRef]
24. Taghiyari, H.R.; Bari, E.; Schmidt, O.; Ghanbary, M.A.T.; Karimi, A.; Paridah, M.T. Effects of nanowollastonite on biological resistance of particleboard made from wood chips and chicken feather against Antrodia vaillantii. *Int. Biodeterior. Biodegrad.* **2014**, *90*, 93–98. [CrossRef]
25. Taghiyari, H.R.; Majidi, R.; Jahangiri, A. Adsorption of nano-wollastonite on cellulose surface: Effects on physical and mechanical properties of medium-density fiberboard (MDF). *Cerne* **2016**, *22*, 215–222. [CrossRef]
26. Taghiyari, H.R.; Samadi, Y.S. Effects of wollastonite nanofibers on fluid flow in medium-density fiberboard. *J. For. Res.* **2015**, *27*, 209–217. [CrossRef]
27. Ghahri, S.; Pizzi, A.; Mohebby, B.; Mirshokraie, A.; Mansouri, H.R. Soy-Based, Tannin-Modified Plywood Adhesives. *J. Adhes.* **2016**, *94*, 1–20. [CrossRef]
28. Jahanshahi, S.; Pizzi, A.; Abdulkhani, A.; Shakeri, A. Analysis and Testing of Bisphenol A—Free Bio-Based Tannin Epoxy-Acrylic Adhesives. *Polymers* **2016**, *8*, 143. [CrossRef]
29. Spina, S.; Zhou, X.; Segovia, C.; Pizzi, A.; Romagnoli, M.; Giovando, S.; Pasch, H.; Rode, K.; Delmotte, L. Phenolic resin adhesives based on chestnut (Castanea sativa) hydrolysable tannins. *J. Adhes. Sci. Technol.* **2013**, *27*, 2103–2111. [CrossRef]
30. Ndiwe, B.; Pizzi, A.; Danwe, R.; Tibi, B.; Konai, N.; Amirou, S. Particleboard bonded with bio-hardeners of tannin adhesives. *Holz als Roh-und Werkst.* **2019**, *77*, 1221–1223. [CrossRef]
31. Pizzi, A. Tannins: Major Sources, Properties and Applications. In *Monomers, Polymers and Composites from Renewable Resources*; Elsevier BV: Amsterdam, The Netherlands, 2008; pp. 179–199.
32. Pizzi, A. Tannins: Prospectives and Actual Industrial Applications. *Biomolecules* **2019**, *9*, 344. [CrossRef] [PubMed]
33. Winandy, J.E.; Muehl, J.H.; Glaeser, J.A.; Schmidt, W. Chicken Feather Fiber as an Additive in MDF Composites. *J. Nat. Fibers* **2007**, *4*, 35–48. [CrossRef]
34. Aranberri, I.; Montes, S.; Azcune, I.; Rekondo, A.; Grande, H.-J. Fully Biodegradable Biocomposites with High Chicken Feather Content. *Polymer* **2017**, *9*, 593. [CrossRef] [PubMed]
35. Aranberri, I.; Montes, S.; Wesołowska, E.; Rekondo, A.; Wrześniewska-Tosik, K.; Grande, H.-J. Improved Thermal Insulating Properties of Renewable Polyol Based Polyurethane Foams Reinforced with Chicken Feathers. *Polymer* **2019**, *11*, 2002. [CrossRef]
36. Aranberri, I.; Montes, S.; Azcune, I.; Rekondo, A.; Grande, H.-J. Flexible Biocomposites with Enhanced Interfacial Compatibility Based on Keratin Fibers and Sulfur-Containing Poly(urea-urethane)s. *Polymer* **2018**, *10*, 1056. [CrossRef]
37. Acda, M.N. Waste chicken feather as reinforcement in cement-bonded composites. *Philipp. J. Sci.* **2010**, *139*, 161–166.
38. Koch, J.W. Physical and mechanical properties of chicken feather materials. Master's Thesis, School of Civil Environmental Engineering, Georgia Institute of Technology, Atlanta, GA, USA, May 2006.
39. Fraser, R.; Parry, D. The molecular structure of reptilian keratin. *Int. J. Boil. Macromol.* **1996**, *19*, 207–211. [CrossRef]
40. Schmidt, W.F. Innovative feather utilization strategies. In *National Poultry Waste Management Symposium Proceedings*; Auburn University: Auburn, AL, USA, 1998; pp. 276–282.
41. ASTM D1037-99. *Standard Test Methods for Evaluating Properties of Wood-Base Fiber and Particle Panel Materials*; ASTM International: West Conshohocken, PA, USA, 1999; Available online: www.astm.org.
42. Taghiyari, H.R.; Enayati, A.; Gholamiyan, H. Effects of nano-silver impregnation on brittleness, physical and mechanical properties of heat-treated hardwoods. *Wood Sci. Technol.* **2012**, *47*, 467–480. [CrossRef]

43. Phuong, L.X.; Shida, S.; Saito, Y. Effects of heat treatment on brittleness of Styrax tonkinensis wood. *J. Wood Sci.* **2007**, *53*, 181–186. [CrossRef]
44. Kaith, B.S.; Mittal, H.; Jindal, R.; Maiti, M.; Kalia, S. Environment Benevolent Biodegradable Polymers: Synthesis, Biodegradability, and Applications. In *Cellulose Fibers: Bio- and Nano-Polymer Composites*; Springer-Verlag: Berlin/Heidelberg, Germany, 2011; pp. 425–451.
45. Hosseinpourpia, R.; Adamopoulos, S.; Mai, C. Effects of acid pre-treatments on the swelling and vapor sorption of thermally modified Scots Pin (Pinus sylvestris L.) wood. *BioResources* **2018**, *13*, 331–345.
46. Esmailpour, A.; Taghiyari, H.R.; Hosseinpourpia, R.; Adamopoulos, S.; Zereshki, K. *Shear Strength of Heat-Treated Solid Wood Bonded with Polyvinyl-Acetate Reinforced by Nanowollastonite. Wood Research 2020*; VUPC a.s.: Bratilsava, Slovakia, 2020; in press.
47. Dinwoodie, J.M. *Timber: Its Nature and Behavior*; Van Nostrand Reinhold: New York, NY, USA, 1981.
48. Fengel, D.; Wegener, G. *Wood: Chemistry, Ultrastructure, Reactions*; Walter de Gruyter: Berlin, Germany, 1984.

© 2020 by the authors. Licensee MDPI, Basel, Switzerland. This article is an open access article distributed under the terms and conditions of the Creative Commons Attribution (CC BY) license (http://creativecommons.org/licenses/by/4.0/).

Article

Hygroscopicity of Waterlogged Archaeological Wood from Xiaobaijiao No.1 Shipwreck Related to Its Deterioration State

Liuyang Han [1,2], Juan Guo [1,2], Kun Wang [3], Philippe Grönquist [4,5], Ren Li [1,2], Xingling Tian [6] and Yafang Yin [1,2,*]

1. Department of Wood Anatomy and Utilization, Research Institute of Wood Industry, Chinese Academy of Forestry, Beijing 100091, China
2. Wood Collections (WOODPEDIA), Chinese Academy of Forestry, Beijing 100091, China
3. College of Material Science and Technology, Beijing Forestry University, Haidian District, Beijing 100083, China
4. Wood Materials Science, ETH Zürich, 8093 Zürich, Switzerland
5. Laboratory for Cellulose & Wood Materials, EMPA, 8600 Dübendorf, Switzerland
6. Heritage Conservation and Restoration Institute, Chinese Academy of Cultural Heritage, Beijing 100029, China
* Correspondence: yafang@caf.ac.cn; Tel.: +86-10-62889468

Received: 20 March 2020; Accepted: 3 April 2020; Published: 6 April 2020

Abstract: Waterlogged archaeological wood (WAW) artifacts, made of natural biodegradable polymers, are important parts of many precious cultural heritages. It is of great importance to understand the hygroscopic behavior of WAW in different deterioration states for the development of optimal drying processes and choices of safe storage in varying conditions. This was investigated in a case-study using two *Hopea* (Giam) and two *Tectona* (Teak) WAW samples collected from the Xiaobaijiao No.1 shipwreck. The deterioration state of WAW was evaluated by the maximum water content (MWC) method and by the cell morphological structure. Both *Hopea* and *Tectona* WAW could be classified into moderately and less decayed WAW. The hygroscopic behavior of moderately and less decayed WAW was then comparatively investigated using Dynamic Vapor Sorption (DVS) measurements alongside two sorption fitting models. Compositional analysis and hydroxyl accessibility measurements of WAW cell walls were shown to correlate with the hygroscopicity of WAW in different deterioration states. It was concluded that moderately decayed WAW possessed higher hygroscopicity and hysteresis than less decayed WAW because of the lower relative content of polysaccharides and the higher relative content of lignin, including the slow hydrolysis of O-acetyl groups of xylan and the partial breakage of β-O-4 interlinks, accompanied by an increased hydroxyl accessibility. This work helps in deciding on which consolidation measures are advised for shipwreck restauration, i.e., pretreatments with specific consolidates during wood drying, particularly for wooden artifacts displayed in museums.

Keywords: morphological structure; sorption behavior; sorption fitting model; compositional analysis; hydroxyl accessibility

1. Introduction

Waterlogged archaeological wooden artifacts counting as valuable cultural heritages are being excavated worldwide [1–3]. In dependence of environmental factors, wood species, period, and processed treatments, WAW excavated even from the same archaeological site or collected either from the surface or the inner part of the same wooden artifact would probably be found in different deterioration states [4–8]. In general, WAW can be divided into severely decayed wood, moderately decayed wood, and less decayed wood [9,10]. The subdivision can be done according to maximum water content, morphological observations, and chemical structure of WAW [11–13]. Even though the water environment remarkably slows down the wood deterioration caused by microbiota, waterlogged wooden artifacts still suffer from a high possibility of deterioration. Hence, the conservation of WAW artifacts generally necessitates a water removal treatment [2,14].

Given that the two main forms of water in wood are free water in cell lumens and cell wall mesopores as well as bound water adsorbed in cell walls [7,15,16], the water-removal treatment incorporates both removal of free water and desorption of bound water. The treatment may lead to cell morphology changes due to surface tension that affect the dimensional stability of archaeological wooden artifacts because of cell wall shrinkage and collapse [15,17–20]. The sorption of bound water relies on the hygroscopic behavior of WAW, which is related to the dimensional instability that causes cracks and distortions in WAW [21]. Previous works have found that buried archaeological wood possesses a higher equilibrium moisture content (EMC) and higher hysteresis coefficients than recent wood [22,23], which could be attributed to the deacetylation of hemicelluloses, the degradation of amorphous celluloses, and a decrease of crystallinity. Moreover, WAW possesses varying deterioration states, each of which features an inhomogeneous deterioration behavior [24] and may require specific water-removal treatment to avoid unnecessary damage caused by cell wall shrinkage. Thus, it is of great importance to understand the hygroscopic response of WAW to different deterioration states in order to develop optimal drying processes, suitable display conditions, and safe storage under varying climatic conditions [25,26]. However, no studies yet illustrate the influence of deterioration state of WAW on its hygroscopicity.

The aim of this work is to compare the hygroscopicity of WAW in different deterioration states and to provide basic knowledge for their preservation, particularly in terms of the selection of drying method and storage conditions. Herein, two hardwood species of WAW, *Hopea* (Giam) and *Tectona* (Teak), were collected from the marine Xiaobaijiao No.1 shipwreck dated as 1821–1850 [27,28]. MWC and observation of cell morphological structure by LM and SEM were adopted to classify the deterioration state of WAW. The hygroscopic behavior was then examined by DVS, including analysis of equilibrium moisture content (EMC) and sorption hysteresis, and by fitting sorption/desorption curves with two frequently used multilayer sorption models; the Guggenheim, Anderson, and De Boer (GAB) and the Generalized D'Arcy and Watt (GDW) models. Furthermore, deuterium exchange measurements were conducted in order to gain information regarding the hydroxyl group accessibility. And finally, the study was complemented by compositional analysis in order to better understand the effect of cell wall degradation on the hygroscopicity of WAW.

2. Materials and Methods

The Xiaobaijiao No.1 shipwreck, now preserved in a waterlogged environment at the Ningbo Base of the Chinese National Center of Underwater Cultural Heritage (Ningbo, China), was a commercial ship in the period from 1821 to 1850 AD. The wreck site (scheme as shown in Figure 1) is located on Yushan Island, China [27]. Considering that an ideal selection of the reference wood specimens is a challenge for the comparison with the broader archaeological wood research, two samples corresponding to each of the two species composing the shipwreck were carefully chosen for comparable results. The locations of four samples (denoted H1, H2, T1, and T2) collected for this study are marked in Figure 1 and the detailed position information is listed in Table 1. Samples H1 and H2 were identified as *Hopea* spp. (Giam), while T1 and T2 were identified as *Tectona* spp. (Teak). As reported by a previous and related publication [28], the maximum water content (MWC) of the samples H1, H2, T1 and T2 were 121.59% ± 16.7%, 264.43% ± 80.55%, 108.51% ± 4.5% and 189.59% ± 65.36%, respectively.

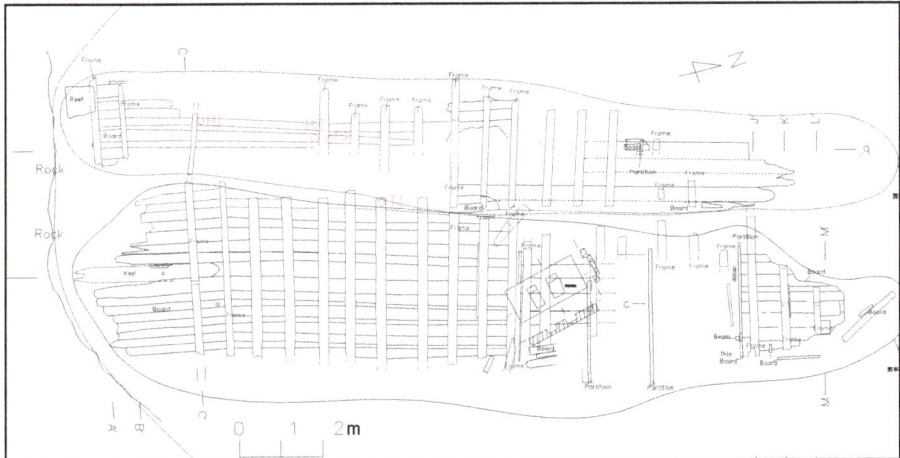

Figure 1. Scheme of the Xiaobaijiao No.1 shipwreck with the sample locations marked in red. *Hopea* WAW: H1, H2; *Tectona* WAW: T1, T2.

Table 1. Detailed position of the samples collected from the Xiaobaijiao No.1 shipwreck.

Sample Name	Sampling Position
H1	The 5th inner layer board of the hull
H2	The 4th inner layer board of the hull
T1	The 6th inner layer board of the hull
T2	The 7th inner layer frame of the hull

2.1. LM

Cross sections of *Hopea* and *Tectona* WAW were prepared by a rotary microtome (RM 2255, Leica, Wetzlar, Germany) with a thickness of 10 μm. A light microscope (BX51, Olympus, Tokyo, Japan) was used to examine microstructure of the specimens.

2.2. SEM

Prior to the SEM examination, all cross-section surfaces of WAW were prepared by a rotary microtome (RM M2255, Leica, Germany). To avoid creating artificial damage to the cell wall structure during cutting with the microtome, the WAW samples were embedded with polyethylene glycol (PEG) 2000 (average molecular mass: 1900–2200 g/mol), which was followed by a rinsing process under flowing water for 30 min to remove the PEG. After mounting the dry samples on aluminum stubs followed by a sputter-coating process with Platinum, the samples were observed using a field emission scanning electron microscope (Quanta 200F FEI, Thermo Fisher Scientific, Waltham, MA, USA) at a voltage of 10 kV.

2.3. DVS

EMC of moderately decayed and less decayed WAW in different relative humidity (RH) states were measured by an automated sorption balance device (DVS Advantage ET85, Surface Measurement Systems Ltd., Wembley, UK). Measurements were mainly conducted according to the protocol found in [29]. Samples were cut into millimeter thick stripes by a razor blade and 30 mg of the waterlogged stripes were initially dried at a partial water vapor pressure of $p/p_0 = 0$ for 600 min. The samples were then exposed to ascending p/p_0 steps ranging from 0 to 0.98 for adsorption and then descending in the same manner for desorption at 25 °C. Equilibrium in each step was defined to be reached at a mass change per time (dm/dt) of less than 0.0005%/min over a 10 min stability window or a maximal time of 1000 min per step. The sorption hysteresis parameters were calculated by the difference of EMC for desorption and adsorption in the same relative humidity.

2.4. Isotherm Models

The obtained isotherms by DVS were fitted with two common sorption models, whose parameters were obtained by least-square fits to the data for each sample. The first of the models, the GAB model, was mainly improved from the commonly used BET isotherm model by increasing the sorbate activity range [30], and was first recommended by the European Project Group COST 90 [31] as a fundamental equation to characterize water sorption in food. It was subsequently introduced for the analysis of wood [7,32,33]. The GAB equation reads as:

$$\text{EMC} = M_m \frac{K_{GAB} \cdot C_{GAB} \cdot RH}{(1 - K_{GAB} \cdot RH) \cdot (1 - K_{GAB} \cdot RH + C_{GAB} \cdot K_{GAB} \cdot RH)} \times 100\% \quad (1)$$

where EMC (%) is the equilibrium moisture content; RH (%) is the air relative humidity; M_m is the monolayer capacity; C_{GAB} (%) is the equilibrium constant related to the monolayer sorption, and K_{GAB} (%) is the equilibrium constant related to the multilayer sorption.

In addition, the internal specific surface area (S_{GAB}) of WAW can be obtained based on the values of Mm provided by the GAB model:

$$S_{GAB} = \frac{M_m \cdot \rho \cdot L \cdot \sigma}{M} \quad (2)$$

where ρ is the density of water, L is the Avogadro number, σ is the average area where water occupies the complete monolayer (0.114 nm^2 was used in this study for the surface area occupied by a single water molecule) and M is the molar mass of water [26,34].

The second isotherm model used, the GDW model, assumes that the Langmuir mechanism governs the monolayer sorption, i.e., only one water molecule can be directly bound to a primary sorption site and that there are three possible scenarios: (a) the number of the secondary sites is lower than the number of primary sites (i.e., primary bound water molecules are not completely converted into the secondary sorption sites, $w < 1$. w is a conversion ratio of primary bound water molecules into the secondary sites); (b) the number of the secondary sites is equal to the primary sites (i.e., each monolayer molecule is converted into the secondary site, $w = 1$); (c) the number of the secondary sites is higher than the primary sites (i.e., each primary bound molecules creates more than one secondary sorption site, $w > 1$) [32]. The GDW model [32,35] equation reads as:

$$\text{EMC} = \frac{m_{GDW} \cdot K_{GDW} \cdot RH}{(1 + K_{GDW} \cdot RH)} \cdot \frac{1 - k_{GDW}(1-w) \cdot RH}{(1 - k_{GDW} \cdot RH)} \times 100\% \quad (3)$$

where m_{GDW} (%) is the maximum amount of water bound to the primary sorption sites, i.e., the monolayer water content. K_{GDW} (%) is a constant of sorption kinetics on the primary sites, and k_{GDW} (%) is a constant of sorption kinetic on the secondary sites.

2.5. Compositional Analysis

The carbohydrates and total lignin of WAW specimens were measured with 3 replicates under the standard procedure according to the National Renewable Energy Laboratory (NREL, Golden, CO, USA) protocol [36,37]. Briefly, the milled specimens were hydrolyzed in 72% H_2SO_4 for 1 h at 30 °C and were then completely hydrolyzed in an autoclave at 121 °C for 1 h. The acid insoluble lignin was determined by weighing the solid, and the monosaccharides in the liquid were detected by high-performance anion exchange chromatography (Dionex ISC 3000, Sunnyvale, CA, USA).

2.6. Hydroxyl Accessibility

The WAW samples measured by DVS were further used to study the hydroxyl accessibility. The samples, initially dried at $p/p_0 = 0$ and 40 °C for 6 h, were exposed to D_2O vapor at $p/p_0 = 0.95$ and 25 °C for 10 h to ensure that the material's accessible hydrogen protium is completely replaced by deuterium [38]. Then, the drying procedure was applied again, and the deuterated dry mass m_D was obtained. The number of available water vapor accessible OH groups (sorption sites) was calculated by equation (3) [29,39]:

$$\text{Number} = \frac{m_D - m_{dry}}{m_{dry} \cdot (M_D - M_H)} \quad (4)$$

where m_{dry} is the dry mass of archaeological wood; M_D is the molar mass of deuterium and M_H is the molar mass of protium.

3. Results and Discussion

3.1. The Deterioration State of Waterlogged Archaeological Wood

According to the MWC values, the most commonly used parameter to classify the deterioration state of waterlogged archaeological wood [9,10], samples H2 and T2 belong to class of moderately decayed wood (185% < MWC < 400%), while H1 and T1 can be regarded as class of less decayed wood (MWC<185%). LM and SEM revealed the morphological structures of the WAW specimens and confirmed these deterioration states [6,9]. Cells in H1 mainly remained intact (Figure 2A) without significant deterioration features visible in the SEM image (Figure 3A). In contrast, the morphological structure of cell walls in H2 displayed pronounced decay patterns (Figure 2C). Parts of the S_2 layers and the S_3 layers of the fiber cell walls were degraded (Figure 3C), which is a sign of erosion bacteria decay pattern [24]. Furthermore, some parts of the cell walls of sample H2 were depleted by microbiological degradation with significant cavities emerging. Similar micro-morphological structure differences were present in T1 and T2. As for less decayed waterlogged *Tectona* (T1), the morphology shown by LM (Figure 2B) and SEM (Figure 3B) also indicated no pronounced sign of degradation. However, obvious features of soft-rot decay [24] were found in T2 by both LM (Figure 2D) and SEM (Figure 3D). Specifically, the S_3 layers of the cell walls of T2 were still intact, while cavities occurred in its S_2 layers. To sum up, the two moderately decayed WAWs were deteriorated to some degree by microorganisms with pronounced alternations of their cell morphologies, while the two less decayed WAWs in this research didn't show any obvious cell morphologies indicating decay. Thus, the deterioration states of *Hopea* and *Tectona* WAW revealed by the morphological method were well consistent with the MWC method in this study.

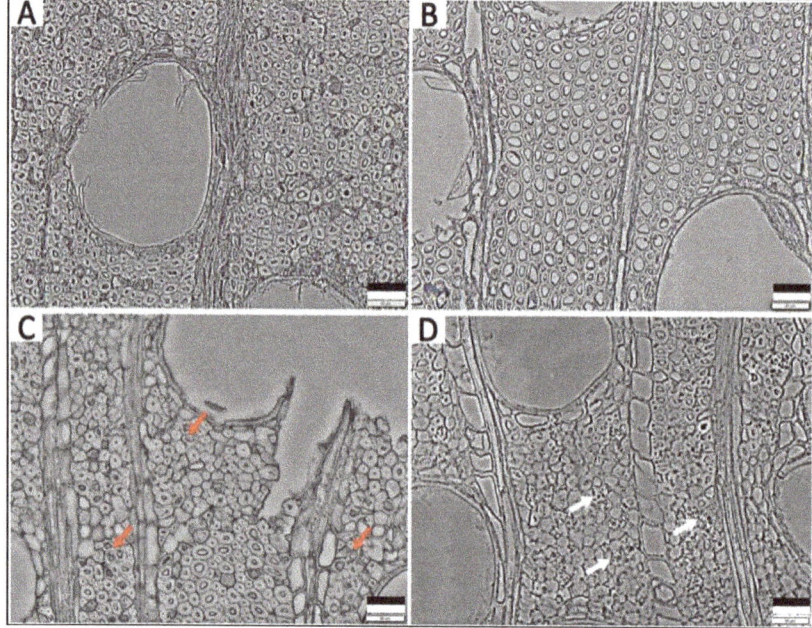

Figure 2. Light microscopy images of *Hopea* waterlogged archaeological wood (WAW): H1 (**A**): Less decayed sample, H2 (**C**): Moderately decayed sample (red arrows display pronounced decay patterns indicating degraded fiber cell walls); *Tectona* WAW: T1 (**B**), Less decayed sample, T2 (**D**): Moderately decayed sample (white arrows show that cavities occurred in its S_2 layers indicate degraded fiber cell walls). Scale bar = 50 μm.

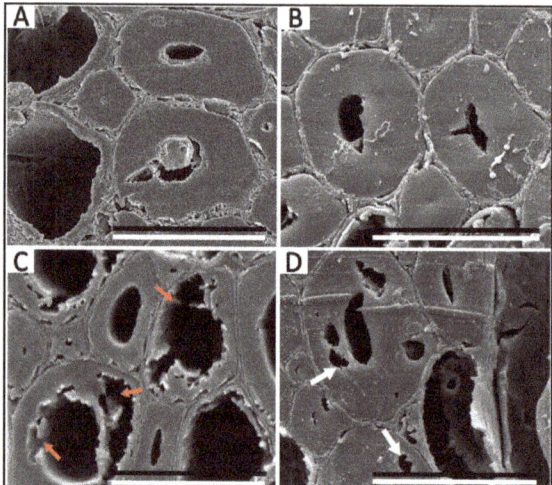

Figure 3. SEM images of *Hopea* WAW: H1 (**A**): Less decayed sample, H2 (**C**): Moderately decayed sample (red arrows indicate decay in the S_2 and in many of the S_3 layers within the fiber cell walls); *Tectona* WAW: T1 (**B**): Less decayed sample, T2 (**D**): Moderately decayed sample (white arrows indicate cavities visible in the S_2 layers). Scale bar = 20 μm.

3.2. Hygroscopicity of WAW in Different Deterioration States

The sorption isotherms of *Hopea* and *Tectona* WAW in different deterioration states are shown in Figure 4 for a single adsorption and desorption cycle per sample. All adsorption and desorption curves display S-shapes, which implies that the sorption isotherms of both moderately and less decayed WAW might be classified as type IV IUPAC isotherms [40–44] (Whether the measured isotherms can be classified as type II or type IV can be debated. Some studies [41,43,44] classified wood sorption isotherms as type II, but this depends on the chosen interpretation of the sorption mechanisms; either exclusively mono/multilayer adsorption, or a mix with capillary condensation in mesopores of the cell wall [40]. Here the authors assume type IV.). The sorption isotherms of all specimens exhibited an upward bend at around 60%–80% RH, which is commonly reported in lignocellulosic materials [26,32,43]. Furthermore, the EMCs of moderately decayed WAW at each RH were all higher than those of less decayed WAW. At the highest relative humidity (98% RH), the EMCs of H2 and T2 reached as high as 25.91% and 27.37%, whereas the EMC of H1 and T1 reached 21.55% and 22.15%, respectively. Furthermore, different relative changes in EMCs (M_m/M_l) in both adsorption and desorption branches were present for both *Hopea* and *Tectona* WAW. The EMCs in both adsorption and desorption of H2 were at least 15% higher than those of H1 (Figure 5A). As for *Tectona*, the EMCs in adsorption for T2 were 19% to 39% higher than those of T1, and the EMCs in desorption for the former were 20%–49% higher than the latter (Figure 5B).

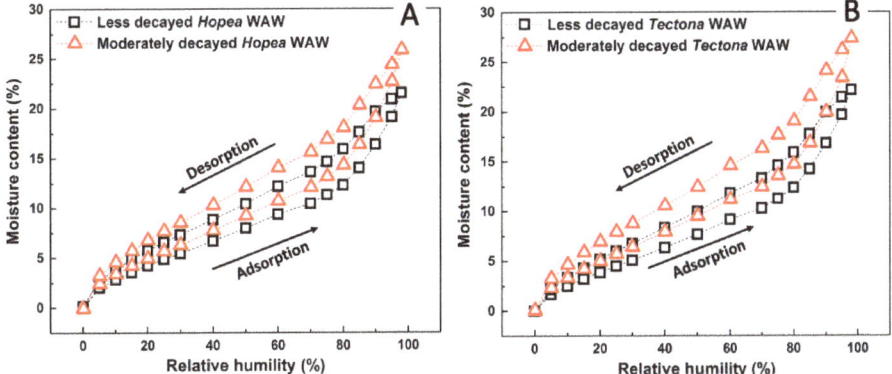

Figure 4. Equilibrium moisture content (EMC) of *Hopea* (**A**) and *Tectona* (**B**) WAW. Water vapor adsorption and desorption curves for the less decayed (H1 and T1, open black squares) and the moderately decayed (H2 and T2, open red triangles) samples.

In addition to the increase of EMCs, sorption hysteresis of moderately decayed WAW was also higher than that of less decayed WAW. The sorption hysteresis is commonly calculated as the difference between adsorption and desorption branches of an isotherm in the range between the highest relative humidity (98% RH) and 0% RH [45]. It is believed to originate from a potential rearrangement of structural components in cell walls [46]. As shown in Figure 6, sorption hysteresis indeed exists in the whole moisture extent from 0% RH to 98% RH. The measurable sorption hysteresis of moderately decayed WAW was found higher than that of less decayed WAW at every humidity condition and for both *Hopea* and *Tectona* WAW. The higher sorption hysteresis of moderately decayed WAW as compared to less decayed WAW might lead to the higher variation in surface moisture content of wood elements under standard changes of relative humidity [46]. Therefore, conservators of archaeological artifacts usually try to consolidate WAW, for example, with lactitol and trehalose, in order to lower the sorption hysteresis [32,47].

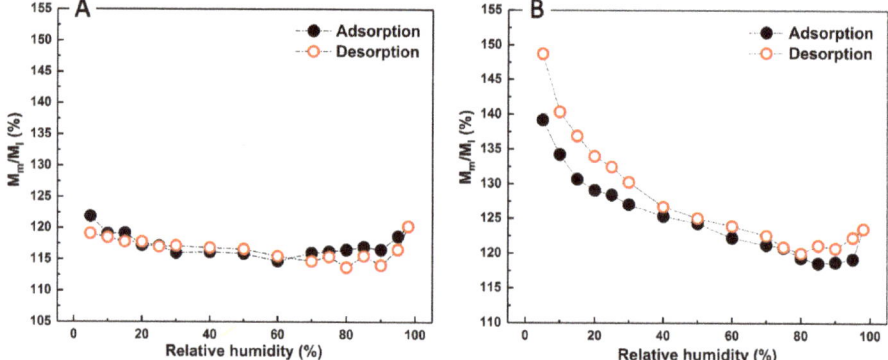

Figure 5. Relative changes in EMCs (M_m/M_l) in adsorption (black solid cycles) and desorption (red open cycles) of *Hopea* (**A**) and *Tectona* (**B**) WAW. M_m = moisture content (MC) of moderately decayed WAW (H2, T2); M_l = MC of less decayed WAW (H1, T1).

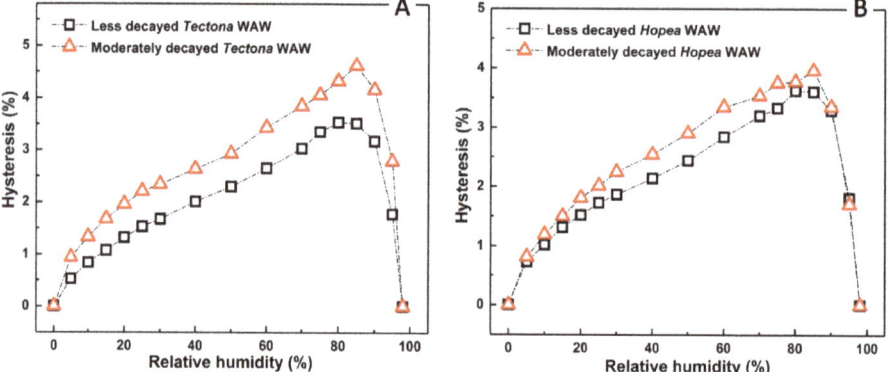

Figure 6. Sorption hysteresis of *Hopea* (**A**) and *Tectona* (**B**) WAW for the less decayed (H1, T1, open black squares) and the moderately decayed (H2, T2, open red triangles) samples.

In order to analyze the sorption process in detail, GAB and GDW sorption models were applied to fit the adsorption and desorption isotherms of *Hopea* and *Tectona* WAW with two deterioration states. The fits were considered to be valid if all the coefficient of determination (R^2) values were above 0.99 [26,48]. The parameters calculated by a least-square fitting were listed in Table 2.

With the analysis by the GAB model, as listed in Table 2, it could be noticed that the C_{GAB} values were approximately an order of magnitude higher than the K_{GAB} values for all samples, indicating much higher heat of sorption of the monolayer as compared to the multilayer [49]. It could also be deduced from the GAB model that the necessary condition for classifying the isotherms as type II (in this context, the authors assume that the sorption isotherm is the result of unrestricted mono/multilayer adsorption up to high p/p_0) was satisfied because the conjunction of the relations $5.57 \leq C_{GAB} < \infty$ and $0.24 < K_{GAB} \leq 1$ was satisfied for the analyzed isotherms [32,50]. The maximum monolayer water content reflected by the M_m coefficient was found 11.74% higher for H2 than the M_m for H1. For *Tectona* WAW it was found 16.74% higher during the adsorption processes. For the desorption processes, the value for H2 was 8.80% higher than that of H1 and in the case of *Tectona*, it was 13.29% higher. The increased maximum monolayer water contents for moderately decayed WAW imply that long-time deterioration increased the number of accessible primary sorption sites compared to less decayed WAW. Besides, the M_m coefficient is proportional to the internal specific surface area

(S_{GAB}) [26,34], from which can be deduced that the internal specific surface area of WAW increases with the deterioration level. It was found that the C_{GAB} coefficient, which represents the total heat of sorption of the monolayer water [26,51], was higher for moderately decayed WAW than that for less decayed WAW. The C_{GAB} coefficients for H2 were 16.48% and 17.91% higher than those of H1, and in the case of *Tectona* they were 40.88% and 49.31% higher for both the adsorption and desorption processes respectively. These results could lead to the interpretation that the monolayer water is bound more strongly to the primary sorption sites for moderately decayed WAW as compared to less decayed WAW.

Table 2. Coefficients of the GAB and GDW models for *Hopea* and *Tectona* WAW in different deterioration states.

Sample	Sorption Phase	GAB Model					GDW Model				
		M_m	K_{GAB}	C_{GAB}	R2	S_{GAB}	m_{GDW}	K_{GDW}	k_{GDW}	w	R2
H1	Adsorption	4.94	0.79	15.59	0.999	187.68	8.70	4.41	0.86	0.39	1
	Desorption	8.30	0.66	10.72	1	315.34	6.69	11.32	0.58	1.97	1
H2	Adsorption	5.52	0.80	18.16	0.999	209.72	9.74	4.86	0.87	0.39	1
	Desorption	9.03	0.68	12.64	1	343.08	9.60	8.23	0.67	1.04	1
T1	Adsorption	4.84	0.80	11.79	1	183.88	8.57	3.71	0.85	0.45	1
	Desorption	8.20	0.67	8.01	1	311.54	6.62	8.38	0.61	1.90	1
T2	Adsorption	5.65	0.81	16.63	0.999	214.66	11.55	3.64	0.88	0.31	1
	Desorption	9.29	0.69	11.96	0.999	352.96	9.20	9.12	0.67	1.22	0.999

Note: M_m (%) is the monolayer capacity, C_{GAB} (%) is the equilibrium constant related to the monolayer sorption, K_{GAB} (%) is the equilibrium constant related to the multilayer sorption, S_{GAB} (m^2/g) is the internal specific surface area, m_{GDW} (%) is the maximum amount of water bound to the primary sorption sites, i.e., the monolayer water content, K_{GDW} (%) is a constant of sorption kinetic on the primary sites, k_{GDW} (%) is a constant of sorption kinetic on the secondary sites, w – conversion ratio of primary bound water molecules into the secondary sites.

Using the GDW model, the maximum content of water bound to primary sites (m_{GDW}) of H2 was 11.95% higher than that of H1, and in the case of *Tectona* WAW 34.77%, during the adsorption processes. During desorption processes, the m$_{GDW}$ of the H2 sample was 43.50% higher than for H1, and in the case of *Tectona*, it was 38.97% higher. This could indicate an increased number of primary sorption sites for water in moderately decayed WAW. The ratio of water molecules bound to primary sites and converted into secondary sites (w) of WAW generally decreased with deterioration state. The values of H1 and H2 were the same. During the desorption processes, for *Tectona*, the value of T2 was 31.11% less than that of T1 during the adsorption. The value of H2 was 47.21% less than that of H1, and the in the case of *Tectona*, it was 35.79% less. The results would imply that each primary bound molecule of moderately decayed WAW created less secondary sorption sites than in the case of less decayed WAW. However, the decreased w of WAW in this study and other related studies [32,46] did not obviously contribute to the increased hygroscopicity because of the significant contribution of the increased number of primary sorption sites.

The GAB and the GDW sorption models both indicated that moderately decayed WAW may possess more sorption sites and display a stronger capability of adsorbing water vapor from the surrounding environment than less decayed WAW.

3.3. The Chemical Deterioration and Increased Hydroxyl Accessibility of Waterlogged Archaeological Wood

The hygroscopicity of wood highly depends on the relative contents of the main components in the cell wall, i.e., cellulose, hemicellulose, and lignin [52]. The results of the compositional analysis of WAW in different deterioration states are shown in Table 3. The amount of cellulose, the most significant component, is reflected by the relative content of glucose. The content of hemicellulose can be assigned to the relative content of xylose, because in hardwoods, xylan dominates the composition of hemicellulose [52]. Finally, the Klason lignin content was used to reflect the content of lignin [53].

Table 3. Chemical compositions of WAW collected from the Xiaobaijiao No.1 shipwreck. Chemical composition 100% means related to investigated components. Statistics for n = 3 measurements. The standard deviations were less than 2%.

Sample	Acid-Insoluble Lignin	Acid-Soluble Lignin	Glucose	Xylose
H1	46.0%	1.0%	45.6%	7.4%
H2	55.8%	1.0%	39.3%	3.9%
T1	45.0%	0.8%	47.6%	6.6%
T2	54.6%	0.8%	38.8%	5.8%

For H2, the relative content of cellulose counted by the proportion of glucose is 39.2% (according to total dry weight), 14% lower than that of H1. Xylose accounted for 3.9%, which was 47.3% lower than in the case of H1. In contrast, the relative content of lignin was 56.6% (55.6% Klason lignin and 1% acid-soluble lignin) for H2, 21% higher than that of H1. Additionally, it was found that the relative content of Klason lignin decreased from 55.6% to 46% from H2 compared to H1. As stated, the increased content of lignin in moderately decayed WAW does not indicate an absolute increase of lignin during the long-term deterioration process, instead, this is mainly a result from the loss of polysaccharides including hemicelluloses and cellulose [54]. For *Tectona* WAW, the compositional analysis presented the same tendency with a 9.6% higher lignin content and a 9.8% lower glucose content as well as a 0.8% lower xylose content for moderately decayed compared to less decayed WAW.

The increase of EMC for the WAW can be understood in terms of the increase of both the number of sorption sites and their accessibility, as reported by a previous and related publication, where it was shown that the changes of chemical and cellulose crystallite structure led to an increased bound water uptake [7]. The hydroxyl accessibility of WAW was obtained in this work to further explore the apparent difference in hygroscopicity between moderately decayed and less decayed WAW.

Figure 7 displays the number of accessible hydroxyl groups obtained by sample deuteration using heavy water. The amount of accessible hydroxyl groups of wood generally ranges from 6.8–10.3 mmol/g according to previous research [38,55]. In this study, the number of accessible hydroxyl groups of H2 was 7.5 ± 0.4 mmol/g, while that of H1 was lower, with a value of 6.8 ± 0.3 mmol/g. For *Tectona* WAW, T2 showed a higher mean value of 7.9 ± 0.5 mmol/g, while T1 showed a lower accessibility of 7.0 ± 0.5 mmol/g. Because the amount of available OH groups is generally believed to directly correlate with the amount of sorption sites, the 10% and 13% higher numbers of hydroxyl groups in moderately decayed WAW could strongly indicate why the EMCs and hysteresis were higher in moderately decayed WAW than for less decayed WAW.

The observed changes in hydroxyl accessibility could be either attributed to alternation of structure or to changes of chemical composition [29,56] upon deterioration. In fact, lignin, hemicellulose, and cellulose possess different amounts of hydroxyl groups [57], and as was reported above, they were each found to degrade to some different extent in WAW. The compositional analysis of different substrates revealed that the deterioration of WAW was mainly related to the decomposition of polysaccharides in the cell wall. Cellulose and hemicelluloses are in general extensively affected by anaerobic microorganisms and by occurring acid or alkali environment conditions [5] during the 170-years deterioration in waterlogged conditions. It is known that slow hydrolysis of O-acetyl groups of xylan always occurs in decayed archaeological hardwood [22,58,59]. As the hemicellulose is the most hydrophilic polymer in wood, its degradation generally results in a reduced hygroscopicity [60]. However, it should be emphasized that the hygroscopicity was higher for WAW than recent wood although the hemicellulose of WAW was severely degraded compared to cellulose and lignin, as illustrated by previous studies [7,32,47]. As shown in Table 3, the relative content of xylose representing hemicellulose decreased a lot. That means that besides hemicellulose, variations in the component structures of cellulose and lignin should also contribute to the increased hygroscopicity of WAW. Besides hemicellulose, cellulose is also considered as a major substance contributing to the hydrophilicity of wood and of other natural biodegradable polymers [61–63]. Although the interior of the cellulose microfibrils is not accessible to water vapor [64], the hydroxyl groups on surface chains are accessible [65–67]. Furthermore, the extent of moisture sorption of cellulose was reported to increase with decreasing degree of crystallinity [61,68]. In WAW, the deterioration of cellulose includes both the degradation of amorphous cellulose and the decrease of relative crystallinity [23,69]. In this research, the observed decrease of relative content of cellulose could demonstrate its degradation, and in addition, changes in amorphous cellulose and crystallinity were reported in the related previous study [69]. Therefore, both aspects might explain the observed increase in hygroscopicity. However, next to the deterioration of cellulose and hemicellulose, the modification of lignin may also lead to an increase of adsorbent functional groups in WAW [7,54,69], responsible for an increased uptake of water vapor. In detail, lignin in archaeological wood generally undergoes modifications such as the partial breakage of β-O-4 interlinks, an alternation of lignin structure, and demethylation/demethoxylation [7,54,70,71]. Although there is few studies on the relationship between WAW and the number of available OH groups, it was reported that the amount of available OH groups of delignified wood increased with the degree of delignification [29]. Both WAW and delignified wood possesses more OH groups because their lignin appears to be less bound to the carbohydrate matrix, enabling a flexible conformation. Additionally, the structure of lignin was partly altered, similar to extracted lignin [29,69]. In the current study, the increase in relative content of lignin of moderately decayed WAW (shown in Table 3) was caused by the loss of cellulose and hemicellulose, which implied that the lignin must have been modified to some degree throughout the deterioration in order to explain the increased hygroscopicity. To sum up, the demonstrated increase of hygroscopicity in more decayed WAW can be attributed to changes of relative components of the cell wall. It could be demonstrated that interpretations regarding hygroscopicity can correlate well between compositional analysis and hydroxyl accessibility measurements.

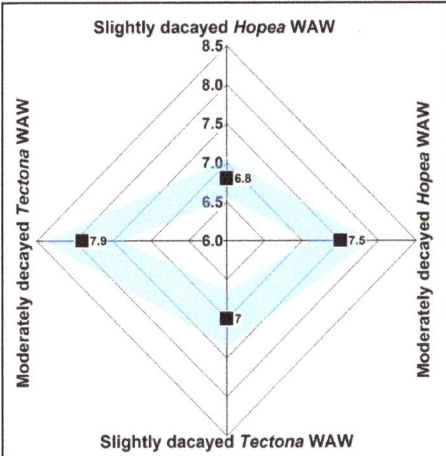

Figure 7. Radar image of the number (mmol/g) of accessible hydroxyl groups of the less decayed (H1, T1) and the moderately decayed (H2, T2) samples. The mean values (black squares) and the standard deviation (blue bands) were calculated from statistics for n = 3 measurements using representative samples.

4. Conclusions

Two hardwood species of waterlogged archaeological wood, *Hopea* and *Tectona*, were collected from a 170-year-old shipwreck, the Xiaobaijiao No.1, in order to investigate the effect of the deterioration on the hygroscopicity. Two deterioration states i.e., moderately decayed and less decayed WAW were confirmed by LM and SEM. In terms of water vapor sorption, moderately decayed WAW showed the higher EMCs and hysteresis than less decayed WAW. The fitted sorption models revealed that, towards

WAW, the monolayer water was bound more strongly to primary sorption sites and the number of primary sorption sites increased with deterioration. The observed increase in hygroscopicity was attributed to a more severe deterioration of the wood chemical components of moderately decayed WAW in contrast to the less decayed WAW. In addition to higher hygroscopicity, moderately decayed WAW showed higher hysteresis than less decayed WAW, particularly above 60% RH, which is a domain close to museum humidity conditions (50%–60% RH for wooden artifacts in China). Therefore, considering the irreversible alternations in chemistry of WAW, the active reduction of sorption sites of WAW could be recommended as a possible strategy for conservation. Differentiated and appropriate conservation treatments would be required for WAW in different deterioration states.

Author Contributions: Conceptualization, L.H., J.G., X.T. and Y.Y.; methodology, L.H. and J.G.; software, L.H., K.W., P.G. and R.L.; validation, L.H., K.W., P.G. and R.L.; formal analysis, L.H. and K.W.; investigation, L.H.; resources, L.H., K.W., P.G., R.L. and X.T.; data curation, L.H., K.W. and P.G.; writing—original draft preparation, L.H. and J.G.; writing—review and editing, L.H., J.G., P.G., Y.Y.; visualization, L.H.; supervision, J.G., X.T. and Y.Y.; project administration, L.H., K.W. and P.G.; funding acquisition, J.G. and L.H. All authors have read and agreed to the published version of the manuscript.

Funding: This work was funded by the Fundamental Research Funds for the Central Non-profit Research Institution of CAF, grant number CAFYBB2018QB006, the Chinese National Natural Science Foundation, grant number 310600450 and the scholarship from China Scholarship Council, grant number CSC201803270037.

Acknowledgments: Tobias Keplinger and Ingo Burgert from the Wood Materials Science group at ETH Zürich are acknowledged for hosting the first author and providing access to equipment and giving comments on the manuscript. The authors would like to thank Xiaomei Jiang and Yonggang Zhang from Research Institute of Wood Industry, Chinese Academy of Forestry, Kunkun Tu and Stéphane Croptier from Wood Materials Science, ETH Zürich and Zheng Jia from Heritage Conservation and Restoration Institute, Chinese Academy of Cultural Heritage for their technical supports.

Conflicts of Interest: The authors declare no conflict of interest.

References

1. Seborg, R.M.; Inverarity, R.B. Preservation of old, waterlogged wood by treatment with polyethylene glycol. *Science* **1962**, *136*, 649–650. [CrossRef] [PubMed]
2. Walsh-Korb, Z.; Avérous, L. Recent developments in the conservation of materials properties of historical wood. *Prog. Mater. Sci.* **2019**, *102*, 167–221.
3. Sandström, M.; Jalilehvand, F.; Persson, I.; Gelius, U.; Frank, P. Deterioration of the seventeenth-century warship Vasa by internal formation of sulphuric acid. *Nature* **2002**, *415*, 893–897.
4. Hoffmann, P.; Singh, A.; Kim, Y.S.; Wi, S.G.; Kim, I.-J.; Schmitt, U. The Bremen Cog of 1380–an electron microscopic study of its degraded wood before and after stabilization with PEG. *Holzforschung* **2004**, *58*, 211–218. [CrossRef]
5. Bjurhager, I.; Halonen, H.; Lindfors, E.L.; Iversen, T.; Almkvist, G.; Gamstedt, E.K.; Berglund, L.A. State of degradation in archeological oak from the 17th century Vasa ship: Substantial strength loss correlates with reduction in (holo)cellulose molecular weight. *Biomacromolecules* **2012**, *13*, 2521–2527. [CrossRef] [PubMed]
6. Macchioni, N.; Pizzo, B.; Capretti, C.; Giachi, G. How an integrated diagnostic approach can help in a correct evaluation of the state of preservation of waterlogged archaeological wooden artefacts. *J. Archaeol. Sci.* **2012**, *39*, 3255–3263. [CrossRef]
7. Guo, J.; Xiao, L.; Han, L.; Wu, H.; Yang, T.; Wu, S.; Yin, Y.; Donaldson, L.A. Deterioration of the cell wall in waterlogged wooden archeological artifacts, 2400 years old. *IAWA J.* **2019**, *1*, 1–25. [CrossRef]
8. Macchioni, N.; Pizzo, B.; Capretti, C. An investigation into preservation of wood from Venice foundations. *Constr. Build. Mater.* **2016**, *111*, 652–661. [CrossRef]
9. Macchioni, N.; Capretti, C.; Sozzi, L.; Pizzo, B. Grading the decay of waterlogged archaeological wood according to anatomical characterisation. The case of the Fiavé site (N-E Italy). *Int. Biodeter. Biodegr.* **2013**, *84*, 54–64. [CrossRef]
10. De Jong, J. Conservation techniques for old waterlogged wood from shipwrecks found in the Netherlands. *Biodeterior. Invest. Tech.* **1977**, *113*, 295–338.

11. Babiński, L.; Izdebska-Mucha, D.; Waliszewska, B. Evaluation of the state of preservation of waterlogged archaeological wood based on its physical properties: Basic density vs. wood substance density. *J. Archaeol. Sci.* **2014**, *46*, 372–383. [CrossRef]
12. Broda, M.; Mazela, B.; Krolikowska-Pataraja, K.; Siuda, J. The state of degradation of waterlogged wood from different environments. *Ann. Wars. Univ. Life Sci. SGGW For. Wood. Technol.* **2015**, *91*, 23–27.
13. Irbe, I.; Bikovens, O.; Fridrihsone, V.; Dzenis, M. Impact of biodeterioration on structure and composition of waterlogged foundation piles from Riga Cathedral (1211 CE), Latvia. *J. Archaeol. Sci. Rep.* **2019**, *23*, 196–202. [CrossRef]
14. Walsh, Z.; Janeček, E.R.; Hodgkinson, J.T.; Sedlmair, J.; Koutsioubas, A.; Spring, D.R.; Welch, M.; Hirschmugl, C.J.; Toprakcioglu, C.; Nitschke, J.R. Multifunctional supramolecular polymer networks as next-generation consolidants for archaeological wood conservation. *Proc. Natl. Acad. Sci. USA* **2014**, *111*, 17743–17748. [CrossRef] [PubMed]
15. Skaar, C. *Wood-Water Relations*; Springer Science & Business Media: Heidelberg, Germany, 2012.
16. Zhan, T.; Jiang, J.; Lu, J.; Zhang, Y.; Chang, J. Frequency-dependent viscoelastic properties of Chinese fir (*Cunninghamia lanceolata*) under hygrothermal conditions. Part 1: Moisture adsorption. *Holzforschung* **2019**, *73*, 727–736. [CrossRef]
17. Choong, E.T.; Achmadi, S.S. Effect of extractives on moisture sorption and shrinkage in tropical woods. *Wood Fiber Sci.* **2007**, *23*, 185–196.
18. Rowell, R.M.; Banks, W.B. *Water Repellency and Dimensional Stability of Wood*; Gen. Tech. Rep. FPL-50; US Department of Agriculture, Forest Service, Forest Products Laboratory: Madison, WI, USA, 1985; 24p.
19. Fackler, K.; Schwanninger, M. Accessibility of hydroxyl groups of brown-rot degraded spruce wood to heavy water. *J. Near Infrared Spec.* **2011**, *19*, 359–368. [CrossRef]
20. Rowell, R. *Protection of Wood Against Biodegradation by Chemical Modification*; Ellis Horwood: Hemel Hempstead, UK, 1993.
21. Thybring, E.E.; Glass, S.V.; Zelinka, S.L. Kinetics of Water Vapor Sorption in Wood Cell Walls: State of the Art and Research Needs. *Forests* **2019**, *10*, 704. [CrossRef]
22. Esteban, L.G.; de Palacios, P.; Fernández, F.G.; Guindeo, A.; Conde, M.; Baonza, V. Sorption and thermodynamic properties of juvenile Pinus sylvestris L. wood after 103 years of submersion. *Holzforschung* **2008**, *62*, 745–751. [CrossRef]
23. Esteban, L.G.; de Palacios, P.; Fernández, F.G.; Martín, J.A.; Génova, M.; Fernández-Golfín, J.I. Sorption and thermodynamic properties of buried juvenile Pinus sylvestris L. wood aged 1,170±40 BP. *Wood Sci. Technol.* **2009**, *43*, 679–690. [CrossRef]
24. Singh, A.P.; Kim, Y.S.; Chavan, R.R. Relationship of wood cell wall ultrastructure to bacterial degradation of wood. *IAWA J.* **2019**, *40*, 845–870. [CrossRef]
25. Rhim, J.W.; Lee, J.H. Thermodynamic Analysis of Water Vapor Sorption Isotherms and Mechanical Properties of Selected Paper-Based Food Packaging Materials. *J. Food Sci.* **2009**, *74*, E502–E511. [CrossRef] [PubMed]
26. Zhang, X.; Li, J.; Yu, Y.; Wang, H. Investigating the water vapor sorption behavior of bamboo with two sorption models. *J. Mater. Sci.* **2018**, *53*, 8241–8249. [CrossRef]
27. Deng, Q. *The Investigation and Excavation of Xiaobaijiao No. 1 Shipwreck Site of Qing Dynasty in East Sea of China, in Early Navigation in the Asia-Pacific Region*; Springer: Berlin/Heidelberg, Germany, 2016; pp. 241–269.
28. Han, L.; Tian, X.; Zhou, H.; Yin, Y.; Guo, J. The Influences of the Anatomical Structure and Deterioration State of Wood from a Qing Dynasty Shipwreck on Wood Color after the Consolidation Treatment. *J. Southwest Forst. Univ. (Nat. Sci.)* **2020**, *40*, 1–7.
29. Grönquist, P.; Frey, M.; Keplinger, T.; Burgert, I. Mesoporosity of Delignified Wood Investigated by Water Vapor Sorption. *ACS Omega* **2019**, *4*, 12425–12431. [CrossRef] [PubMed]
30. Timmermann, E.O. Multilayer sorption parameters: BET or GAB values? *Colloids Surf. A Physicochem. Eng. Asp.* **2003**, *220*, 235–260. [CrossRef]
31. Wolf, W.; Spiess, W.; Jung, G. *Standardization of Isotherm Measurements (COST-Project 90 and 90 bis), in Properties of Water in Foods*; Springer: Berlin/Heidelberg, Germany, 1985; pp. 661–679.
32. Majka, J.; Babiński, L.; Olek, W. Sorption isotherms of waterlogged subfossil Scots pine wood impregnated with a lactitol and trehalose mixture. *Holzforschung* **2017**, *71*, 813–819. [CrossRef]
33. Basu, S.; Shivhare, U.; Mujumdar, A. Models for sorption isotherms for foods: A review. *Dry. Technol.* **2006**, *24*, 917–930. [CrossRef]

34. Kozłowska, A.; Kozłowski, R. Analysis of water adsorption by wood using the Guggenheim-Anderson-de Boer equation. *Eur. J. Wood Prod.* **2012**, *70*, 445–451.
35. Furmaniak, S.; Terzyk, A.P.; Gauden, P.A.; Rychlicki, G. Applicability of the generalised D'Arcy and Watt model to description of water sorption on pineapple and other foodstuffs. *J. Food Eng.* **2007**, *79*, 718–723. [CrossRef]
36. Sluiter, A.; Hames, B.; Ruiz, R.; Scarlata, C.; Sluiter, J.; Templeton, D.; Crocker, D. Determination of structural carbohydrates and lignin in biomass. *Lab. Anal. Proced.* **2008**, *1617*, 1–16.
37. Han, L.; Wang, K.; Wang, W.; Guo, J.; Zhou, H. Nanomechanical and Topochemical Changes in Elm Wood from Ancient Timber Constructions in Relation to Natural Aging. *Materials* **2019**, *12*, 786. [CrossRef] [PubMed]
38. Thybring, E.E.; Thygesen, L.G.; Burgert, I. Hydroxyl accessibility in wood cell walls as affected by drying and re-wetting procedures. *Cellulose* **2017**, *24*, 2375–2384. [CrossRef]
39. Popescu, C.-M.; Hill, C.A.S.; Curling, S.; Ormondroyd, G.; Xie, Y. The water vapour sorption behaviour of acetylated birch wood: How acetylation affects the sorption isotherm and accessible hydroxyl content. *J. Mater. Sci.* **2013**, *49*, 2362–2371. [CrossRef]
40. Thommes, M.; Kaneko, K.; Neimark, A.V.; Olivier, J.P.; Rodriguez-Reinoso, F.; Rouquerol, J.; Sing, K.S. Physisorption of gases, with special reference to the evaluation of surface area and pore size distribution (IUPAC Technical Report). *Pure Appl. Chem.* **2015**, *87*, 1051–1069. [CrossRef]
41. Broda, M.; Majka, J.; Olek, W.; Mazela, B. Dimensional stability and hygroscopic properties of waterlogged archaeological wood treated with alkoxysilanes. *Int. Biodeter. Biodegr.* **2018**, *133*, 34–41. [CrossRef]
42. Brunauer, S.; Deming, L.S.; Deming, W.E.; Teller, E. On a theory of the van der Waals adsorption of gases. *J. Am. Chem. Soc.* **1940**, *62*, 1723–1732. [CrossRef]
43. Popescu, C.-M.; Hill, C.A.; Kennedy, C. Variation in the sorption properties of historic parchment evaluated by dynamic water vapour sorption. *J. Cult.* **2016**, *17*, 87–94. [CrossRef]
44. Engelund, E.T.; Thygesen, L.G.; Svensson, S.; Hill, C.A.S. A critical discussion of the physics of wood–water interactions. *Wood Sci. Technol.* **2012**, *47*, 141–161. [CrossRef]
45. Fredriksson, M.; Thybring, E.E. Scanning or desorption isotherms? Characterising sorption hysteresis of wood. *Cellulose* **2018**, *25*, 4477–4485. [CrossRef]
46. Olek, W.; Majka, J.; Stempin, A.; Sikora, M.; Zborowska, M. Hygroscopic properties of PEG treated archaeological wood from the rampart of the 10th century stronghold as exposed in the Archaeological Reserve Genius loci in Poznań (Poland). *J. Cult.* **2016**, *18*, 299–305. [CrossRef]
47. Broda, M.; Curling, S.F.; Spear, M.J.; Hill, C.A.S. Effect of methyltrimethoxysilane impregnation on the cell wall porosity and water vapour sorption of archaeological waterlogged oak. *Wood Sci. Technol.* **2019**, *53*, 703–726. [CrossRef]
48. Esteban, L.G.; Simón, C.; Fernández, F.G.; de Palacios, P.; Martín-Sampedro, R.; Eugenio, M.E.; Hosseinpourpia, R. Juvenile and mature wood of Abies pinsapo Boissier: Sorption and thermodynamic properties. *Wood Sci. Technol.* **2015**, *49*, 725–738. [CrossRef]
49. De Oliveira, G.H.H.; Corrêa, P.C.; de Oliveira, A.P.L.R.; Reis, R.C.d.; Devilla, I.A. Application of GAB model for water desorption isotherms and thermodynamic analysis of sugar beet seeds. *J. Food Process Eng.* **2017**, *40*, e12278. [CrossRef]
50. Lewicki, P.P. The applicability of the GAB model to food water sorption isotherms. *Int. J. Food Sci.* **1997**, *32*, 553–557. [CrossRef]
51. Maskan, M.; Göğüş, F. The fitting of various models to water sorption isotherms of pistachio nut paste. *J. Food Eng.* **1997**, *33*, 227–237. [CrossRef]
52. Salmén, L.; Burgert, I. Cell wall features with regard to mechanical performance. A review COST Action E35 2004–2008: Wood machining—Micromechanics and fracture. *Holzforschung* **2009**, *63*, 121–129. [CrossRef]
53. Pettersen, R.C. *The Chemical Composition of Wood*; ACS Publications: Washington, DC, USA, 1984.
54. Xia, Y.; Chen, T.Y.; Wen, J.L.; Zhao, Y.L.; Qiu, J.; Sun, R.C. Multi-analysis of chemical transformations of lignin macromolecules from waterlogged archaeological wood. *Int. J. Biol. Macromol.* **2017**, *109*, 407–416. [CrossRef]
55. Altgen, M.; Willems, W.; Hosseinpourpia, R.; Rautkari, L. Hydroxyl accessibility and dimensional changes of Scots pine sapwood affected by alterations in the cell wall ultrastructure during heat-treatment. *Polym. Degrad. Stabil.* **2018**, *152*, 244–252. [CrossRef]

56. Hill, C.A. *Wood Modification: Chemical, Thermal and Other Processes*; John Wiley & Sons: Hoboken, NJ, USA, 2007; Volume 5.
57. Zelinka, S.L. Preserving ancient artifacts for the next millennia. *Proc. Natl. Acad. Sci. USA* **2014**, *111*, 17700–17701. [CrossRef]
58. Popescu, C.-M.; Dobele, G.; Rossinskaja, G.; Dizhbite, T.; Vasile, C. Degradation of lime wood painting supports. *J. Anal. Appl. Pyrol.* **2007**, *79*, 71–77. [CrossRef]
59. Popescu, C.-M.; Hill, C.A.S. The water vapour adsorption–desorption behaviour of naturally aged Tilia cordata Mill. wood. *Polym. Degrad. Stabil.* **2013**, *98*, 1804–1813. [CrossRef]
60. Rautkari, L.; Hill, C.A.; Curling, S.; Jalaludin, Z.; Ormondroyd, G. What is the role of the accessibility of wood hydroxyl groups in controlling moisture content? *J. Mater. Sci.* **2013**, *48*, 6352–6356. [CrossRef]
61. Guo, X.; Liu, L.; Hu, Y.; Wu, Y. Water vapor sorption properties of TEMPO oxidized and sulfuric acid treated cellulose nanocrystal films. *Carbohydr. Polym.* **2018**, *197*, 524–530. [CrossRef] [PubMed]
62. Sharma, P.R.; Joshi, R.; Sharma, S.K.; Hsiao, B.S. A simple approach to prepare carboxycellulose nanofibers from untreated biomass. *Biomacromolecules* **2017**, *18*, 2333–2342. [CrossRef] [PubMed]
63. Klemm, D.; Heublein, B.; Fink, H.P.; Bohn, A. Cellulose: Fascinating biopolymer and sustainable raw material. *Angew. Chem. Int. Ed.* **2005**, *44*, 3358–3393. [CrossRef] [PubMed]
64. Hofstetter, K.; Hinterstoisser, B.; Salmén, L. Moisture uptake in native cellulose–the roles of different hydrogen bonds: A dynamic FT-IR study using Deuterium exchange. *Cellulose* **2006**, *13*, 131–145. [CrossRef]
65. Fernandes, A.N.; Thomas, L.H.; Altaner, C.M.; Callow, P.; Forsyth, V.T.; Apperley, D.C.; Kennedy, C.J.; Jarvis, M.C. Nanostructure of cellulose microfibrils in spruce wood. *Proc. Natl. Acad. Sci. USA* **2011**, *108*, 1195–1203. [CrossRef]
66. Šturcová, A.; His, I.; Apperley, D.C.; Sugiyama, J.; Jarvis, M.C. Structural details of crystalline cellulose from higher plants. *Biomacromolecules* **2004**, *5*, 1333–1339. [CrossRef]
67. Eyley, S.; Thielemans, W. Surface modification of cellulose nanocrystals. *Nanoscale* **2014**, *6*, 7764–7779. [CrossRef]
68. Mihranyan, A.; Llagostera, A.P.; Karmhag, R.; Strømme, M.; Ek, R. Moisture sorption by cellulose powders of varying crystallinity. *Int. J. Pharm.* **2004**, *269*, 433–442. [CrossRef] [PubMed]
69. Han, L.; Tian, X.; Keplinger, T.; Zhou, H.; Li, R.; Svedström, K.; Burgert, I.; Yin, Y.; Guo, J. Even Visually Intact Cell Walls in Waterlogged Archaeological Wood Are Chemically Deteriorated and Mechanically Fragile: A Case of a 170 Year-Old Shipwreck. *Molecules* **2020**, *25*, 1113. [CrossRef] [PubMed]
70. Colombini, M.P.; Lucejko, J.J.; Modugno, F.; Orlandi, M.; Tolppa, E.L.; Zoia, L. A multi-analytical study of degradation of lignin in archaeological waterlogged wood. *Talanta* **2009**, *80*, 61–70. [CrossRef] [PubMed]
71. Pedersen, N.B.; Gierlinger, N.; Thygesen, L.G. Bacterial and abiotic decay in waterlogged archaeological Picea abies (L.) Karst studied by confocal Raman imaging and ATR-FTIR spectroscopy. *Holzforschung* **2015**, *69*, 103–112. [CrossRef]

© 2020 by the authors. Licensee MDPI, Basel, Switzerland. This article is an open access article distributed under the terms and conditions of the Creative Commons Attribution (CC BY) license (http://creativecommons.org/licenses/by/4.0/).

Article

Study on the Properties of Partially Transparent Wood under Different Delignification Processes

Yan Wu [1,2,*], Jichun Zhou [1,2], Qiongtao Huang [3], Feng Yang [4,*], Yajing Wang [1,2] and Jing Wang [1,2]

1. College of Furnishings and Industrial Design, Nanjing Forestry University, Nanjing 210037, China; 15250988513@163.com (J.Z.); lionzaka@163.com (Y.W.); wangjing_9711@163.com (J.W.)
2. Co-Innovation Center of Efficient Processing and Utilization of Forest Resources, Nanjing Forestry University, Nanjing 210037, China
3. Department of Research and Development Center, Yihua Lifestyle Technology Co., Ltd., Shantou 515834, China; huangqt@yihua.com
4. Fashion Accessory Art and Engineering College, Beijing Institute of Fashion Technology, Beijing 100029, China
* Correspondence: wuyan@njfu.edu.cn (Y.W.); yangfeng@bift.edu.cn (F.Y.)

Received: 29 February 2020; Accepted: 14 March 2020; Published: 15 March 2020

Abstract: Two common tree species of Betula alnoides (*Betula*) and New Zealand pine (*Pinups radiata D. Don*) were selected as the raw materials to prepare for the partially transparent wood (TW) in this study. Although the sample is transparent in a broad sense, it has color and pattern, so it is not absolutely colorless and transparent, and is therefore called partially transparent. For ease of interpretation, the following "partially transparent wood" is referred to as "transparent wood (TW)". The wood template (FW) was prepared by removing part of the lignin with the acid delignification method, and then the transparent wood was obtained by impregnating the wood template with a refractive-index-matched resin. The goal of this study is to achieve transparency of the wood (the light transmittance of the prepared transparent wood should be improved as much as possible) by exploring the partial delignification process of different tree species on the basis of retaining the aesthetics, texture and mechanical strength of the original wood. Therefore, in the process of removing partial lignin by the acid delignification method, the orthogonal test method was used to explore the better process conditions for the preparation of transparent wood. The tests of color difference, light transmittance, porosity, microstructure, chemical groups, mechanical strength were carried out on the wood templates and transparent wood under different experimental conditions. In addition, through the three major elements (lignin, cellulose, hemicellulose) test and orthogonal range analysis method, the influence of each process factor on the lignin removal of each tree species was obtained. It was finally obtained that the two tree species acquired the highest light transmittance at the experimental level 9 (process parameters: $NaClO_2$ concentration 1 wt%, 90 °C, 1.5 h); and the transparent wood retained most of the color and texture of the original wood under partial delignification up to 4.84–11.07%, while the mechanical strength with 57.76% improved and light transmittance with 14.14% higher than these properties of the original wood at most. In addition, the wood template and resin have a good synergy effect from multifaceted analysis, which showed that this kind of transparent wood has the potential to become the functional decorative material.

Keywords: transparent wood; orthogonal test; partial delignification; light transmittance

1. Introduction

Transparent wood is usually prepared by impregnating delignified wood template with a refractive-index-matched resin [1]. In recent years, transparent wood, an emerging achievement in wood modification, has attracted attention and research due to its many advantages such as light weight, light transmission, environmental protection, and high mechanical properties. Most of the experiments focus on electronic equipment, optical devices and energy-saving buildings, etc. For example, Li et al. [2] used 2–5 cm thick translucent wood composite material (light transmittance 40%) as the wall material of the house model. The wall material is made by the method of H_2O_2 vapor delignification and epoxy resin impregnation, which can effectively capture the external environment light. In addition, unlike the transparent wood roof, the light intensity inside the translucent wood house model is more conducive to people's daily life. In 2019, Wang et al. [3] used photochromic materials to infiltrate modified wood templates to obtain photochromic transparent wood. This kind of transparent wood appears a bright purple to colorless color change under light and shows about 65% good optical transmittance and 90% high optical haze, which is very important for the application of Smart windows and anti-counterfeiting materials. In the same year, Li et al. [4] successfully assembled the perovskite solar cells treated at low temperature (<150 °C) directly on the transparent wood substrate for the first time, with the power conversion efficiency up to 16.8%, which confirmed that the transparent wood is suitable as a substrate for solar cell modules and has potential in energy-efficient building applications. Celine Montanari et al. [5] also put forward the view that transparent wood should be endowed with multiple functions. Using polyethylene glycol as the base material, PCMs (phase-change materials) with stable shape was embedded in delignified wood templates to obtain functional transparent wood. The application of the heat storage and reversible light transmittance of the transparent wood in thermal energy storage was discussed.

In fact, the development and application of transparent wood reflects the broad development prospects of wood composite materials [6], and the numerical improvement of light transmittance is no longer a single pursuit. On the contrary, people are paying more and more attention to the functionality of transparent wood and its corresponding application fields. At present, most of the transparent or translucent wood tends to be glass, colorless and no texture, and lacks the visual and tactile natural characteristics of wood [7]. Based on the inspiration given by this cutting-edge subject, from another perspective, if wood still has a certain degree of light transmission on the basis of retaining most of the natural color and texture, then in the home industry, transparent wood can also become a highly functional decorative material. This concept increases the application of wood modification in some fields [8].

The process of preparing transparent wood mainly involves the acquisition of delignified wood template and the impregnation of refractive-index-matched resin. It is not difficult to find that delignification is usually an important step in the preparation of transparent wood. Regarding lignin, it accounts for about 30% of the wood mass fraction, and cross-linking with different polysaccharides in the wood increases the mechanical strength of the wood. Components such as lignin and wood extracts cause the wood to undergo strong light scattering and light absorption in the visible light range [9], so that the wood shows color and texture. Because the purpose of this experiment is to prepare transparent wood with color and texture (not absolutely transparent), it is necessary to retain a part of the lignin to achieve the effect of color development, that is, to study the process of removing part of the lignin. In fact, it takes a long time and a lot of chemicals to remove all the lignin [10]. In addition, the removal of all lignin will weaken the wood structure, mainly for the wood template, such as pine, poplar and other low-density tree species are easily broken after delignification, so it is a great challenge for preparing the transparent wood of large-scale or low-density tree species [11]. Therefore, it is of great significance to explore the partial delignification process for the preparation of transparent wood. Although the sample to be prepared is transparent in a broad sense, it has color and pattern, and thus is not absolutely colorless and transparent. Therefore, it was called partially transparent in the title. For ease of interpretation, the following "partially transparent wood" is referred to as "transparent wood (TW)".

It is known in pre-experiments and references that the amount of lignin removal has a direct impact on the performance of transparent wood, but there is almost no systematic research on the specific impact of removing part of the lignin process on the performance of transparent wood, including the related qualitative and quantitative analysis, and most of the literature on transparent wood removes almost all lignin or chromogenic substances. Therefore, the purpose of this experiment is to study the process technology of partial delignification and the performance of the corresponding transparent wood. The above has certain novelty. It is hoped that the transparency of the transparent wood prepared from different tree species can be improved as much as possible on the basis of retaining the aesthetics, texture and mechanical strength of the wood. In this experiment, the orthogonal test method was used to explore the best technological conditions of removing part of the lignin in the process of preparing transparent wood.

2. Experimental Section

2.1. Materials

Here, common tree species (one example of dark tree species and one light tree species) are selected: Betula alnoides (*Betula*) and New Zealand pine (*Pinups radiata D. Don*), which are representative tree species in hardwood and coniferous wood respectively. The wood veneers of Betula alnoides and New Zealand pine that are both from Yihua Lifestyle Technology Co., Ltd., China., and these wood veneers are produced in the actual production line. In the experiment, the wood samples were cut to the size of 20 mm (length)×20 mm (width)×0.5 mm (thickness). The physical properties of the two tree species such as air-dry density relative, moisture content and thickness are shown in Table 1. Chemical reagents (analytical grade) used are as follows: ethanol absolute, acetic acid (CH_3COOH) and sodium hydroxide (NaOH) were all produced by Nanjing Chemical Reagent Co., Ltd. Methyl methacrylate (MMA) and sodium hypochlorite ($NaClO_2$) were supplied by Shanghai Macklin Biochemical Co., Ltd. Azobisisobutyronitrile (AIBN) was supplied by Tianjin Benchmark Chemical Reagent Co., Ltd.

Table 1. Physical properties of two tree species.

Wood Species	Air-Dry Density Relative (g/cm^3)	Moisture Content (%)	Thickness (mm)
Betula alnoides (A)	0.65	9.96	0.50
New Zealand pine (B)	0.31	10.01	0.50

2.2. Experimental Methods

In the experiment, acid delignification was applied, and chlorite method was specifically used to prepare the wood template. After obtaining the wood template, a refractive-index-matched resin was impregnated to make the transparent wood. The orthogonal test method is a scientific and effective method for selecting the optimal scheme in multi-factor experiments. With the least number of tests and the most appropriate test method, an optimal test condition and the best scheme will be obtained [12]. Therefore, in the process of removing part of the lignin by the acid method, the orthogonal test method was selected in the experiment to explore the better process conditions for the preparation of transparent wood. With regard to the design of orthogonal experiments, combined with the reading reference of relevant literature and the experience of preliminary experiments, the following three representative factors are selected in the experiment: $NaClO_2$ concentration (X), reaction temperature (Y), reaction time (Z). The formulation of specific experimental factor levels is shown in Table 2.

Table 2. Factor level of the orthogonal experiment.

	X/wt%	Y/°C	Z/min
1	0.4	70	45
2	0.7	80	90
3	1	90	135

According to the principle of orthogonal experiment design method [13], the orthogonal table L_9 (3^4) is used to arrange the three factors/three levels of tests (L_9 means that nine experiments are needed, at most four factors can be observed, and each factor is three levels). Better production conditions can be determined from these nine sets of data [14]. Table 3 is the proposed orthogonal test table. Therefore, each tree species corresponds to nine different wood templates.

Table 3. Orthogonal test table on the preparation of wood templates.

	XYZ	X (wt%)	Y (°C)	Z (min)
1	111	0.4	70	45
2	122	0.4	80	90
3	133	0.4	90	135
4	212	0.7	70	90
5	223	0.7	80	135
6	231	0.7	90	45
7	313	1	70	135
8	321	1	80	45
9	332	1	90	90

2.3. Preparation of Transparent Wood

The following preparation process is shown in Figure 1.

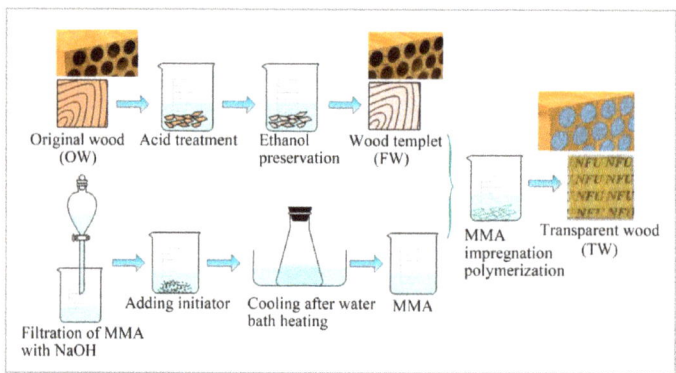

Figure 1. The main preparation process of transparent wood (TW).

2.3.1. Preparation of Wood Templates

Betula alnoides and New Zealand pine were dried at 103 °C for a few minutes until they were completely dry, and then stored in a drying dish. Subsequently, a certain concentration of $NaClO_2$ solution was prepared, and the pH value was adjusted to 4.6 by CH_3COOH. The samples of the two tree species were heat-treated according to the values in the orthogonal experiment table, and wood templates under different experimental conditions were obtained. The above treated wood templates were washed with deionized water and stored in the ethanol absolute solution. The ethanol absolute solution can displace the residual water in the wood, thus greatly improving the permeability of wood templates.

2.3.2. Preparation of the Polymer

The polymerization inhibitor inside the pure MMA monomer was removed by NaOH solution, and then the MMA was prepolymerized in the water bath at 75 °C, in which 0.38 wt% AIBN was used as the initiator. After 15 minutes, the prepolymerized MMA was cooled to room temperature in the ice water bath to terminate the reaction.

2.3.3. Obtainment of the Transparent Wood

Take out the wood templates stored in the ethanol absolute solution, and let the wood templates and the prepolymerized MMA fully infiltrate for 0.5–1 h under vacuum. Then, the resin-infiltrated wood template was sandwiched in a mold formed by two glass slides and covered them with aluminum foil paper. Finally, the samples were put into an oven at 70 °C for 5 h to complete the further polymerization reaction [15]. The above preparation process was repeated several times to obtain transparent wood under different experimental conditions [16]. In order to facilitate the following experimental test analysis, Betula alnoides is abbreviated as A and New Zealand pine is abbreviated as B; untreated wood samples are collectively referred to as OW, wood templates after removing some lignin are collectively referred to as FW, and the transparent wood after impregnation is collectively referred to as TW. Taking Betula alnoides as an example, untreated wood is OW-A, and the wood templates treated with $NaClO_2$ solution (one-to-one corresponding to the setting level of orthogonal test) are labeled FW-A-1, FW-A-2, FW-A-3, FW-A-4, FW-A-5, FW-A-6, FW-A-7, FW-A-8, FW-A-9; the transparent wood made from the corresponding wood templates are named TW-A-1, TW-A-2, TW-A-3, TW-A-4, TW-A-5, TW-A-6, TW-A-7, TW-A-8, TW-A-9. New Zealand pine (B) is marked in the same way as Betula alnoides, except that A is replaced by B.

3. Performance Testing

3.1. Three Elements Test

The experiment explores the influence of partial delignification process on the properties of transparent wood, and the main components of wood were analyzed qualitatively and quantitatively at each orthogonal test level through the test of three elements (lignin, cellulose, and hemicellulose), in order to refine and compare the corresponding sample characteristics under different process parameters. In the experiment, the contents of three elements in Betula alnoides and New Zealand pine raw samples OW, wood templates samples FW were measured by the National Renewable Energy Laboratory (NREL) method [17].

3.2. Color Difference Test

The color reader (PANTONE) was used to measure the color difference between Betula alnoides and New Zealand pine OW, TW samples, and the changes and causes of wood color before and after the experiment were compared and analyzed [18]. The color of each tree species sample is kept as consistent as possible to reduce the test error. Each sample is tested for at least 3-5 replicates and the average value is obtained. The values of the color difference test for the samples are respectively expressed by three parameters L, a and b: L value reflects the brightness, the higher the value is, the higher the brightness is; a value reflects the red-green degree, the positive value is red, the negative value is green; b value reflects the orange-blue degree, the positive value is orange, the negative value is blue; the larger the absolute value of a and b is, the darker the color is [19].

3.3. Light Transmittance Test

The light transmittance of two tree species OW and TW were measured by the ultraviolet-visible spectrophotometer (Shanghai youke UV1900PC) at 350–800 nm wavelength. Samples under the same experimental conditions were selected for more than two repeated tests to reduce experimental errors.

3.4. SEM Test

The micro morphology of two tree species OW, DW and TW samples were observed by a FEI Quanta 200 scanning electron microscope (SEM). In the experiment, the wood samples were tangential sections obtained by cutting longitudinally along the trunk. In order to observe the change of wood duct or tracheid during the experiment, the samples were cut along the thickness direction by ultra-thin microtome.

3.5. Specific Surface Area and Pore Size Distribution Test

We used an automatic specific surface area and pore size distribution tester (ASAP2020) to test Surface area and Adsorption average pore width of OW (original wood) and FW (wood templates). The relationship between the partial removal of lignin and the distribution of pores in the samples was discussed, as well as the specific effect of changes in pores in these wood templates on light transmittance.

3.6. Fourier Infrared Test

The infrared absorption spectrum was obtained by Fourier transform infrared spectroscopy (FTIR), and then the group characteristics and changes of OW, FW and TW were compared and analyzed by the characteristic absorption peaks presented in the figure, so as to carry out the qualitative analysis of related chemical components.

3.7. Mechanical Performance Test

The mechanical properties of wood before and after the experiment were tested and calculated. A computer-controlled electronic universal testing machine (SANS-CMT6104) was used to measure the mechanical tensile properties of OW and TW. The upper and lower clamps of the testing machine first fixedly clamped the sample, and set no additional load at this time (to reduce the experimental error), then the lower clamp was fixed, the upper clamp stretched the sample in the direction of the wood grain until it broke, and the upward stretching speed was set to 5 mm/min.

The following formulas are used in mechanical tests:

$$\sigma = \frac{F}{S} \tag{1}$$

$$S = b \times h \tag{2}$$

In the formula, σ is the tensile strength; F is the maximum force borne by the specimen when it is broken; S is the original cross-sectional area in the tensile direction of the sample; b is the initial width of the tensile section of the sample, and h is the initial thickness of the tensile section [20].

4. Results and Discussion

4.1. Three Elements Analysis

According to the National Renewable Energy Laboratory (NREL) method and the calculation of fixed formula template (included in NREL), the three major elements content of natural wood samples and delignified wood templates were obtained. The samples tested here need to be processed into 20–80 mesh wood flour and dried until absolutely dry, and it should be noted that the total lignin content here is the sum of acid-insoluble lignin and acid-soluble lignin content [21,22]. The contents of lignin, cellulose, and hemicellulose in Betula alnoides and New Zealand pine OW (original wood) and FW (wood templates) samples are as shown in Figure 2. Generally, wood is mainly composed of cellulose, hemicellulose and lignin, accounting for more than 90% of the total amount of wood, in which cellulose and hemicellulose are colorless substances with simple structure; relatively speaking, the structure of lignin is more complex, which is also one of the main factors of wood coloration [23].

The abscissa 1–9 in Figure 2 respectively corresponds to 9 levels of the orthogonal test (0 is the original wood sample), and each level records the specific content of the remaining three elements of the wood sample after partial delignification. It can be seen from Figure 2 that in the process of removing lignin, the contents of the three major elements are all decreased to varying degrees, indicating that the removal of lignin will have a certain impact on the contents of cellulose and hemicellulose. Because the discussion is centered on the removal of part of the lignin, the core of the process exploration is also closely related to the lignin. Therefore, for the two species of Betula alnoides and New Zealand pine, the multi-factor orthogonal range analysis method [24] is combined with the lignin content after the experiment. As a result, Table 4 was obtained. Through the orthogonal range analysis method, the influence of various process parameters on the partial delignification of each tree species in the multi-factor experiment can be obtained, which is meaningful for the further discussion of the process: it can be obtained from Table 4, for Betula alnoides, the influence order of each factor on the lignin content is Y > X > Z, that is, the reaction temperature > NaClO$_2$ concentration > reaction time; for New Zealand pine, the order of influence of various factors on lignin content is Z > X > Y, that is, reaction time > NaClO$_2$ concentration > reaction temperature.

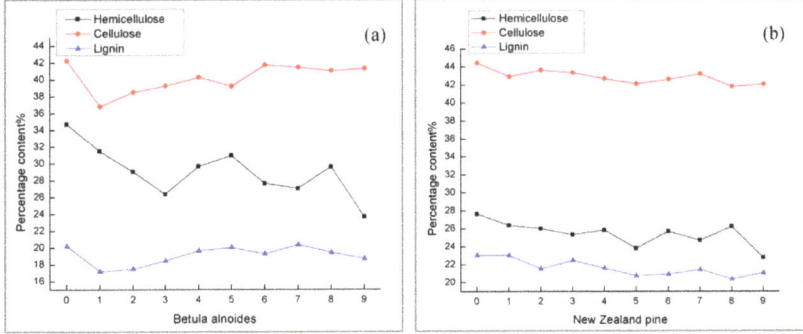

Figure 2. The contents of lignin, cellulose, and hemicellulose in (**a**) Betula alnoides and (**b**) New Zealand pine OW and FW samples.

Table 4. Orthogonal test results and range analysis.

	Betula Alnoides		
Lignin content/%	X	Y	Z
Mean value1	28.95	29.39	29.57
Mean value2	29.42	29.86	27.45
Mean value3	26.77	25.88	28.12
Range	2.65	3.98	2.12
	New Zealand Pine		
Lignin content/%	X	Y	Z
Mean value1	25.91	25.64	26.10
Mean value2	25.11	25.34	24.88
Mean value3	24.58	24.61	24.61
Range	1.33	1.03	1.49

4.2. Color Difference Analysis

For the analysis of color difference values of the above tree species OW, FW and TW, although the specific values are different, the change trend is the same. Therefore, Figure 3 (0 is ow, 1–9 corresponds to nine levels of orthogonal test) is drawn with Betula alnoides as the representative. For the wood template FW, it can be seen from Figure 3 that the remaining lignin content is positively correlated with a and b values, and negatively correlated with L values. It is known that the lignin is one of the main causes of wood color, and the higher the content of lignin, the more orange it is [25]. The positive value of b represents orange, and the larger the value is, the darker the color is. Therefore, the change trend of lignin content and b value is consistent with the theory, that is, the more the residual lignin content of FW is, the greater the b value is. Because the reagent used to remove partial lignin is sodium hypochlorite, which has certain bleachability, the whiteness of the wood template increases, the brightness also increases, and the L value increases. For the TW of the two tree species, the filling of the transparent resin improves the lightness of the wood, and the L value is higher than that of the original wood OW, and lower than that of FW. The a values are slightly reduced, and the b values are increased compared to FW. It can be seen that the filling of the resin not only makes the wood have a certain degree of light transmission, but also retains most of the color and texture of the wood.

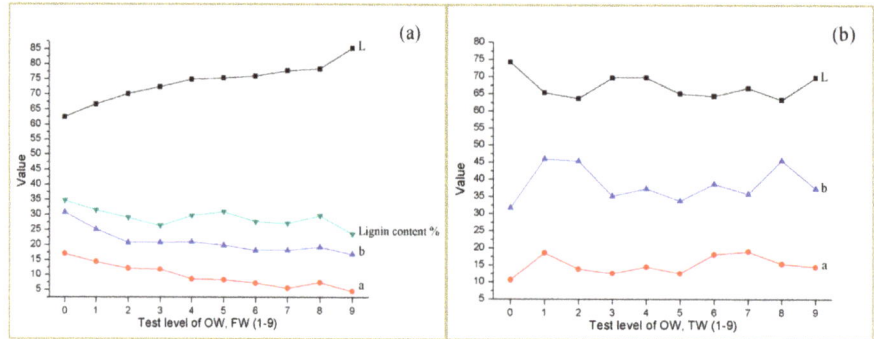

Figure 3. The color difference of OW, (**a**) FW and (**b**) TW.

Figure 4 shows the samples of level 5 and 9 in the orthogonal test, because the amount of delignification in level 5 tends to average compared with the whole, and the content of delignification in level 9 is the highest, which is very representative. The sample with certain light transmittance is placed on the paper printed with "NFU" under the sunlight; and under the condition of specific light source, the light transmittance shows the unique texture and color of wood, which is very novel and beautiful, showing the feasibility of the process and the potential of the material as a functional decoration material.

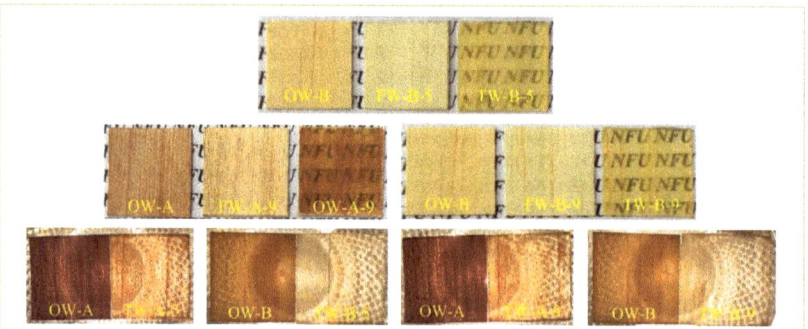

Figure 4. Contrast photos of OW-A, FW-A-5, TW-A-5, OW-B, FW-B-5, TW-B-5, OW-A, FW-A-9, TW-A-9; OW-B, FW-B-9, TW-B-9.

4.3. Light Transmittance Analysis

Figure 5a,b show the light transmittance of Betula alnoides and New Zealand pine OW and TW samples respectively. The numbers 1–9 correspond to the 9 levels of orthogonal test (0 is the original wood sample). The light transmittance of the original wood of the two tree species is very low in the visible light wavelength range, which is mainly due to the light absorption of the color-producing components such as lignin, and also includes the light scattering caused by the porous structure of wood [26]. It can be seen from Figure 5 that the OW light transmittance of the darker Betula alnoides with a higher density is slightly lower than that of the lighter New Zealand pine with a lower density. After the experiment, the light transmittance of TW of the two tree species has been improved compared to the OW, and the light transmittance of New Zealand pine has increased more than Betula alnoides: For Betula alnoides, TW-A-9 has a maximum increase of 11.39% at 800 nm; for New Zealand pine, TW-B-9 has a maximum increase of 14.14% at 800 nm. At the same time, it can be found that the light transmittance values corresponding to 1–9 are negatively correlated with the remaining lignin content, that is, the lower the remaining lignin content is, the more the light transmittance increases. The reason for this is that the more lignin is removed from the wood template, the more pores there are, and the more transparent resin is infiltrated, so the higher the light transmittance of the sample is [27]. In this test, the performance is explored by the process, and the process is deduced from the performance. The goal is to hope that the light transmittance of the wood made of different tree species can be improved as much as possible while retaining the beauty, texture, and strength of the solid wood. Here, it can be concluded that the optimal experimental level for Betula alnoides is 9; the optimal experimental level for New Zealand pine is 9.

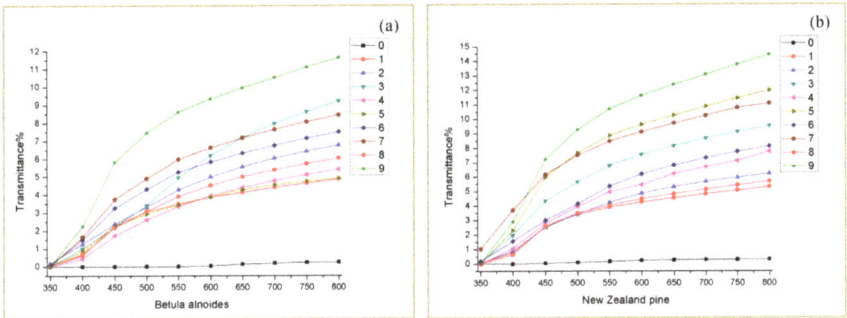

Figure 5. The light transmittance of OW and TW for two tree species (**a**,**b**).

4.4. SEM Analysis

Taking orthogonal test levels 5 and 9 as examples for analysis, since the amount of lignin removed by level 5 tends to be average compared with the whole, and the content of lignin removed by level 9 is the highest, which are all very representative, the observed sample is OW-A, FW-A-5, FW-A-9, TW-A-5, TW-A-9; OW-B, FW-B-5, FW-B-9, TW-B-5, TW-B-9. From the analysis of the three major elements (lignin, cellulose, and hemicellulose), it can be seen that most of the lignin is still retained, so the difference between the wood templates FW of the same tree species is not large. Figure 6a,f present the micro-morphology of OW-A and OW-B respectively, showing the natural porous structure of the wood. It is known that lignin is mainly concentrated in the cell corner intercellular layer, as is the case for birch and New Zealand pine. In the process of removing part lignin, cracks appear in the triangle area where three cells intersect due to the decrease of lignin [28], as shown in Figure 6b,c,g,h. From FW to TW of two tree species, after impregnation polymerization, it was observed that PMMA not only fully filled the cracks but also penetrated into the cells of wood. The honeycomb porous structure of the wood was almost eliminated, and the cell walls were more closely bonded, which indicated that the wood template and PMMA had a good synergistic effect, and played a role in replacing part of the lignin to connect the cellulose skeleton to a certain extent.

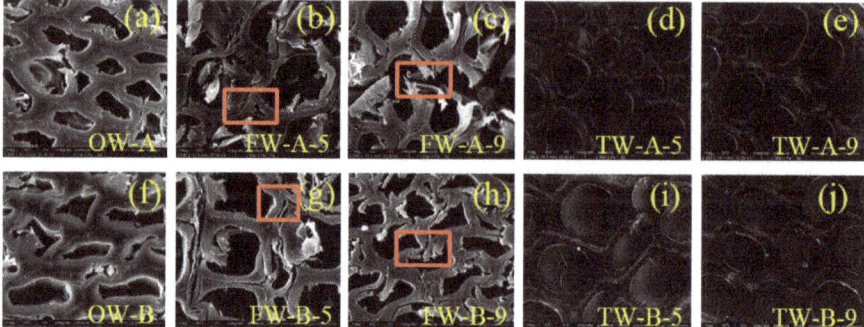

Figure 6. The micro morphology of (**a**) OW-A, (**b**) FW-A-5, (**c**) FW-A-9, (**d**) TW-A-5, (**e**) TW-A-9; (**f**) OW-B, (**g**) FW-B-5, (**h**) FW-B-9, (**i**) TW-B-5, (**j**) TW-B-9.

4.5. Specific Surface Area and Pore Size Distribution Analysis

The BET specific surface area refers to the total area of a unit mass of material [29]. It can be obtained from Figure 7a,b that the BET specific surface area value and the average adsorption pore size value have a negative correlation as a whole. At the same time, the more the delignification content is, the larger the specific surface area is, and the smaller the average adsorption pore diameter is. This is because the original adsorption pore diameter is mostly from the larger conduit hole in the wood. When part of the lignin is removed, there will be a lot of nano-sized micro pore diameters in the wood template, so the average adsorption pore size value decreases, and the specific surface area increases. It is not difficult to find that both the Betula alnoides and New Zealand pine have the highest specific surface area value at experimental level 9. This is because the wood template (level 9) has more pores. Owing to the increase of wood template pores, the value of specific surface area increases and PMMA has better penetration effect. Therefore, for both tree species, orthogonal test levels 9 has the highest light transmittance, corresponding to the above.

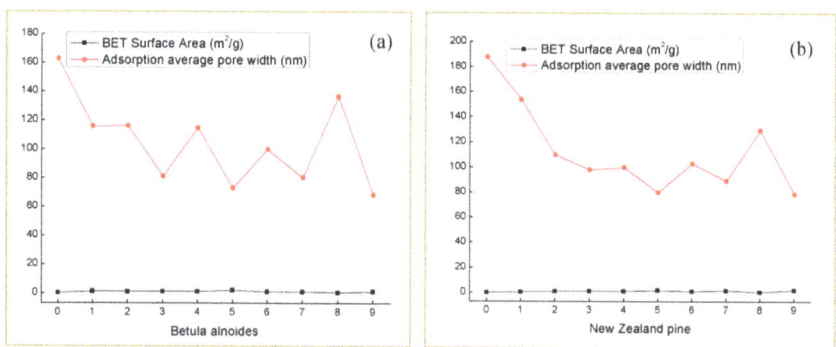

Figure 7. Surface area and adsorption average pore width of OW and FW for two tree species (**a**,**b**).

4.6. Fourier Infrared Analysis

Figure 8a,b shows the infrared spectrum of OW and FW for two tree species. The characteristic absorption peaks of OW-A and OW-B include 3417 cm^{-1} (O–H stretching vibration in cellulose), 2920 cm^{-1} (C–H stretching vibration), 1735 cm^{-1} (Acetyl site of hemicellulose), and 1504 cm^{-1} (stretching vibration of aromatic skeleton group of lignin), 1165 cm^{-1} (C–O–C stretching vibration in cellulose) [30]. Although Betula alnoides and New Zealand pine are two different tree species, their main components are similar as wood, so the main stretching vibrations of the spectrum are basically the same. FW1-9 (including two tree species) correspond to the 9 levels of the orthogonal test respectively. It is found that the peak intensity of FW weakens at 3417 cm^{-1}, 1735 cm^{-1}, and 1504 cm^{-1}, which proves that the content of lignin in the samples was indeed decreasing [31]. Partial lignin has been removed, and the content of cellulose and hemicellulose has also been slightly reduced, which is in line with the test results of the three major elements. Figure 8c,d are the infrared spectra of TW from Betula alnoides and New Zealand pine. After resin impregnation polymerization, TW not only has part of the characteristic absorption peak of the wood, but also has the characteristic peak of PMMA (2992^{-1} and 2950 cm^{-1} for C–H, 1740 cm^{-1} for C=O, and 1191 and 1145 cm^{-1} for C–O) [32], which shows that PMMA and wood template have good synergistic effect.

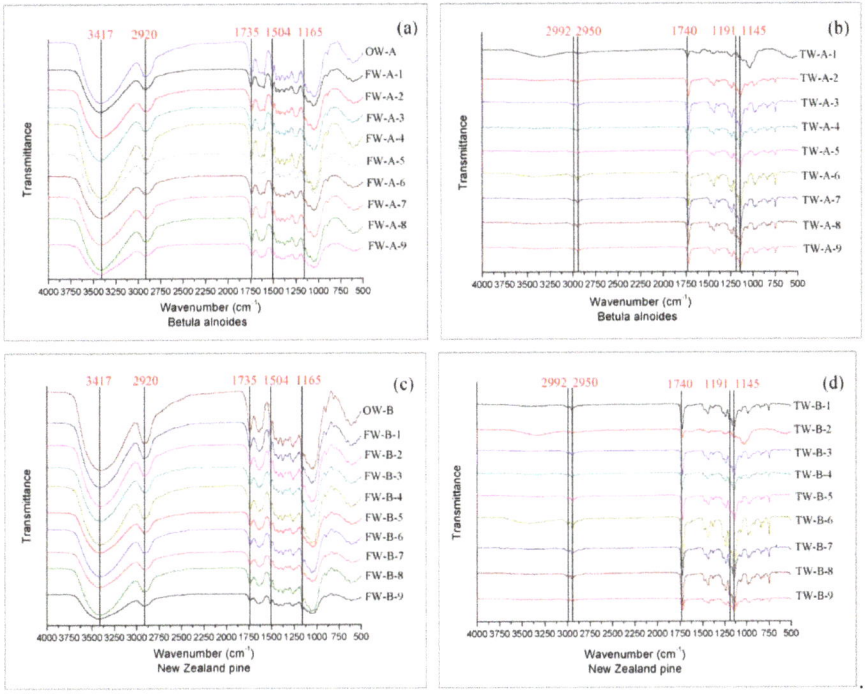

Figure 8. The infrared spectrum of OW, FW and TW for two tree species corresponding to (**a**–**d**).

4.7. Mechanical Performance Analysis

Under the load of the longitudinal tension, the samples went through the elastic deformation stage, the yield stage, the plastic deformation stage and instantaneous fracture stage. With the moment of fracture, the load value dropped sharply. Therefore, the force that the sample bore at the moment of fracture was the load value F corresponding to the highest point of the curve. Both OW and TW (1–9 corresponding to 9 levels of orthogonal test) of Betula alnoides and Pinus New Zealand were tested to compare the mechanical properties of the original wood and the impregnated wood samples. After each sample was tested, the fracture load F (taking the average of multiple tests) was obtained, and then the tensile strength σ value was calculated according to the formula, as shown in Figure 9, where the σ value represents the maximum bearing capacity of the sample under static tensile condition. It can be seen from Figure 9 that the tensile strength and ductility of all TW of Betula alnoides and New Zealand pine are higher than that of OW: the maximum increase rate is 57.58% for Betula alnoides and 57.76% for New Zealand pine. This shows that PMMA has a good synergy with the partially delignified wood templates from a macro perspective, so the overall mechanical properties have been improved.

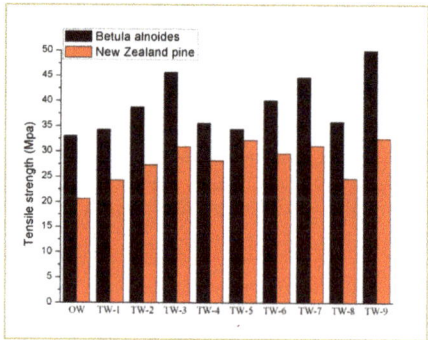

Figure 9. The tensile strength of OW and TW for two tree species.

5. Conclusions

Through the study on the performance of the transparent wood under the process of removing part of the lignin, it is concluded that this kind of transparent wood has a certain degree of light transmittance while retaining most of the wood color and texture. The light transmittance of some tree species is increased by 14.14%, and the mechanical strength is increased by 57.76%, demonstrating that it has good development prospects in the home furnishing industry and will become a highly functional decorative material. At the same time, it can be concluded that the amount of lignin removal has a direct impact on the performance of transparent wood. In the test, combined with the orthogonal test method, the qualitative and quantitative analysis is carried out, and a good process plan is selected. In addition, although the specific values of the two selected tree species are different in the test, the overall trend of change is basically the same, which shows that this experiment is applicable to a variety of wood and is practical, and worthy of detailed further discussion.

Author Contributions: Conceptualization, Y.W.; Data curation, J.Z. and Y.W.; Funding acquisition, Y.W., Q.H. and F.Y.; Investigation, J.Z. and J.W.; Methodology, J.Z.; Writing—original draft, J.Z.; Writing—review & editing, Y.W. and F.Y. All authors have read and agreed to the published version of the manuscript.

Funding: The authors gratefully acknowledgement the financial support from the project funded by Yihua Lifestyle Technology Co., Ltd. Projects funded (YH-NL-201507 and YH-JS-JSKF-201904003), the Special Scientific Research Fund of Construction of High-level teachers Project of Beijing Institute of Fashion Technology (BIFTQG201805) and the Project of Science and Technology Plan of Beijing Municipal Education Commission (KM202010012001).

Acknowledgments: The authors gratefully acknowledgement the financial support from P.R. China and would like to thank Shang Huang for helping us prepare for the samples.

Conflicts of Interest: The authors declare no conflict of interest.

References

1. Zhu, M.W.; Song, J.W.; Li, T.; Gong, A.; Wang, Y.B.; Dai, J.Q.; Yao, Y.G.; Luo, W.; Henderson, D.; Hu, L.B. Highly Anisotropic, Highly Transparent Wood Composites. *Adv. Mater.* **2016**, *28*, 5181–5187. [CrossRef] [PubMed]
2. Li, Y.Y.; Guo, X.L.; He, Y.M.; Zheng, R.B. House model with 2–5 cm thick translucent wood walls and its indoor light performance. *Eur. J. Wood Wood Prod.* **2019**, *77*, 843–851. [CrossRef]
3. Wang, L.H.; Liu, Y.J.; Zhan, X.Y.; Luo, D.; Sun, X.W. Photochromic transparent wood for photo-switchable smart window applications. *J. Mater. Chem. C* **2019**, *7*, 8649–8654. [CrossRef]
4. Li, Y.Y.; Cheng, M.; Jungstedt, E.; Xu, B.; Sun, L.C.; Berglund, L. Optically Transparent Wood Substrate for Perovskite Solar Cells. *ACS Sustain. Chem. Eng.* **2019**, *7*, 6061–6067. [CrossRef]
5. Montanari, C.; Li, Y.Y.; Chen, H.; Yan, M.; Berglund, L.A. Transparent Wood for Thermal Energy Storage and Reversible Optical Transmittance. *ACS Appl. Mater. Interfaces* **2019**, *11*, 20465–20472. [CrossRef]

6. Zhao, Z.Y.; Miao, Y.F.; Yang, Z.Q.; Wang, H.; Sang, R.J.; Fu, Y.C.; Huang, C.X.; Wu, Z.H.; Zhang, M.; Sun, S.J.; et al. Effects of Sulfuric Acid on the Curing Behavior and Bonding Performance of Tannin-Sucrose Adhesive. *Polymers* **2018**, *10*, 651. [CrossRef]
7. Xu, W.; Fang, X.Y.; Han, J.T.; Wu, Z.H.; Zhang, J.L. Effect of coating thickness on sound absorption property of four wood species commonly used for piano soundboards. *Wood Fiber Sci.* **2020**, *52*, 28–43. [CrossRef]
8. Wu, Y.; Zhou, J.C.; Huang, Q.T.; Yang, F.; Wang, Y.J.; Liang, X.M.; Li, J.Z. Study on the Colorimetry Properties of Transparent Wood Prepared from Six Wood Species. *ACS Omega* **2020**, *5*, 1782–1788. [CrossRef]
9. Fang, G.Z.; Gui, Z. Analysis of Wood Chemical Composition of 20 Tree Species. *China Pulp Pap.* **2002**, *21*, 79.
10. Huang, C.X.; Wang, X.; Liang, C.; Jiang, X.; Yang, G.; Xu, J.; Yong, Q. A sustainable process for procuring biologically active fractions of high-purity xylooligosaccharides and water-soluble lignin from Moso bamboo prehydrolyzate. *Biotechnol. Biofuels* **2019**, *12*, 189. [CrossRef]
11. Li, Y.; Fu, Q.; Rojas, R.; Yan, M.; Lawoko, M.; Berglund, L. A New Perspective on Transparent Wood: Lignin-retaining Transparent Wood. *ChemSusChem* **2017**, *10*, 3445–3451. [CrossRef] [PubMed]
12. Yang, K.H. Application of Orthogonal Test in the Design of Liqueur Flavors. *Liquor Mak.* **2018**, *45*, 79–80.
13. Lan, W.T.; Wu, A.X.; Yu, P. Development of a new controlled low strength filling material from the activation of copper slag: Influencing factors and mechanism analysis. *J. Clean. Prod.* **2019**. [CrossRef]
14. Yang, R. Discussion on the application of orthogonal test method in production practice. *J. Zhengzhou Coal Manag. Inst.* **2000**, *15*, 71–74.
15. Li, Y.Y.; Fu, Q.L.; Yu, S.; Yan, M.; Berglund, L. Optically Transparent Wood from a Nanoporous Cellulosic Template: Combining Functional and Structural Performance. *Biomacromolecules* **2016**, *17*, 1359. [CrossRef]
16. Zhao, Z.; Sakai, S.; Wu, D.; Chen, Z.; Zhu, N.; Huang, C.; Sun, S.; Zhang, M.; Umemura, K.; Yong, Q. Further Exploration of Sucrose-Citric Acid Adhesive: Investigation of Optimal Hot-Pressing Conditions for Plywood and Curing Behavior. *Polymers* **2019**, *11*, 1996. [CrossRef]
17. Sluiter, A.; Hames, B.; Ruiz, R.; Scarlata, C.; Sluiter, J.; Templeton, D.; Crocker, D. Determination of structural carbohydrates and lignin in biomass. In *Laboratory Analytical Procedure (LAP), Technical Report NREL/TP-510-42618*; United States Department of Energy: Washington, DC, USA, 2011.
18. Wu, J.M.; Wu, Y.; Yang, F.; Tang, C.Y.; Huang, Q.T.; Zhang, J.L. Impact of Delignification on Morphological, Optical and Mechanical Properties of Transparent Wood. *Compos Part A-Appls.* **2019**, *117*, 324–331. [CrossRef]
19. Lin, S.S.; Zhang, Q.; Yu, J.; Duan, H.J. Analysis of Relationship between Chromatic Aberration Value and Pigment Content of Peel in Different Peach Varieties. *Acta Agric. Jiangxi* **2018**, *30*, 36.
20. Wu, Y.; Wu, J.M.; Wang, S.Q.; Feng, X.H.; Chen, H.; Tang, Q.W.; Zhang, H.Q. Measurement of mechanical properties of multilayer waterborne coatings on wood by nanoindentation. *Holzforschung* **2019**, *73*, 1–7. [CrossRef]
21. Huang, C.X.; Lin, W.; Lai, C.; Li, X.; Jin, Y.; Yong, Q. Coupling the post-extraction process to remove residual lignin and alter the recalcitrant structures for improving the enzymatic digestibility of acid-pretreated bamboo residues. *Bioresour. Technol.* **2019**, *285*, 121355. [CrossRef]
22. Gu, J.; Pei, W.H.; Tang, S.; Yan, F.; Peng, Z.W.; Huang, C.X.; Yang, J.L.; Yong, Q. Procuring biologically active galactomannans from spent coffee ground (SCG) by autohydrolysis and enzymatic hydrolysis. *Int. J. Biol. Macromol.* **2020**, *149*, 572–580. [CrossRef] [PubMed]
23. Wang, Y.; Fu, S.Y. Research Progress in Transparent Wood. *China Pulp Pap.* **2018**, *37*, 68–72.
24. Mei, L.F.; Wang, Z.H.; Yan, D.B.; Chen, S.X. Research on Galvanized Steel Laser Welding Quality of Vehicle Body based on Orthogonality Range Analysis Method. *Appl. Laser* **2017**, *37*, 85–90.
25. Jia, C.; Li, T.; Chen, C.J.; Dai, J.Q.; Kierzewski, I.M.; Song, J.W.; Li, Y.J.; Yang, C.P.; Wang, C.W.; Hu, L.B. Scalable, anisotropic transparent paper directly from wood for light management in solar cells. *Nano Energy* **2017**, *36*, 366–373. [CrossRef]
26. Gan, W.T.; Xiao, S.L.; Gao, L.K.; Gao, R.N.; Li, J.; Zhan, X.X. Luminescent and transparent wood composites fabricated by PMMA and γ-Fe_2O_3@YVO_4:Eu^{3+} nanoparticles impregnation. *ACS Sustain. Chem. Eng.* **2017**, *5*, 3855–3862. [CrossRef]
27. Wu, Y.; Tang, C.; Wu, J.; Huang, Q.T. Research Progress of Transparent Wood: A Review. *For. Eng.* **2018**, *3*, 12–18.
28. Qin, J.K.; Bai, T.; Shao, Y.L.; Zhao, X.; Li, S.; Hu, Y.C. Fabrication and Characterization of Multilayer Transparent Wood of Different Species. *J. Beijing For. Univ.* **2018**, *40*, 113–120.

29. Wu, L.W.; Wang, C.F.; Li, N.; Zhao, S.; Gao, M. Uncertainty evaluation of measurement of specific surface area of Al_2O_3 carrier by BET method. *Ind. Catal.* **2016**, *24*, 79–80.
30. Rana, R.; Langenfeld-Heyser, R.; Finkeldey, R.; Polle, A. FTIR Spectroscopy, Chemical and Histochemical Characterisation of Wood and Lignin of Five Tropical Timber Wood Species of the Family of Dipterocarpaceae. *Wood Sci. Technol.* **2010**, *44*, 225–242. [CrossRef]
31. Yaddanapudi, H.S.; Hickerson, N.; Saini, S.; Tiwari, A. Fabrication and characterization of transparent wood for next generation smart building applications. *Vacuum* **2017**, *146*, 649–654. [CrossRef]
32. Kavale, M.S.; Mahadik, D.B.; Parale, V.G.; Wagh, P.B.; Gupta, S.C.; Rao, A.V.; Barshilia, H.C. Optically Transparent, Superhydrophobic Methyltrimethoxysilane Based Silica Coatings without Silylating Reagent. *Appl. Surf. Sci.* **2011**, *258*, 156–162. [CrossRef]

© 2020 by the authors. Licensee MDPI, Basel, Switzerland. This article is an open access article distributed under the terms and conditions of the Creative Commons Attribution (CC BY) license (http://creativecommons.org/licenses/by/4.0/).

Article

Effect of Veneer-Drying Temperature on Selected Properties and Formaldehyde Emission of Birch Plywood

Pavlo Bekhta [1],*, Ján Sedliačik [2] and Nataliya Bekhta [1]

[1] Department of Wood-Based Composites, Cellulose, and Paper, Ukrainian National Forestry University, 79057 Lviv, Ukraine; natalijabekhta@gmail.com
[2] Department of Furniture and Wood Products, Technical University in Zvolen, 96001 Zvolen, Slovakia; jan.sedliacik@tuzvo.sk
* Correspondence: bekhta@nltu.edu.ua; Tel.: +38-032-2384499

Received: 6 February 2020; Accepted: 2 March 2020; Published: 5 March 2020

Abstract: In this study, the effect of the veneer-drying process at elevated temperatures on selected properties and formaldehyde emission of plywood panels was determined. We assume that during the veneer drying at high temperatures, more formaldehyde is released from it, and therefore, a lower formaldehyde emission can be expected from the finished plywood. Prior to bonding, birch veneers were dried at 160 °C (control) and 185 °C in an industrial veneer steam dryer (SD) and at 180 °C, 240 °C and 280 °C in an industrial veneer gas dryer (GD). Two types of adhesives were used: urea–formaldehyde (UF) and phenol–formaldehyde (PF) resins. Bonding quality, bending strength and modulus of elasticity in bending, water absorption and thickness swelling of plywood samples were determined. The formaldehyde emission level of samples was also measured. It was concluded from the study that the effects of veneer-drying temperatures on the bonding strength and physical and mechanical properties of plywood panels were significant. Veneer-drying temperatures of 185 °C/SD, 180 °C/GD and 240 °C/GD negatively affected the bending strength and the modulus of elasticity along and across the fibres for both UF and PF plywood samples. Bonding strength mean values obtained from all test panels were above the required value (1.0 MPa) indicated in EN 314-2 standard. The lowest formaldehyde emissions for the UF and PF plywood samples were observed in the samples from veneer dried in a steam dryer at 185 °C/SD.

Keywords: birch plywood; veneer-drying temperature; formaldehyde emission; bending strength; modulus of elasticity; bonding strength; thickness swelling; water absorption

1. Introduction

Plywood and other wood-composite materials such as particle boards, medium- (MDF) and high density fibreboard (HDF) are becoming more popular and are widely used for the manufacture of furniture, cabinets, engineering floors, housing and various construction products. These materials are mainly bonded with thermosetting formaldehyde-type adhesives such as urea–formaldehyde (UF) resin, melamine–urea–formaldehyde (MUF) resin, phenol–formaldehyde (PF) resin, etc. UF adhesives are widely used in the production of wood-based panels because of their excellent adhesion to lignocellulose, excellent internal cohesion, easy processing and application, the lack of colour in the finished product and low cost. However, their poor resistance to external factors, especially moisture, and their tendency to release formaldehyde vapours are significant limitations. Wood-composite materials are the most important sources of formaldehyde emission (FE), a harmful gas when released inside buildings. Because of this, one of the disadvantages of plywood and wood composite materials is their toxicity—the release of formaldehyde. The FE from these materials is indisputable [1]. Adhesives based on UF and PF are carcinogenic in very high concentrations. Therefore, formaldehyde was reclassified in 2004 by the International Agency for Research on Cancer (IARC) as 'carcinogenic to humans (Group 1)' [2], compelling companies to reduce FE to lower levels.

Methods for reducing FE in wood-composite materials have been widely discussed over the years, and many of them have been described in [3]. However, a decrease in the formaldehyde content of the resins may impair the bonding strength, and adhesives without formaldehyde can increase the cost of the plywood [4,5].

It is known that natural wood contains and emits volatile organic compounds, including formaldehyde [3,6,7]. Therefore, the issue of FE no longer focuses solely on the adhesive systems used to bond plywood panels. This may also be due to the influence of certain wood species. Typically, coniferous species tend to have a higher content of formaldehyde than hardwood species [7].

It has been reported that heat treatment contributes to the FE from solid wood [8]. Lignin and hemicellulose have the potential for FE, and lignin appears to have more potential in this respect than cellulose and hemicellulose. Moreover, thermohydrolytic processing of wood can lead to FE from polysaccharides [9]. Furthermore, the FE levels depend on numerous factors such as wood species, moisture content (MC), outside temperature, and storing time [9–14].

Wood as a natural material contains formaldehyde [6]. The FE from solid wood increases at elevated temperatures and prolonged heating times [9,15], even in the absence of wood resin [16]. In the technological process of manufacturing plywood, the veneer can be subjected to high temperatures in the drying and pressing processes.

Veneer drying is one of the most important stages in the production of wood-based panels, such as plywood and laminated veneer lumber (LVL). As is well known, the purpose of the drying process is to reduce the MC of the veneer to obtain the appropriate values for bonding. Prior to the bonding process, the MC of all veneer sheets should be below 7% [17]. Best bonding results are obtained in plywood panels with veneers having 4–6% MC [11]. The amount of moisture in the wood combined with the water in the adhesives will significantly affect the wetting, spreading, penetration and even curing of the adhesives [18]. Moreover, Aydin et al. [11] reported that the FE from poplar and spruce plywood decreased with increasing veneer MC.

Baldwin [19] stated that the veneer-drying process accounts for some 70% of the thermal energy consumed in plywood production and approximately 60% of the mill's total energy requirement. The drying temperatures vary mainly from 90 to 160 °C and are normal in the plywood industry. The use of high drying temperatures reduces the drying time of the veneer and increases productivity. The small thickness of veneers allows fast drying. Reducing drying time and energy consumption offers great potential for economic benefits for the woodworking industry [20]. It has been concluded that the practice of high drying temperature could save energy and drying time by 44% and 25%, respectively, compared with normal drying temperature [21]. However, drying to very low MC and at a very high temperature or at a moderate temperature for a long period inactivates the veneer surface, causing poor veneer wetting and therefore poor bonding [22].

In addition, the drying temperature affects both the physical and chemical properties of the veneer surface and therefore the durability and physical–mechanical properties of the plywood decreases [23]. Many studies have been conducted on the effect of veneer-drying temperature on the ability of veneer surfaces to bond [24,25], on the relationship between surface inactivation and bonding strength [22] and optimal conditions for surface preparation [26]. In addition, in recent years, many studies have been conducted on the effect of heat treatment on the properties of plywood panels made from thermally treated veneer [23,27–34].

Some physical and mechanical properties of plywood panels manufactured from alder veneer sheets dried at 20, 110, 150, and 180 °C were determined by Aydin and Colakoglu [23]. These authors concluded that shear strength values of plywood panels decreased and formaldehyde emission increased clearly with increasing veneer drying temperature; no clear difference was found for bending strength of panels manufactured from veneers dried at 110, 150, and 180 °C. Nazerian et al. [27] showed that heat treatment of beech, maple and poplar veneers at 120 or 180 °C for 4 h in a small heating unit under atmospheric pressure alters the physical and chemical properties significantly, but the strength properties begin to deteriorate. In other work [28] poplar veneers were subjected to heat treatment at 80, 130 and 180 °C for 1, 3, and 5 h in a small heating unit under atmospheric pressure. Based on the findings in that study, the bending strength and modulus of elasticity of LVL panels decreased with increasing in temperature and time. Jamalirad et al. [29,30] found that water absorption, thickness swelling, shear and bending strength of plywood samples improve at higher drying temperature 180 °C for 2 h. As we can see, in many of these studies, veneers were subjected to heat treatment at high temperatures for a long time (1–5 h).

In practice, steam and gas dryers are used for veneer drying [35]. In the steam dryer the drying agent is heated by steam and in the gas dryer the drying agent is heated by flue gases from combustion of different types of fuel. Moreover, in the gas dryers (GD), the veneer is dried under severe conditions; the drying temperatures are high (200–280 °C), which significantly shortens the drying time, although there is a risk of extra cracks since the veneer becomes brittle. In the steam dryers (SD), the veneer dries under milder conditions than in gas dryers. Typical drying temperatures are of the order of 160 °C in the steam dryers. There is some information in the literature on veneer drying at temperatures of 100–180 °C, but we did not find any information on veneer drying at high temperatures (180–280 °C), including in gas and steam dryers, and how these high temperatures act for a short period of time on the properties and formaldehyde emission of plywood panels. On the other hand, as mentioned above, heat treatment promotes the release of formaldehyde from solid wood [8].

We hypothesize that during the veneer drying at high temperatures, more formaldehyde is released from it, and therefore, a lower FE can be expected from the finished plywood. Therefore, the purpose of this study was to determine the effects of high temperature drying of veneers on some physical and mechanical properties and formaldehyde emission of plywood panels.

2. Materials and Methods

2.1. Materials

Birch (*Betula verrucosa* Ehrh.) wood veneers obtained by rotary cutting at industrial conditions were used in the study. Rotary cut veneer sheets with 30 cm by 30 cm dimensions and 1.5 mm thickness were classified into five groups after the peeling process and they were dried at 160 °C (control) and 185 °C in an industrial veneer steam dryer (SD) and at 180 °C, 240 °C and 280 °C in an industrial veneer gas dryer (GD). These drying temperatures were obtained from local producer of rotary-cut veneer and plywood panels. The veneer drying was performed in industrial conditions. Each veneer sheet was dried to 4–6% MC.

Two types of formaldehyde-based adhesives were used in this study: commercial UF and PF resins. UF resin solution used in the manufacturing was composed of 100 parts of UF resin by weight, 15 parts of wheat flour by weight, and 4 parts of hardener by weight. The PF resin was used for plywood panel manufacturing without any filler or additive. The formulations of the adhesive mixtures are given in Table 1.

Table 1. Formulations of adhesive mixtures used for the manufacturing of plywood panels.

	Ingredients of Adhesive	Parts by Weight
Urea–formaldehyde	UF resin (with 66% solid)	100
	Wheat flour	15
	Hardener	4
Phenol–formaldehyde	PF resin (with 47% solid)	100

2.2. Plywood Panel Manufacturing

Plywood was processed according to an experimental design, where the factors were the types of adhesive (UF and PF), types of veneer dryer (SD and GD) and temperatures of veneer drying (160 °C, 185 °C, 180 °C, 240 °C and 280 °C) (Table 2). Three-layer and five-layer plywood panels of 300 mm × 300 mm were made for the strength and formaldehyde release tests, respectively. Four plywood panels were made for each experimental condition—two panels for the strength test and two panels for the formaldehyde release test. Four control panels were also made for UF and PF adhesives. Totally 48 panels were manufactured.

Table 2. Conditions of manufacturing three-layer plywood panels.

Test No.	Adhesive Type	Type of Drier	Temperature of Veneer Drying (°C)	Adhesive Spread Rate (g/m^2)	Pressing Pressure (MPa)	Pressing Temperature (°C)	Pressing Time (min)
1	UF	SD	160	160	1.8	105	5/6 *
2		SD	185				
3		GD	180				
4		GD	240				
5		GD	280				
6	PF	SD	160	160	1.8	145	5/6 *
7		SD	185				
8		GD	180				
9		GD	240				
10		GD	280				

* Pressing time for five-layer plywood panels.

Plywood panels were made in an electrically heated hydraulic laboratory press. The specific pressing pressure of 1.8 MPa and temperature of 105 °C for UF adhesive and 145 °C for PF adhesive were used, and 5 min (for a three-layer panel) or 6 min (for a five-layer panel) pressing time (during the last 30 s of the press cycle the pressure was continuously reduced to 0 MPa). The glue spread was 160 g·m^{-2} based on wet mass. The adhesive mixture was applied onto one side of every uneven ply. The plies were assembled perpendicularly to each other (veneer sheets were laid up tight/loose) to form plywood of three/five plies. Glue was applied onto the veneer surface with a hand roller spreader.

2.3. Panel Testing

During the experiment, all plywood samples were conditioned prior to testing for 2 weeks at 20 ± 2 °C and 65 ± 5% relative humidity. The panels were cut to extract test samples according to the standard requirements. The shear strength was measured according to EN 314-1 [36] and EN 314-2 [37] methods after pretreatment for intended use in interior conditions for UF, and exterior conditions for PF. For the shear strength test, one-half of the samples were tested in dry conditions and the other half in wet conditions. UF plywood test pieces were immersed in water at 20 ± 3 °C for 24 h; PF plywood test pieces were immersed for 4 h in boiling water, then dried in the ventilated oven for 16 h at the temperature of 60 ± 3 °C, then immersed in boiling water for 4 h, followed by cooling in water at 20 ± 3 °C for at least 1 h. Ten samples (a total of 480 samples) were used for each variant both dry and wet shear strength mechanical testing.

Bending strength (MOR) and modulus of elasticity (MOE) tests were carried out for plywood panels manufactured according to EN 310 [38] standard. Twelve samples (a total of 288 samples) were used for the evaluation of plywood MOR and MOE. MOR and MOE in bending were carried out in parallel (∥) and perpendicular (⊥) directions, depending on the surface layer.

Dimensional stability in the form of thickness swelling (TS) and water absorption (WA) of the samples were determined according to a water-soaking test based on EN 317 [39], using test pieces of dimension 50 × 50 mm. They were immersed in distilled water for four different periods of 2, 24, 48 and 72 h. After this time the test pieces were removed from the water, weighed, and the thickness was measured. The samples were weighed to the nearest 0.001 g and measured to the nearest 0.01 mm immediately. Six replicate samples were tested for each type of plywood panel. The per cent change from the original thickness represents the TS, and the per cent weight change from the original weight represents the WA.

2.4. Formaldehyde Release from Plywood

The formaldehyde release level was measured by means of the desiccator method according to EN ISO 12460-4 [40] standard. The test principle is a determination of the quantity of formaldehyde emitted from plywood samples absorbed in a specified volume of distilled water during 24 h in the glass desiccator.

The test pieces were prepared from each type of plywood with dimensions (150 ± 1) mm × (50 ± 1) mm (length × width) with a total of 1735 cm^2 (10 pieces). They were then placed in a desiccator with an enclosed volume of 11 litres with a glass crystallizing dish containing 300 mL of distilled water (Figure 1). Samples were removed from the desiccator after 24 h and the obtained formaldehyde solution was prepared for spectroscopic analysis. To determine the formaldehyde content, 25 mL of tested water solution from the desiccator was mixed with 25 mL of acetylacetone-ammonium acetate solution in a 100 mL flask. The stoppered flasks were then heated in a water bath at 65 ± 2 °C for 10 min and subsequently cooled to ambient temperature for 60 ± 5 min. The absorbance of samples was measured on a UviLine SI 5000 spectrophotometer (SI Analytics, College Station, TX, USA) at 412 nm. The formaldehyde content was determined using the calibration curve that was prepared from the standard formaldehyde solutions. The emission tests were carried out in duplicate.

Figure 1. Plywood samples in the glass desiccator.

2.5. Analysis of Variance

Analysis of variance (ANOVA) at a 0.05 significance level was carried out using IBM SPSS Statistics software (IBM Corp., Armonk, NY, USA) to estimate the relative importance of the effects of the study variables such as types of adhesive and veneer-drying temperature and their interactions on the properties of plywood panels.

3. Results and Discussion

3.1. Shear Strength of Plywood Samples

Average values of the bonding strength of plywood samples made of birch veneer, dried at high temperatures are given in Table 3. Drying veneer at high temperatures in both steam and gas dryers provides high values of bonding strength of the plywood samples, glued with UF and PF adhesives, compared with control samples. The Duncan test with a 95% confidence level was used to compare the mean values of variance sources and the results for statistical evaluation are presented in Table 3. ANOVA analysis showed that the effect of type of adhesive used, the form of the drying agent and the temperature of the veneer-drying on the bonding strength of plywood samples is statistically significant ($p \leq 0.05$). Bonding shear strength mean values obtained from the samples of all plywood panels were also above the limit value (1.0 MPa) indicated in EN 314-2 [37] standard. The limit value in the standard indicates the minimum standard requirement for bonding strength. Therefore, the plywood panels tested in this study have met the standard requirement for the bonding strength.

Table 3. Shear strength of plywood panels.

Drying Temperature (°C)	Moisture Content (%)	Shear Strength (MPa)			
		UF		PF	
		Dry Test	Wet Test	Dry Test	Wet Test
160 °C/SD (control)	6.6	2.27 (0.19) A	1.81 (0.17) A	2.97 (0.24) B	1.78 (0.19) B
185 °C/SD	5.5	3.09 (0.22) C	2.52 (0.12) B	3.23 (0.12) C	1.92 (0.16) B
180 °C/GD	6.2	3.02 (0.28) C	2.67 (0.12) C	2.57 (0.22) A	1.85 (0.10) B
240 °C/GD	4.7	3.08 (0.20) C	2.69 (0.18) C	2.60 (0.21) A	1.93 (0.23) B
280 °C/GD	4.8	2.75 (0.15) B	2.48 (0.21) B	2.56 (0.18) A	1.61 (0.11) A

Values in parentheses are standard deviations. Different letters denote a significant difference. The means followed by the same letter do not statistically differ from each other ($p \leq 0.05$).

The highest values of bonding strength (2.67 and 2.69 MPa) are observed in plywood samples glued with UF adhesive using veneer dried at temperatures 180 °C/GD and 240 °C/GD. The difference between the bonding strength values for these temperatures is insignificant, and therefore, in practical terms, it is more economically advantageous to use a lower veneer-drying temperature of 180 °C/GD. In contrast, the lowest value of bonding strength (2.48 MPa), not considering the control sample, was recorded for UF plywood samples from veneer, dried at the highest used temperature of 280 °C/GD. For PF plywood samples, the smallest value of bonding strength (1.61 MPa), as in the case of UF plywood samples, was found for plywood samples from veneer dried at the highest applied temperature of 280 °C/GD. This result may be because: (1) excessive veneer drying may cause some drying defects, especially surface cracks [41], which may lead to loss of strength; (2) the thermal processes have caused poor wetting of the wood with adhesive [24,42], and the poor wettability is considered as an indicator of poor bonding strength [43]; (3) at high temperatures during veneer drying extractives migrate to the surface where they concentrate and physically block adhesive contact with wood [22]; (4) the surface roughness increased with increasing veneer-drying temperature; a rough veneer can contribute to the excessive penetration of the adhesive into the wood, increasing the actual (true) surface area while reducing the proportion of the adhesive to that surface [44], as a consequence, there is insufficient adhesive to adhere, resulting in a decrease of bonding quality [45].

Veneer-drying temperatures of 185 °C/SD, 180 °C/GD and 240 °C/GD showed the same effect on the bonding strength; the strength values did not differ significantly. In addition, it can be noted that for UF plywood samples, the bonding strength values for all the investigated temperatures are higher than the value for the control sample. For PF plywood samples, the bonding strength values for the investigated temperatures, except the temperature of 280 °C/GD, are slightly higher than the bonding strength value for the control sample, but the difference between these values is not significant.

That is, the veneer-drying temperature of 185 °C/SD, 180 °C/GD and 240 °C/GD did not affect the bonding strength of PF plywood samples. In general, UF plywood samples showed higher values of bonding strength than PF plywood samples. This can be explained by the fact that higher drying temperatures (180–280 °C) initiate degradation of hemicelluloses and formation of acetic acid and formic acid leading to lower pH of veneer surfaces [46–48], which can affect the curing of acid-catalysed UF resin and results in improving the shear strength and dimensional stability of plywood samples [29]. Özşahin and Aydin [49] also found that bonding shear strength values of panels manufactured with PF glue were lower than those of panels manufactured with UF glue. They explained this by different pretreatments of plywood samples bonded with UF and PF glues.

Comparing the effect of the veneer-drying temperature on the bonding strength for one type of dryer, it can be noted that the higher temperature 185 °C/SD in the steam dryer had a positive effect on the bonding strength of UF and PF plywood samples compared with the lower temperature of 160 °C/SD in the same dryer. Among the temperatures of 180 °C/GD, 240 °C/GD and 280 °C/GD of gas dryers, the best bonding strength values for UF and PF plywood samples were obtained at the veneer-drying temperatures of 180 °C/GD and 240 °C/GD.

The results obtained by Theander et al. [50] showed that the drying temperatures (180 °C) and degradation of hemicelluloses cause the production of sugar monomers, which play an important role in the bonding quality and in increasing the bonding strength. Fengel and Wegener [51] stated that lignin becomes softer above 160 °C. When lignin softens, it enters the micropores of the wood and makes the woody tissue more homogeneous. This significantly reduces the tendency of wood to swell during soaking and increases bonding strength.

Demirkir et al. [33] reported that the bonding strength values of plywood panels with PF resin increased, as the veneer-drying temperature increased. Jamalirad et al. [30] also showed that the mechanical properties of plywood with increasing drying temperature up to 180 °C for 2 h do not negatively affect shear strength and MOR of plywood samples. Opposing results were obtained in [23], which found that shear strength values of plywood panels decreased clearly with increasing veneer-drying temperature.

3.2. Formaldehyde Emission of Plywood Samples

FE of plywood samples is expressed as the arithmetic average of the two tests and shown in Figure 2. As can be seen from Figure 2 the values of FE of plywood samples made using veneer dried at high temperatures were lower, except the temperature of 280 °C/GD for PF plywood samples, than the value of FE for control sample. The lowest FE of 0.31 and 0.09 mg/L for the UF and PF plywood samples, respectively, were observed in the samples from veneer dried in a steam dryer at 185 °C/SD. At this temperature, the values of FE are less than 2.2 times and 1.6 times, respectively, for UF and PF plywood samples than those values in the control samples. As can be seen from Figure 2, the veneer-drying temperatures of 180 °C/GD and 240 °C/GD in the gas dryer are more acceptable in terms of reducing FE in plywood samples than the highest veneer-drying temperature of 280 °C/GD. The drying temperature of 280 °C/GD can be considered unacceptable for veneer drying in terms of the bonding strength and FE of plywood samples.

Similar results were obtained by Murata et al. [31] who showed that heating veneer sheets in the temperature range of 150 °C to 170 °C effectively reduced the FE of plywood. Opposing results were obtained by Aydin and Colakoglu [23] who found that FE values of plywood panels increased with increasing veneer-drying temperature.

Hasegawa [8] showed that solid wood dried at high temperatures emits more formaldehyde (HCHO) than that dried at low temperatures. The border of drying temperature, which significantly increases the HCHO emission rate, is dependent on the wood species and its components, e.g., lignin [52, 53].

It has been reported in the literature that the amount of organic compounds emitted during thermal drying increases with increasing temperature. The main mechanisms of emission during drying of wood are direct evaporation, steam distillation and thermal degradation. In thermal degradation, high molecular weight organic compounds are cleaved into low molecular weight organic compounds with increasing temperature [51,54].

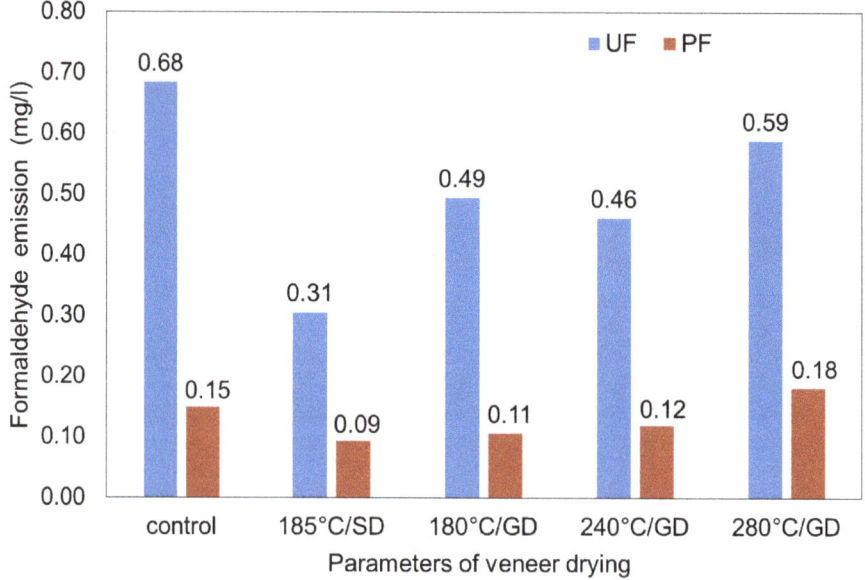

Figure 2. Formaldehyde emission of plywood panels.

3.3. Bending Strength and Modulus of Elasticity of Plywood Samples

The average values of the MOR and MOE of plywood samples made of birch veneer, dried at high temperatures are shown in Table 4. ANOVA analysis showed that the type of adhesive used, the type of drying agent and the veneer-drying temperature significantly ($p \leq 0.05$) affect the MOR and MOE of plywood samples.

Mean values obtained for MOR and MOE of plywood panels were higher than the limit values for structural purpose solid wood panels (35 MPa for MOR (∥) and 5 MPa for MOR (⊥), 8500 MPa for MOE (∥) and 470 MPa for MOE (⊥)) indicated in EN 13,353 [55] for panels having thickness up to 20 mm.

The MOR and MOE values of UF/PF plywood samples from veneer dried at elevated temperatures were (Table 4):

- along the fibres, between 118.1 and 139.7 MPa/131.0–172.8 MPa and 9301.5–10988.3 MPa/ 9930.4–13924.5 MPa, respectively;
- across the fibres, between 24.2 and 27.8 MPa/25.5–31.2 MPa and 981.9–1351.6 MPa/ 1197.3–1560.7 MPa, respectively.

Veneer-drying temperatures of 185 °C/SD, 180 °C/GD and 240 °C/GD had a negative effect on the MOR and MOE along and across the fibres for both UF and PF plywood samples. In addition, the values of the MOR and MOE of PF plywood samples are higher than the similar values for UF plywood samples. UF plywood samples made from veneer dried at temperatures of 185 °C/SD and 180 °C/GD have lower values of MOR along and across the fibres than the reference samples. In contrast, UF plywood samples from veneer dried in a gas dryer at temperatures of 240 °C/GD and 280 °C/GD have higher values of MOR along the fibres than control samples. The highest values of MOR along and across the fibres (139.7 and 27.8 MPa, respectively) are observed in UF samples made from veneer dried at the highest drying temperature of 280 °C/GD. PF plywood samples made from veneer dried at temperatures of 185 °C/SD, 180 °C/GD and 240 °C/GD have lower values of MOR along and across the fibres than the reference samples. As in the case of UF samples, the highest value of MOR along and across the fibres are observed in PF samples made from veneer dried at temperature of 280 °C/GD.

Table 4. Duncan's test results for selected mechanical properties of plywood panels.

Drying Temperature (°C)	Bending Strength (MPa)				Modulus of Elasticity (MPa)			
	UF		PF		UF		PF	
	MOR (∥)	MOR (⊥)	MOR (∥)	MOR (⊥)	MOE (∥)	MOE (⊥)	MOE (∥)	MOE (⊥)
160 °C/SD (control)	130.5 (7.1) B	26.2 (2.6) AB	161.0 (9.9) C	31.1 (5.1) C	11,275.5 (905.9) B	1229.6 (92.3) C	14,888.3 (976.9) E	1363.6 (135.6) B
185 °C/SD	118.1 (8.4) A	25.1 (1.8) A	145.8 (11.3) B	29.6 (2.7) BC	9301.5 (898.2) A	1135.3 (64.2) B	11,625.3 (471.3) B	1277.5 (69.6) AB
180 °C/GD	126.3 (4.9) B	24.2 (1.4) A	146.6 (9.9) B	27.0 (3.2) AB	10,852.7 (786.5) B	981.9 (22.0) A	12,686.1 (858.0) C	1232.8 (117.2) A
240 °C/GD	133.6 (10.4) BC	25.5 (2.3) A	131.0 (6.6) A	25.5 (2.3) A	9738.2 (901.5) A	1111.4 (107.6) B	9930.4 (620.3) A	1197.3 (118.1) A
280 °C/GD	139.7 (7.6) C	27.8 (3.0) B	172.8 (11.4) D	31.2 (4.5) C	10,988.3 (743.4) B	1351.6 (99.8) D	13,924.5 (1015.4) D	1560.7 (111.1) C

Values in parentheses are standard deviations. Different letters denote a significant difference. The means followed by the same letter do not statistically differ from each other ($p \leq 0.05$).

High veneer-drying temperatures adversely affected the MOE of plywood samples. The values of the MOE for UF and PF plywood samples, with the exception of MOE (⊥) at 280 °C/GD, are lower than the values of the MOE for control samples. In [23] no clear difference was found for MOR of plywood panels manufactured from veneers dried at elevated temperatures. Kol and Seker [34] found that heat treatment had an enhancement effect on MOE and adverse effect on MOR of LVL panels. Nazerian et al. [27] showed that the MOR and MOE of the LVL manufactured from heat-treated poplar veneers decreased with increasing temperature of treatment. Higher weight decrease was obtained from the LVL with veneers treated at 180 °C for 5 h. The optimum drying temperature values were obtained (between 160 and 165 °C) in Scots pine plywood and (between 190 and 196 °C) in alder plywood, for best shear strength, MOR and MOE values [56].

3.4. Water Absorption and Thickness Swelling of Plywood Samples

Statistical analysis using ANOVA showed that the type of adhesive, the type of drying agent and the veneer-drying temperature significantly ($p \leq 0.05$) affected the WA and TS of the plywood samples. The values of TS for the UF plywood samples made from veneer, dried at high temperatures in the steam and gas dryers are less than the TS values of the control samples (Figure 3). The least TS is demonstrated by UF and PF plywood samples from the veneer dried at 240 °C/GD, while the smallest WA is demonstrated by the UF and PF samples from the veneer dried at 240 °C/GD and 280 °C/GD, respectively. With regard to the influence of high temperatures on WA of plywood samples, the picture here is not so straightforward. The WA is greater at drying temperatures of 185 °C/SD and 180 °C/GD, and at drying temperatures of 240 °C/GD and 280 °C/GD the WA is less than the WA of the control samples.

Drying of veneer at high temperatures reduces the WA and TS of PF plywood samples (Figure 4). The TS values for all PF plywood samples are smaller than for the control samples. The WA values are also smaller for all investigated temperatures, except for a temperature of 185 °C/SD, than for the control samples. The smallest values of WA and TS for PF plywood samples are observed at 280 °C/GD and 240 °C/GD, respectively.

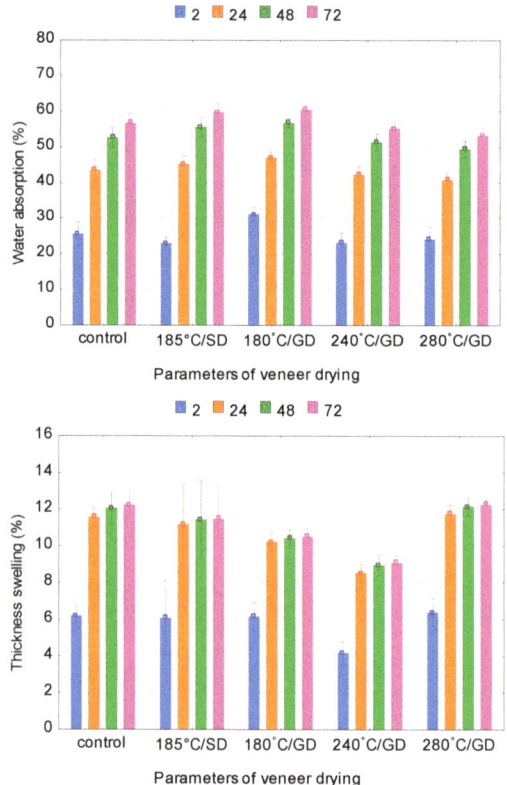

Figure 3. Water absorption and thickness swelling of urea–formaldehyde (UF) plywood samples.

It is known that heat treatment causes a decrease in affinity of the treated material to water. One of the indicators that characterize the affinity of wood to water is the contact angle. Measurements of the contact angle in previous work showed that the contact angle increases on thermally treated veneers, plywood and OSB panels [57–59] with the increase in temperature and duration of treatment, indicating a decrease in the affinity of the treated material to water.

Kol and Seker [34] found that heat treatment had an enhancement effect on TS of LVL panel. Nazerian et al. [27] showed that increase of the temperature and treatment time resulted in better dimensional stability for the LVLs manufactured from heat-treated poplar veneers. Jamalirad et al. [29] showed that WA and TS of plywood samples were improved with increasing drying temperature up to 180 °C for 2 h.

It is obvious that with the increase in amount of the heat-treated veneer sheets in a plywood packet the MC decreases. A decrease in MC is expected because the amount of free hydroxyl groups is reduced during heat treatment and the possibility of moisture absorption from the environment is reduced. The equilibrium moisture content (EMC) of the plywood panels decreased with increasing veneer-drying temperature [23]. Several authors [23,46] showed that the EMC values of alder, beech and spruce plywood panels decreased significantly with high-temperature veneer drying.

The improvement in the dimensional stability of the plywood samples from veneer dried at high temperatures is mainly due to the decrease in hygroscopicity due to the chemical changes at high temperatures. Theoretically, existing OH groups in hemicelluloses have the most significant effect on the physical properties of wood. High temperature reduces the EMC of wood [60], the veneer and, as a consequence, TS of plywood panels made from veneer [27]. Zhang et al. [61] showed that changing the character of wood from hydrophilic to more hydrophobic by extracting hemicellulose can also potentially improve the dimensional stability of wood and wood composites.

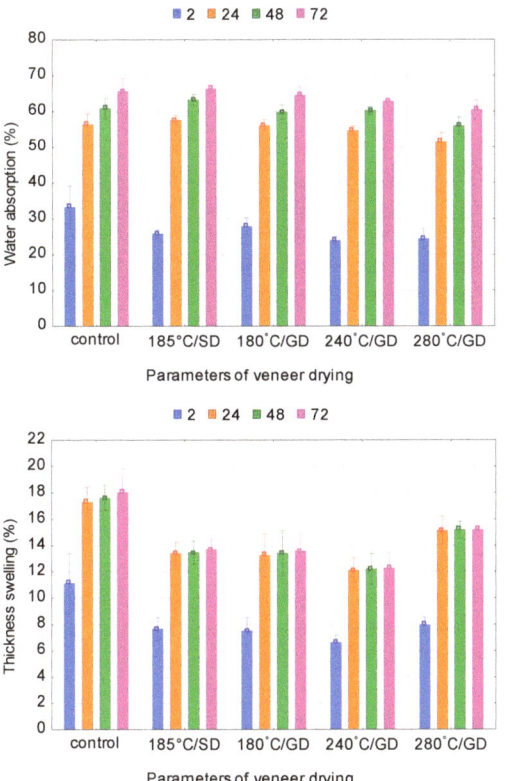

Figure 4. Water absorption and thickness swelling of phenol–formaldehyde (PF) plywood samples.

4. Conclusions

In this study, the birch veneer dried at elevated temperatures was successfully used for the bonding of plywood. The results of this study confirmed our assumption that veneer drying at elevated temperatures reduces the FE of plywood. The lowest FE for the UF and PF plywood samples were observed in the samples from veneer dried in a steam dryer at 185 °C/SD. It was also concluded from the study that the effect of veneer-drying temperatures on the bonding strength and physical and mechanical properties of plywood panels was significant. Veneer-drying temperatures of 185 °C/SD, 180 °C/GD and 240 °C/GD had a negative effect on the MOR and MOE along and across the fibres for both UF and PF plywood samples. The highest values of MOR along and across the fibres are observed in UF and PF plywood samples made from veneer dried at the highest drying temperature of 280 °C/GD. Bonding strength means values obtained from all test panels were above the required value (1.0 MPa) according to EN 314-2 standard. The effect of veneer drying at elevated temperatures on the dimensional stability of plywood samples was not as evident as in the case of bonding strength and mechanical properties of samples. The dimensional stability of the PF plywood samples was better than the dimensional stability of the UF samples.

Based on the findings of this study, an optimum veneer-drying temperature of 185 °C/SD could be recommended in industrial application for maintaining a balance between the formaldehyde emissions and bonding strength of the UF- and PF-bonded plywood panels.

Author Contributions: Conceptualization, P.B.; Methodology, P.B., J.S. and N.B.; investigation, P.B., J.S. and N.B.; Writing—original draft preparation, P.B.; Writing—review and editing, P.B., J.S. and N.B. All authors have read and agreed to the published version of the manuscript.

Funding: This work was supported by the Slovak Academic Information Agency and by the Slovak Research and Development Agency under the contracts No. APVV-14-0506, APVV-17-0456, APVV-17-0583 and APVV-18-0378; and ITMS project code: 313011T720 "LignoPro".

Conflicts of Interest: The authors declare no conflict of interest.

References

1. Kelly, T.J.; Smith, D.L.; Satola, J. Specific emission rates of formaldehyde from materials and consumer products found in California homes. *Environ. Sci. Technol.* **1999**, *33*, 81–88. [CrossRef]
2. Formaldehyde, 2–Butoxyethanol and 1–tert–Butoxypropan–2–ol. In *Monographs on the Evaluation of Carcinogenic Risk to Humans*; World Health Organization–International Agency for Research on Cancer: Lyon, France, 2006; Volume 88, 478p.
3. Roffael, E. Volatile organic compounds and formaldehyde in nature, wood and wood based panels. *Holz Roh Werkst.* **2006**, *64*, 144–149. [CrossRef]
4. Wang, S.Y.; Yang, T.H.; Lin, L.T.; Lin, C.J.; Tsai, M.J. Properties of low-formaldehyde-emission particleboard made from recycled wood-waste chips sprayed with PMDI/PF resin. *Build. Environ.* **2007**, *42*, 2472–2479. [CrossRef]
5. Amazio, P.; Avella, M.; Errico, M.E.; Gentile, G.; Balducci, F.; Gnaccarini, A.; Moratalla, J.; Belanche, M. Low formaldehyde emission particleboard panels realized through a new acrylic binder. *J. Appl. Polym. Sci.* **2011**, *122*, 2779–2788. [CrossRef]
6. Meyer, B.; Boehme, C. Formaldehyde emission from solid wood. *For. Prod. J.* **1997**, *47*, 45–48.
7. Weigl, M.; Wimmer, R.; Sykacek, E.; Steinwender, M. Wood-borne formaldehyde varying with species, wood grade, and cambial age. *For. Prod. J.* **2009**, *59*, 88.
8. Hasegawa, A. The effect of drying temperature on chemical-substance emission from solid wood (in Japanese). *J. Environ. Eng. AIJ* **2009**, *73*, 1267–1273. [CrossRef]
9. Schäfer, M.; Roffael, E. On the formaldehyde release of wood. *Holz Roh Werkst.* **2000**, *58*, 259–264. [CrossRef]
10. Martinez, E.; Belanche, M.I. Influence of veneer wood species on plywood formaldehyde emission and content. *Holz Roh Werkst.* **2000**, *58*, 31–34. [CrossRef]
11. Aydin, I.; Colakoglu, G.; Colak, S.; Demirkir, C. Effects of moisture content on formaldehyde emission and mechanical properties of plywood. *Build. Environ.* **2006**, *41*, 1311–1316. [CrossRef]
12. Bohm, M.; Salem, M.Z.M.; Srba, J. Formaldehyde emission monitoring from a variety of solid wood, plywood, blockboard and flooring products manufactured for building and furnishing materials. *J. Hazard. Mater.* **2012**, *221–222*, 68–79. [CrossRef] [PubMed]
13. Salem, M.Z.M.; Bohm, M.; Srba, J.; Berankova, J. Evaluation of formaldehyde emission from different types of wood-based panels and flooring materials using different standard test methods. *Build. Environ.* **2012**, *49*, 86–96. [CrossRef]
14. Salem, M.Z.M.; Bohm, M. Understanding of Formaldehyde Emissions from Solid Wood: An Overview. *BioResources* **2013**, *8*, 4775–4790. [CrossRef]
15. McDonald, A.G.; Gifford, J.S.; Steward, D.; Dare, P.H.; Riley, S.; Simpson, I. Air emission from timber drying: High temperature drying and re-drying of CCA treated timber. *Holz Roh Werkst.* **2004**, *62*, 291–302. [CrossRef]
16. Jiang, T.; Gardner, D.J.; Baumann, M.G.D. Volatile organic compound emissions arising from the hot-pressing of mixed hardwood particleboard. *For. Prod. J.* **2002**, *52*, 66–77.
17. Lutz, J.F. *Wood Veneer: Log Selection, Cutting, and Drying*; Technical Bulletin No. 1577; U.S. Department of Agriculture: Washington, DC, USA, 1978.
18. Vick, C.B. Adhesive bonding of wood materials. In *Wood handbook-Wood as an Engineering Material*; Gen. Tech. Rep. FPL–GTR–113; U.S. Department of Agriculture, Forest Service, Forest Products Laboratory: Madison, WI, USA, 1999; Chapter 9; 463p.
19. Baldwin, R.F. Veneer drying and preparation. In *Plywood and Veneer-Based Products*; Miller Freeman Books: San Francisco, CA, USA, 1995.

20. Rice, R.W. Mass Transfer, Creep and Stress Development during the Drying of Red Oak. Ph.D. Thesis, Virginia Polytechnique Institute and State University, Blacksburg, VA, USA, 1988.
21. Theppaya, T.; Prasertsan, S. Optimization of rubber wood drying by response surface method and multiple contour plots. *Dry. Technol.* **2004**, *22*, 1637–1660. [CrossRef]
22. Frihart, C.R.; Hunt, C.G. Adhesives with wood materials, bond formation and performance. In *Wood Handbook—Wood as an Engineering Material*; General Technical Report FPL-GTR-190; U.S. Department of Agriculture, Forest Service, Forest Products Laboratory: Madison, WI, USA, 2010; Chapter 10.
23. Aydin, I.; Colakoglu, G. Formaldehyde Emission, Surface Roughness, and Some Properties of Plywood as Function of Veneer Drying Temperature. *Dry. Technol.* **2005**, *23*, 1107–1117. [CrossRef]
24. Christiansen, A.W. How overdrying wood reduces its bonding to phenolformaldehyde adhesives: A critical review of the literature, part I—Physical responses. *Wood Fiber Sci.* **1990**, *22*, 441–459.
25. Lehtinen, M. *Effects of Manufacturing Temperatures on the Properties of Plywood*; TRT Report No 92; Helsinki University of Technology, Laboratory of Structural Engineering and Building Physics: Helsinki, Finland, 1998.
26. River, B.H.; Gillespie, R.H.; Vick, C.B. Wood as an adherent. In *Treatise on Adhesion and Adhesives*; Minford, J.D., Ed.; Marcel Dekker, Inc.: New York, NY, USA, 1991; Volume 7, 230p.
27. Nazerian, M.; Dahmardeh Ghalehno, M.; Bayat Kashkooli, A. Effect of wood species, amount of juvenile wood and heat treatment on mechanical and physical properties of laminated veneer lumber. *J. Appl. Sci.* **2011**, *11*, 980–987. [CrossRef]
28. Nazerian, M.; Ghalehno, M.D. Physical and Mechanical Properties of Laminated Veneer Lumber Manufactured by Poplar Veneer. *J. Agric. Sci. Technol. A* **2011**, *1*, 1040–1045. [CrossRef]
29. Jamalirad, L.; Doosthoseini, K.; Koch, G.; Mirshokraie, S.A. Physical and mechanical properties of plywood manufactured from treated red-heart beech (*Fagus Orientalis* L.) wood veneers. *BioResources* **2011**, *6*, 3973–3986.
30. Jamalirad, L.; Doosthoseini, K.; Koch, G.; Mirshokraie, S.A.; Welling, J. Investigation on bonding quality of beech wood (*Fagus orientalis* L.) veneer during high temperature drying and aging. *Eur. J. Wood Prod.* **2012**, *70*, 497–506. [CrossRef]
31. Murata, K.; Watanabe, Y.; Nakano, T. Effect of Thermal Treatment of Veneer on Formaldehyde Emission of Poplar Plywood. *Materials* **2013**, *6*, 410–420. [CrossRef] [PubMed]
32. Zdravkovic, V.; Lovric, A.; Stankovic, B. Dimensional Stability of Plywood Panels made from Thermally Modified Poplar Veneers in the Conditions of Variable Air Humidity. *Drvna Ind.* **2013**, *64*, 175–181. [CrossRef]
33. Demirkir, C.; Colakoglu, G.; Colak, S.; Aydin, I.; Candan, Z. Influence of Aging Procedure on Bonding Strength and Thermal Conductivity of Plywood Panels. *Acta Phys. Pol. A* **2016**, *19*, 1230–1234. [CrossRef]
34. Kol, H.S.; Seker, B. Properties of laminated veneer lumber manufactured from heat treated veneer. *Pro Ligno* **2016**, *12*, 3–8.
35. Bekhta, P.A. *Veneer Production*; Osnova: Kyjiv, Ukraine, 2003; 256p.
36. EN 314-1. *Plywood. Bonding Quality. Part 1: Test Methods*; European Committee for Standardization: Brussels, Belgium, 2004.
37. EN 314-2. *Plywood. Bonding Quality. Part 2: Requirements*; European Committee for Standardization: Brussels, Belgium, 1993.
38. EN 310. *Wood-Based Panels. Determination of Modulus of Elasticity in Bending and of Bending Strength*; European Committee for Standardization: Brussels, Belgium, 1993.
39. EN 317. *Particleboards and Fiberboards. Determination of Swelling in Thickness after Immersion in Water*; European Committee for Standardization (CEN): Brussels, Belgium, 1993.
40. EN ISO 12460-4. *Wood Based Panels. Determination of Formaldehyde Release. Part 4: Desiccator Method (ISO 12460-4:2016)*; European Committee for Standardization: Brussels, Belgium, 2016.
41. Denig, J.; Wengert, E.M.; Simpson, W.T. *Drying Hardwood Lumber*; Gen. Tech. Rep. FPL–GTR–118; US Department of Agriculture, Forest Service, Forest Products Laboratory: Madison, WI, USA, 2000.
42. Christiansen, A.W. Effect of overdrying of yellow-poplar veneer on physical properties and bonding. *Holz Roh Werkst.* **1994**, *52*, 139–149. [CrossRef]
43. Dougal, E.F.; Krahmer, R.L.; Wellons, J.D.; Kanarek, P. Glueline characteristics and bond durability of southeast asian species after solvent extraction and planing of veneers. *For. Prod. J.* **1980**, *30*, 48–53.
44. Neese, J.L.; Reeb, J.E.; Funck, J.W. Relating traditional surface roughness measures to glue-bond quality in plywood. *For. Prod. J.* **2004**, *54*, 67–73.

45. Collett, B.M. A review of the surface and interfacial adhesion in wood science and related fields. *Wood Sci. Technol.* **1972**, *6*, 1–42. [CrossRef]
46. Aydin, I. Activation of wood surface for glue bonds by mechanical pre-treatment and its effects on some properties of veneer surface and plywood panels. *Appl. Sur. Sci.* **2004**, *233*, 268–274. [CrossRef]
47. Boonstra, M.J.; Van Acker, J.; Pizzi, A. Anatomical and molecular reasons for property changes of wood after full-scale industrial heat-treatment. In Proceedings of the Third European Conference on Wood Modification, Cardiff, UK, 15–16 October 2007; pp. 343–358.
48. Tumen, I.; Aydemir, D.; Gunduz, G.; Uner, B.; Cetin, H. Changes in the Chemical Structure of Thermally Treated Wood. *Bioresources* **2010**, *5*, 1936–1944.
49. Özşahin, Ş.; Aydin, İ. Prediction of the optimum veneer drying temperature for good bonding in plywood manufacturing by means of artificial neural network. *Wood Sci. Technol.* **2014**, *48*, 59–70. [CrossRef]
50. Theander, O.; Bjurman, J.; Boutelje, J.B. Increase in the content of low-molecular carbohydrates at lumber surfaces during drying and correlations with nitrogen content, yellowing and mould growth. *Wood Sci. Technol.* **1993**, *27*, 381–389. [CrossRef]
51. Fengel, D.; Wegener, G. *Wood: Chemistry, Ultrastructure, Reactions*; Walter de Gruyter Prees: Berlin, Germany, 1989.
52. Salthammer, T.; Boehme, C.; Meyer, B.; Siwinski, N. Release of primary compounds and reaction products from oriented strand board (OSB). In Proceedings of the Healthy Buildings Conference, Singapore, 7–11 December 2003; pp. 160–165.
53. Tudor, E.M.; Barbu, M.C.; Petutschnigg, A.; Réh, R.; Krišťák, Ľ. Analysis of Larch-Bark Capacity for Formaldehyde Removal in Wood Adhesives. *Int. J. Environ. Res. Public Health* **2020**, *17*, 764. [CrossRef] [PubMed]
54. Spets, J.P.; Ahtila, P. Reduction of organic emissions by using a multistage drying system for wood-based biomasses. *Dry. Technol.* **2004**, *22*, 541–561. [CrossRef]
55. EN 13353. *Solid Wood Panels (SWP)–Requirements*; European Committee for Standardization: Brussels, Belgium, 2008.
56. Özşahin, Ş.; Demir, A.; Aydin, İ. Optimization of Veneer Drying Temperature for the Best Mechanical Properties of Plywood via Artificial Neural Network. *Anatol. Environ. Anim. Sci.* **2019**, *4*, 589–597. [CrossRef]
57. Unsal, O.; Candan, Z.; Buyuksarı, U. Effects of thermal modification on surface characteristics of OSB panels. *Wood Res.* **2010**, *55*, 51–58.
58. Zdravković, V.; Lovrić, A. Influence of thermal treatments on wettability and water spread on the surface of poplar veneer. In Proceedings of the First Serbian Forestry Congress, Belgrade, Serbia, 11–13 November 2010.
59. Candan, Z.; Buyuksarı, U.; Korkut, S.; Unsal, O.; Cakicier, N. Wettability and surface roughness of thermally modified plywood panels. *Ind. Crop. Prod.* **2012**, *36*, 434–436. [CrossRef]
60. Andor, T.; Lagaňa, R. Selected properties of thermally treated ash wood. *Acta Fac. Xylologiae Zvolen* **2018**, *60*, 51–60. [CrossRef]
61. Zhang, Y.; Hosseinaei, O.; Wang, S.; Zhou, Z. Influence of hemicellulose extraction on water uptake behavior of wood strands. *Wood Fiber Sci.* **2011**, *43*, 1–7.

© 2020 by the authors. Licensee MDPI, Basel, Switzerland. This article is an open access article distributed under the terms and conditions of the Creative Commons Attribution (CC BY) license (http://creativecommons.org/licenses/by/4.0/).

Article

Physical and Mechanical Properties of Ammonia-Treated Black Locust Wood

Mariana Domnica Stanciu [1,*], Daniela Sova [1], Adriana Savin [2], Nicolae Ilias [3,4] and Galina A. Gorbacheva [5]

1. Department of Mechanical Engineering, Faculty of Mechanical Engineering, Transilvania University of Brașov, B-dul Eroilor 29, 500036 Brașov, Romania; sova.d@unitbv.ro
2. National Institute of Research and Development for Technical Physics, B-dul Mangeron 47, 700050 Iasi, Romania; asavin@phys-iasi.ro
3. Faculty of Mining, University of Petrosani, 20 Universității Street, 332006 Petroșani, Romania; iliasnic@yahoo.com
4. Technical Sciences Academy of Romania, B-dul Dacia 26, 030167 Bucuresti, Romania
5. Faculty of Forestry, Forest Harvesting, Wood Processing Technologies and Landscape Architecture, Mytishchi Branch of Bauman Moscow State Technical University, 1st Institutskaya street, 141005 Mytischi, Russia; gorbacheva-g@yandex.ru
* Correspondence: mariana.stanciu@unitbv.ro

Received: 3 January 2020; Accepted: 30 January 2020; Published: 7 February 2020

Abstract: Because of the uneven colour of black locust wood, different technologies are used to change the colour, the bestknown being chemical and thermal treatments. Some of them affect the mechanical properties of wood, such as elasticity modulus, strength, durability. This study aims to compare the physical and mechanical properties of black locust wood control samples and treated wood samples with ammonia hydroxide, in terms of density profile, colour values (CIE L^*, a^*, b^*), mechanical properties of samples subjected to static bending, viscous-elastic properties (storage modulus (E'), loss modulus (E'') and damping ($\tan\delta$)). Two types of ammonia-fuming treatment were applied on samples: first treatment T1-5% concentration of ammonia hydroxide for 30 days; second treatment T2-10% concentration for 60 days. The results highlighted the following aspects: the overall colour change in the case of the second treatment is 27% in comparison with 7% recorded for the control samples; the lightness and yellowness values are the most affected by the second ammonia treatment of black locust wood. The density increased with almost 20% due to ammonium fuming (10% concentration/60 days); in case of static bending, the elastic modulus (MOE) tends to decrease with increasing the exposure time to ammonium, but the modulus of rupture (MOR) increases with almost 17% and the breaking force increases too, with almost 41%. In the case of dynamic mechanical analysis, the temperature leads to different viscous-elastic behaviour of each type of samples.

Keywords: black locust wood; ammonia treated wood; colour change; dynamic mechanical analysis

1. Introduction

The physical and mechanical properties of wood are very important for its use. The black locust wood (Latin name *Robinia pseudoacacia*) has different applications, like fence posts, boatbuilding, flooring, furniture, mine timbers, railroad ties, turned objects, fingerboards and back plate of guitars, some parts of musical instruments, veneer, wooden garden benches, pergolas etc. [1]. Due to the anatomical and chemical structure, the black locust wood is characterized by high natural durability, a fact for which it is used mainly for outdoor products. For the product diversification, harmonizing the colour of the wood with the architecture of the space and increasing the resistance to environmental factors, the colour of the black locust wood is modified by various treatments, but the most significant are chemical treatments. Chemical and thermal modifications can lead to improved wood characteristics, such as its durability, but also to the elastic properties modification, either by reducing or improving the mechanical strength. The surface modification of wood by different technologies improves the UV stability and change the surface energy of wood [2]. According to the macro and microscopic structure, natural wood is a composite material consisting of fibers and the matrix(lignin) [3–5]. In this context, Weigl et al. reported the influence of ammonia treatment on different wood species related to the modification of their physical properties [6]. They also noticed that one of the effect of ammonia treatment is the increase of wood affinity to water, but also the color modification and stabilization are enhanced [6]. The most receptive to ammonia-fumed coloring is black locust wood. In [7,8], the authors reported that the colour of ammonia-fumed oak is more resistant than the heat treated oak after UV exposure. Also, in [8] the authors studied the influence of wood species, time of exposure to ammonia gas on colour change of wood, using Fourier Transform Infrared (FTIR) spectroscopic analysis. The most significant chemical modification is related to the reaction of ammonia with carboxylate groups from wood with formation of ammonium salt, the reaction with aldehyde and ketonic groups resulting imines and the reaction with ester functional groups leading to formation of amides [9]. In [10], the authors studied the density profile and thermal properties of untreated beech wood and treated with ammonia and compressed beech wood. Many researchers investigated the effect of ammonia treatment on oak wood, both from colour modification and mechanical properties points of view [6,7,9,10]. The uneven colour of the natural black locust wood is an aesthetic disadvantage of this species and therefore, the steaming of the black locust wood and fuming with ammonia gas for the uniformity of the colour, is a widespread practice in some countries. In [11,12], the authors reported the significant effect of black locust wood treated with ammonia on shear strength in case of different glues, in comparison with untreated wood. Kačík et al. [12] evaluated the modification of chemical components of black locust wood after hot-water pretreatment, highlighting that the most affected by the treatment are lignin, hemicelluloses and holocelluloses [12].

Even if there are numerous studies on the colour changes of wood exposed to ammonia gas, there is some lack of information about the correlation between viscous-elastic properties of black locust wood, density profile and colour modification parameters. The chemical reactions have led to mechanical properties modification. The aim of the study is to analyze the physical and mechanical properties of black locust wood, both untreated and treated specimens, with solution of ammonium hydroxide, knowing that the darker wood colour in ammonia fuming is accomplished through chemical reactions between ammonia gas and wood compounds, which can affect the properties of black locust wood.

2. Materials and Methods

2.1. Materials

Three kinds of samples were tested: control samples-without treatment, cut in radial–longitudinal direction (S_RT_0) and cut in semi-radial–longitudinal direction ($S_{SR}T_0$); samples submitted to treatment T1 (surface treatment), cut in radial–longitudinal direction (S_RT_1) and cut in semi-radial–longitudinal direction ($S_{SR}T_1$); samples submitted to treatment T2 (in-depth treatment), cut in radial–longitudinal direction (S_RT_2) and cut in semi-radial–longitudinal direction ($S_{SR}T_2$). 24 samples of black locust wood were investigated, four from each type. The treatment of wood samples consisted in the exposure of wood to fumes of ammonium hydroxide (T1-5% concentration and 30 days; T2-10% concentration and 60 days). Ammonium hydroxide reacts with the wood tannins. After the treatment, the samples were dried in an oven chamber to obtain the moisture content range between 7% and 8%, specific for some parts of musical instruments construction. The physical and geometric characteristics of the wood samples are indicated in Table 1. The wood specimens with sizes $50 \times 10 \times 5$ mm^3 were subjected to bending, the 6 N force being applied in the middle of the distance between supports (40 mm). In Figure 1, the specimens tested for dynamic mechanical analysis (DMA) and colour change assessment are presented.

Table 1. Physical characteristics of wood samples prepared for dynamic mechanical analysis (DMA) test.

Wood Sample Codes		Dimensions			Mass	Apparent Density	Average Density	Standard Deviation
		Width b (mm)	Length L (mm)	Thickness h (mm)	m (g)	ρ (g/cm^3)	ρ_{av} (g/cm^3)	STDV
$S_R T_0$ MC = 7%	$S_R T_0 1$.	10.6	50	4.88	1.58	0.611	0.640	0.026
	$S_R T_0 2$.	10.5	50	4.9	1.63	0.634		
	$S_R T_0 3$.	10.56	49.95	4.92	1.73	0.667		
	$S_R T_0 4$.	10.56	49.98	4.88	1.67	0.648		
$S_{SR} T_0$ MC = 7%	$S_{SR} T_0 1$.	10.64	50.04	4.88	1.93	0.743	0.716	0.031
	$S_{SR} T_0 2$.	10.64	50.05	4.86	1.92	0.742		
	$S_{SR} T_0 3$.	10.6	50.03	4.86	1.79	0.695		
	$S_{SR} T_0 4$.	10.6	50.01	4.88	1.77	0.684		
$S_R T_1$ MC = 7.5%	$S_R T_1 1$.	10.35	49.9	4.97	1.94	0.756	0.735	0.021
	$S_R T_1 2$.	10.59	50	5.02	1.88	0.707		
	$S_R T_1 3$.	10.45	50.01	5.09	1.94	0.729		
	$S_R T_1 4$.	10.45	49.98	4.94	1.93	0.748		
$S_{SR} T_1$ MC = 8%	$S_{SR} T_1 1$.	10.49	49.96	4.92	1.63	0.632	0.704	0.085
	$S_{SR} T_1 2$.	10.45	49.97	4.86	1.61	0.634		
	$S_{SR} T_1 3$.	10.46	49.97	4.91	2.06	0.803		
	$S_{SR} T_1 4$.	10.48	49.97	4.94	1.94	0.750		
$S_R T_2$ MC = 8%	$S_R T_2 1$.	10.49	49.93	4.85	1.87	0.736	0.794	0.039
	$S_R T_2 2$.	10.47	50.04	4.88	2.07	0.810		
	$S_R T_2 3$.	10.58	50	4.85	2.10	0.819		
	$S_R T_2 4$.	10.57	49.93	4.76	2.04	0.812		
$S_{SR} T_2$ MC = 8%	$S_{SR} T_2 1$.	10.56	50	4.94	2.18	0.836	0.803	0.031
	$S_{SR} T_2 2$.	10.55	50	4.96	2.08	0.795		
	$S_{SR} T_2 3$.	10.52	50.01	4.92	2.11	0.815		
	$S_{SR} T_2 4$.	10.52	49.99	4.98	2.00	0.764		

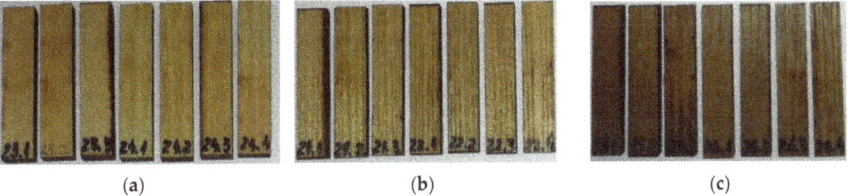

Figure 1. The wood samples: (**a**) Control sample (without treatment); (**b**)T1-samples subjected to 1st treatment (5% ammonium hydroxide); (**c**) T2-samples subjected to 2nd treatment (10% ammonium hydroxide).

2.2. Methods

The experimental investigations consisted of density profile analysis (DPA) on samples subjected to the same treatment as the other ones subjected to colour measurement (CM), a static three-point bending testand dynamic mechanical analysis (DMA). Finally, from the samples subjected to DMA, there were prepared some specimens which were covered with thin gold particles for scanning electron microscopy.

2.2.1. Density Profile Analysis (DPA)

The density profile of black locust samples on transversal direction and radial longitudinal direction was determined using an X-ray Density Profile Analyzer DPX300. The samples with the dimensions 50 × 50 × 30 mm^3 were automatically weighed and tested by the equipment devices. Then, each specimen was introduced in the X-ray device train where the equipment measured the density profile using the X-ray flux. Figure 2 presents the steps followed during the DPA.

(a)

(b)

Figure 2. X-ray Density Profile Analyzer DPX300: (a) weighting the samples; (b) the samples in the X-ray device train.

2.2.2. Colour Measurement (CM)

In order to find out the colour variation of black locust wood affected by chemical treatment with ammonium hydroxide, the chroma meter CR-400 Konica Minolta was used. Measuring results were colour values using the $L^*a^*b^*$ colour system, where L^* describes the lightness, and a^* and b^* describe the chromatic coordinates on the green–red and blue–yellow axes. The overall colour change ΔE^* was calculated with relations (1)–(3) [7]:

$$\Delta E^*_{T0-T1} = \sqrt{\left(L_0^* - L_1^*\right)^2 + \left(a_0^* - a_1^*\right)^2 + \left(b_0^* - b_1^*\right)^2}, \tag{1}$$

$$\Delta E^*_{T1-T2} = \sqrt{\left(L_1^* - L_2^*\right)^2 + \left(a_1^* - a_2^*\right)^2 + \left(b_1^* - b_2^*\right)^2}, \tag{2}$$

$$\Delta E^*_{T0-T2} = \sqrt{\left(L_0^* - L_2^*\right)^2 + \left(a_0^* - a_2^*\right)^2 + \left(b_0^* - b_2^*\right)^2}. \tag{3}$$

where ΔE^*_{T0-T1} represents the overall colour change between samples subjected to treatment T1 and untreated samples (T0); ΔE^*_{T1-T2}—the overall colour change between samples subjected to treatment T2 and samples subjected to treatment T1; ΔE^*_{T0-T2}—the overall colour change between samples subjected to treatment T2 and untreated samples (T0); L_0^*, a_0^*, b_0^*—colour system in terms of lightness, greenness and yellowness of control (untreated) samples; L_1^*, a_1^*, b_1^*—lightness, greenness and yellowness of treated samples subjected to the first treatment (5% ammonia); L_2^*, a_2^*, b_2^*—lightness, greenness and yellowness of treated samples subjected to the second treatment (10% ammonia).

2.2.3. Bending Tests

(1). Static Three-Points Bending Test

The first step was the three-points bending test performed on three types of samples, ten samples for each type, by using the universal machine LS100 Lloyd's Instrument with the load capacity of 100 kN. The aim of this test was to determine the breaking force, elasticity modulus of bending (MOE) and modulus of rupture (MOR) of black locust wood samples for untreated, T1-treated and T2-treated specimens with similar dimensions (length × width × thickness: 150 mm × 10 mm × 6.5 mm) like those of the samples subjected to DMA. The moisture content of wood (MC) was 6–8%, the environmental temperature T = 22 ± 1 °C and the relative humidity of air RH = 50% ± 5%. In Figure 3a, the principles of the three-point bending test can be observed and in Figure 3b, the breaking of black locust wood sample during the bending test. The speed of loading was set at 5 mm/min and the span between supports was 64 mm.

(a)   (b)

Figure 3. Three-point bending test: (a) the principle of loading; (b) the breaking of samples.

(2). Dynamic Mechanical Analysis (DMA)

The method of dynamic mechanical analysis (DMA) consists in applying an oscillating force at different frequencies (f = 1, 3.3, 5, 10, 50 Hz) in two cases. Firstly, isothermal conditions were used (temperature was kept constant at 30 °C during the test). Each sample was subjected to five DMA procedures with different load frequencies. The second analysis consists in the variation of temperature between 30 and 120 °C for 45 min, being repeated for different loading frequencies at the same values as in the first tests. The device returns the response of the material as a function of temperature and frequency that depends on the viscous-elastic nature of the material. The storage modulus (E'), the loss modulus (E'') and the complex modulus (E^*) are calculated from the material response to the sine wave. The ratio of the loss modulus and the storage modulus is called damping, denoted by tanδ, which represents the capacity of the material to store strain energy. This type of analysis predicts the flow behavior of wood in different environmental conditions. In Figure 4, the principles of samples loading and the main components of the equipment are shown. Figure 4a presents the direction of wood grain related to loading: the force is applied perpendicularly to the longitudinal axis (denoted L or x) producing a radial or semiradial bending moment (denoted by M_R and M_{SR}). In Figure 4b, the principles of the three-point bending test are shown and in Figure 4c, the main parts of the DMA equipment and the position of the wood sample within the device.

2.2.4. Scanning Electron Microscope Hitachi S3400N

The microscopic views of black locust wood were captured with a Hitachi S3400N scanning electron microscope (SEM). Before SEM analysis, the small pieces of wood samples were coated with a thin layer of gold (Au), which is a conducting material, as can be seen in Figure 5. The SEM tests were performed in semi-vacuum conditions.

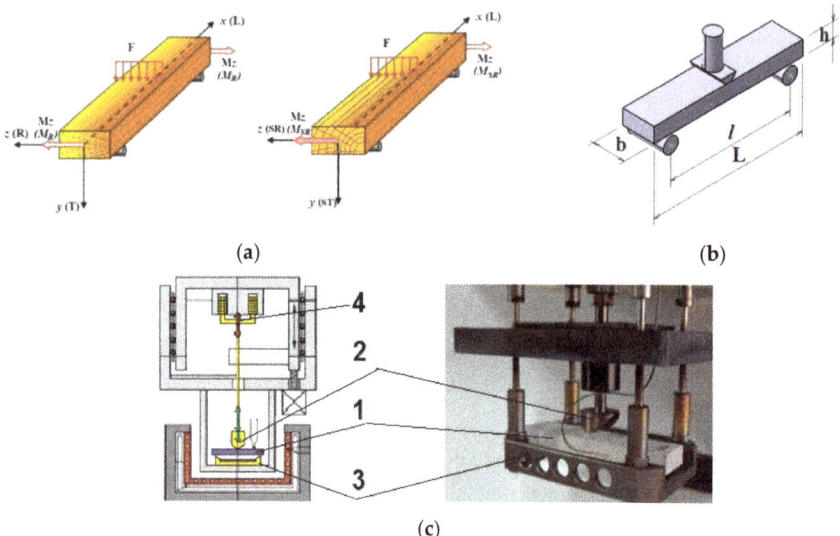

Figure 4. The DMA experimental set-up: (**a**) the types of wood samples (L—longitudinal direction, along wood grain; R—radial direction; T—tangential direction; SR—semi-radial direction; ST—semi-tangential direction); (**b**) the sample position on the equipment supports; (**c**) the main parts of DMA 242C Netzsch equipment: 1–specimen, 2–force application device, 3–specimen fixing/support devices, 4–electronic system for cyclic force application.

Figure 5. The gold-coated samples prepared for scanning electron microscope (SEM).

3. Results

3.1. X-ray Density Profile Analyzer DPX300

Density is correlated with the physical characteristics of the wood species and with their mechanical properties. Figure 6 presents the density profile of black locust samples in longitudinal–radial direction (Figure 6a) and in transversal direction (radial–tangential plane) (Figure 6b). The average values of density determined at 8%–10% moisture content, ranged between 700 and 780 kg/m^3, values that are similar to those reported in literature [13–15]. It is known that black locust wood is characterized by a complex and uneven structure, with clearly visible annual rings. Since black locust wood is a deciduous ring porous species with wide early wood vessels, heavily clogged by tyloses [16], the density profile varies on the annual ring width depending on the areas with early and late wood, as can be seen in Figure 6b. There is a slight increase in density of ammonia-treated black locust wood in comparison with untreated black locust wood, but the values do not exceed the average values recorded in the literature for black locust wood. Density determined by using the X-ray method is almost 7% higher than the calculated one (from specific gravity relation as ratio between mass and volume of samples), as can be seen in Figure 7.

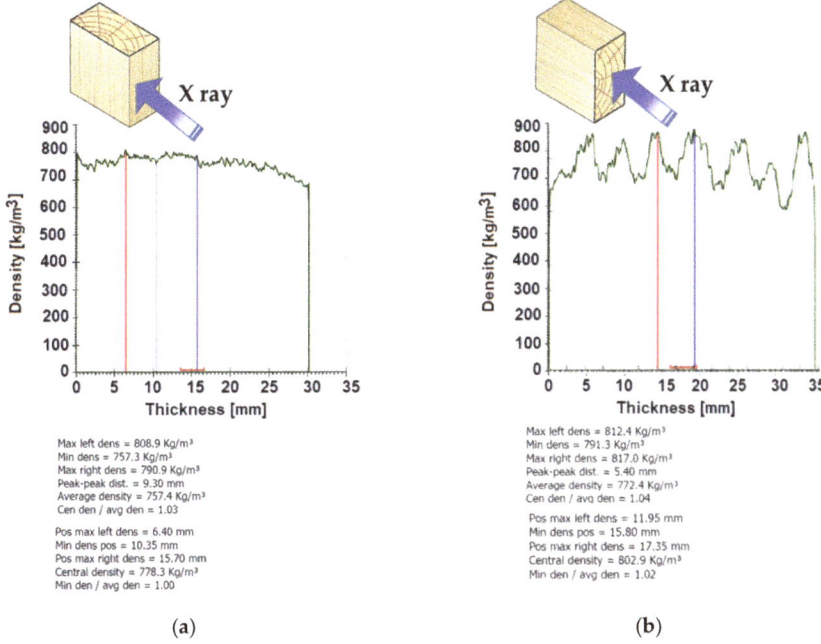

Figure 6. Variation of density profile of black locust wood samples: (**a**) longitudinal radial direction; (**b**) transversal direction of wood.

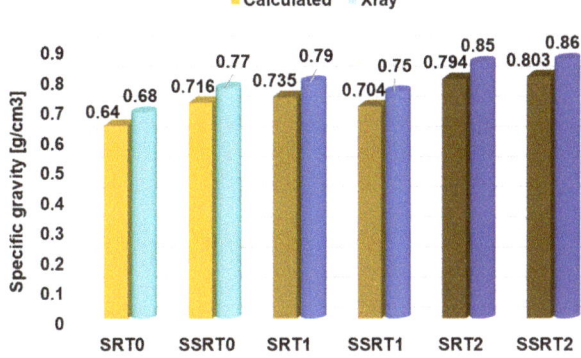

Figure 7. Comparison of conventional density (calculated) and density determined with X-ray equipment.

3.2. Colour Measurement

Figure 8a shows the influence of ammonia treatment on black locust wood from lightness L^*, chroma from green to red (a^*) (Figure 8b) and chroma from a blue to yellow (b^*) point of view (Figure 8c). The major colour changes were recorded in case of lightness: after the first treatment, the lightness decreased with 10% and after the second treatment, the differences were about 37% in comparison with the control samples. The overall colour change ΔE^*_{T0-T2} of the samples after the second treatment was around 27.073 units, as compared to the overall colour change ΔE^*_{T0-T1} obtained after the first treatment, whose value was 7.108 units. The overall colour change value ΔE^*_{T1-T2} obtained after the second treatment in relation to the first ammonia exposure was 20.379 units, which means that the period of exposure to ammonia (60 days) and the concentration of ammonia (10%) have the greatest influence on the colour change.

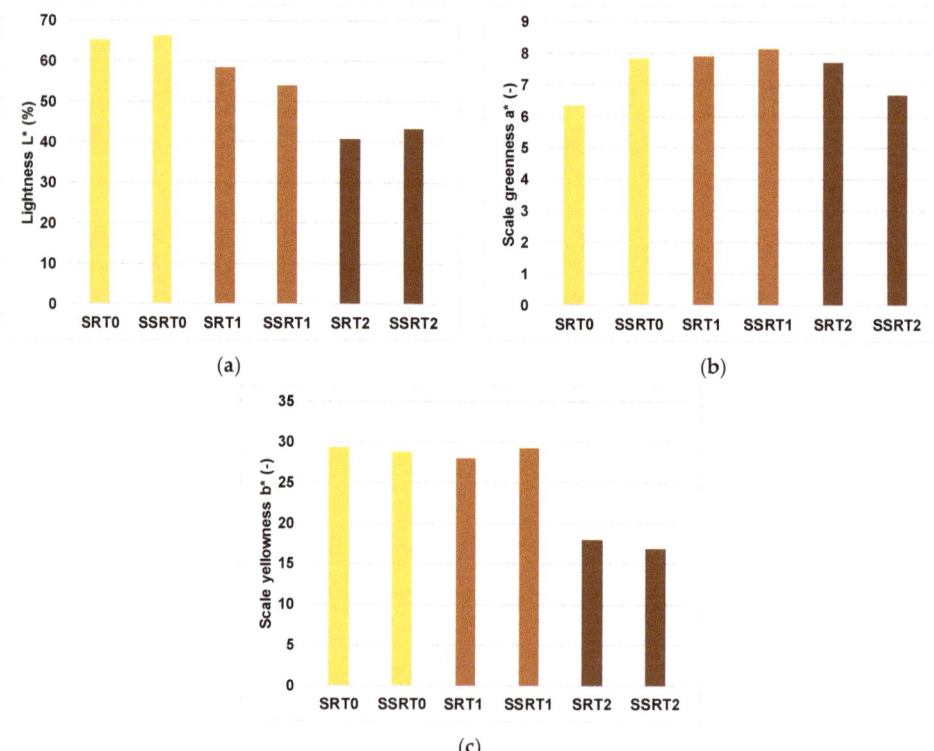

Figure 8. The colour measurement: (a)the lightness; (b) the scale greenness a^*; (c) the scale yellowness b^*.

3.3. Bending Test

3.3.1. Static Three-Point Bending Test

The static three-point bending test has revealed that the MOE decreases with almost 11% in case of treated samples, as compared to control samples, but the breaking force increases with 21% (for samples T1) and with 41% for the second treatment. Also, MOR increases for ammonia treated samples with 15%–17% in comparison with control samples. Figure 9 shows the characteristic curves for each type of tested samples. In Table 2, the average values and standard deviation of the main mechanical properties of control and treated samples are presented. The similar percentile changes were obtained by Weigl et al. [6], Čermák and Dejmal [10], who considered the reduction of mechanical properties values not significant. On the other hand, Rousek et al. [17] studied the possibilities of mechanical properties improvement in case of beech wood modification with ammonia gas, reporting a tendency to increase the mechanical properties of beech wood treated with ammonia, especially the MOR to compression.

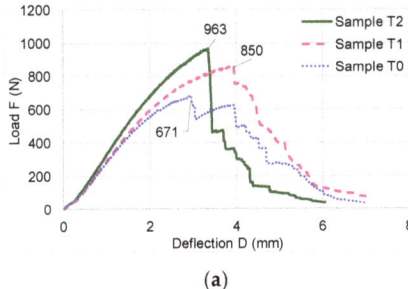 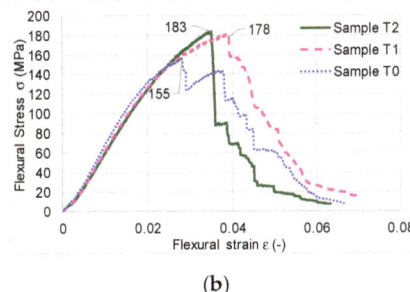

Figure 9. The characteristic curves to static three-point bending test (**a**) load versus deflection variation; (**b**) stress versus flexural strain variation.

Table 2. Mechanical properties obtained from the static flexural test, due to the presence of ammonia in the black locust wood samples.

Average Values (10 Samples for Each Type)	Samples					
	T0	STDEV	T1	STDEV	T2	STDEV
MOE of Bending (MPa)	7190	236.44	6469	214.38	6338	268.53
Differences MOE (%)	0		−10.03		−11.85	
Break Force F (N)	679.19	84.64	860.93	103.80	963.50	110.74
Differences F (%)	0		+26.75		+41.86	
MOR of Bending (MPa)	156.51	24.63	180.51	25.14	183.41	28.53
Differences MOR (%)	0		+15.33		+17.18	
Maximum Deflection (mm)	7.01	0.512	7.00	0.653	6.05	0.721

3.3.2. Dynamic Mechanical Analysis (DMA)

The experimental investigations have resulted in numerous data about the viscous-elastic behavior of different tested species. Thus, the first analysis consists in applying the load at different frequencies (1, 3.33, 5, 10, 50 Hz) at constant temperature (T = 30 °C), being determined the storage modulus values E', loss modulus values E'' and the damping $\tan\delta$. During the cyclic loading, wood tends to store increasingly more energy due to internal friction occurring in wood, as shown in Figure 10, but the damping capacity of wood decreases over time, regardless of species. Generally, the ability of wood samples to store energy increases with increasing the time of loading. This trend is similar with the strain hardening of steel, which is the process of making a metal harder and stronger through plastic deformation. The second analysis consists of temperature scanning at different frequencies.

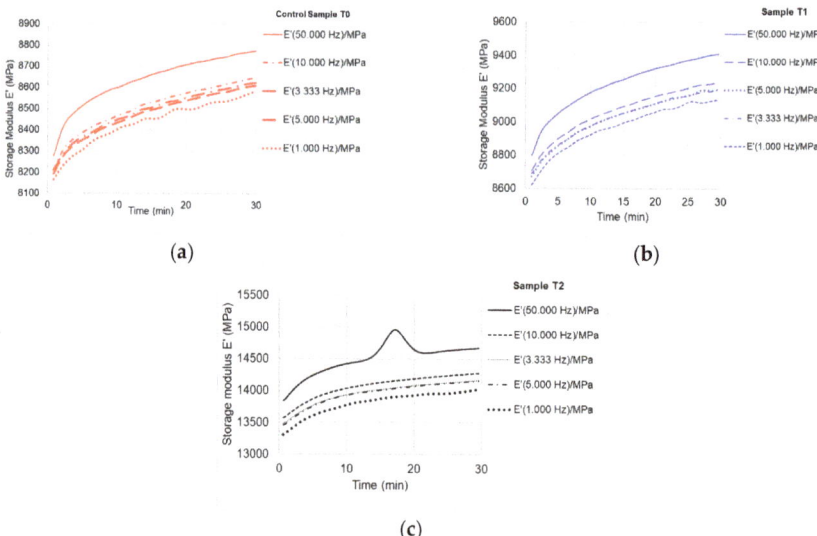

Figure 10. Variation of storage modulus E' with loading frequencies: (**a**) control sample; (**b**) sample T1 (first treatment); (**c**) sample T2 (second treatment).

(1) Isothermal Conditions (T = 30 °C)

It can be seen that the elastic (storage) modulus E' tends to increase by 6%–7%, thus increasing the time of loading. By increasing the frequencies (from 1 to 50 Hz), the storage modulus increases by almost 2.4% (Figure 10). The treatment of the dried black locust samples with ammonia leads to the increase of the storage modulus: after the first treatment, the increase is 4.8%–5% and after the second treatment, the increase is 50%, as compared to the control samples (Figure 10a–c). This phenomenon is explained by the effect of ammonia evaporation that leads to the increase of wood stiffness [16,17]. It is worth to mention that the black locust wood samples treated with ammonia were conditioned and dried to 6%–8% moisture content. Thus, the plasticization effect of ammonia vapor was eliminated. An improvement of mechanical properties of ammonia treated samples in case of beech wood was reported by [18] who noticed the differences between the treatment with ammonia gas (dry wood) and ammonia with water (wet wood).

The values of storage modulus ranged between 8800 MPa for untreated samples, 9400 MPa for the first treatment of black locust samples and 14,000 MPa for the second treatment (Figure 11). Similar values are reported by other authors [16,19–21]. The ratio between loss modulus and storage modulus represents damping (tanδ), which is a sensitive indicator of the mechanical or thermal conditions during the mechanical energy input dispersed as heat by internal friction caused by chain motion. The damping tends to decrease by increasing the loading time at a constant temperature of 30 °C. Because wood is a natural polymer and has a stratified structure of early and late wood, damping occurs gradually, in stages. The energy dissipates progressively, the cellular voids leading to the damping of the internal wood friction. This phenomenon is observed in form of variation curves of damping in Figure 12. In case of the samples T2 (Figure 12c) it can be noticed that the slope of the curve shows a decreasing linear variation over time, which can be influenced by chemical modification of wood by formation of different chemical groups.

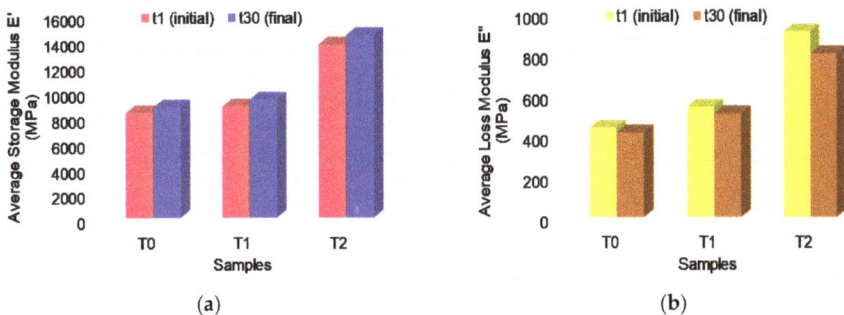

Figure 11. Comparison between dynamic properties of testes samples: (**a**) storage modulus E'; (**b**) loss modulus E''.

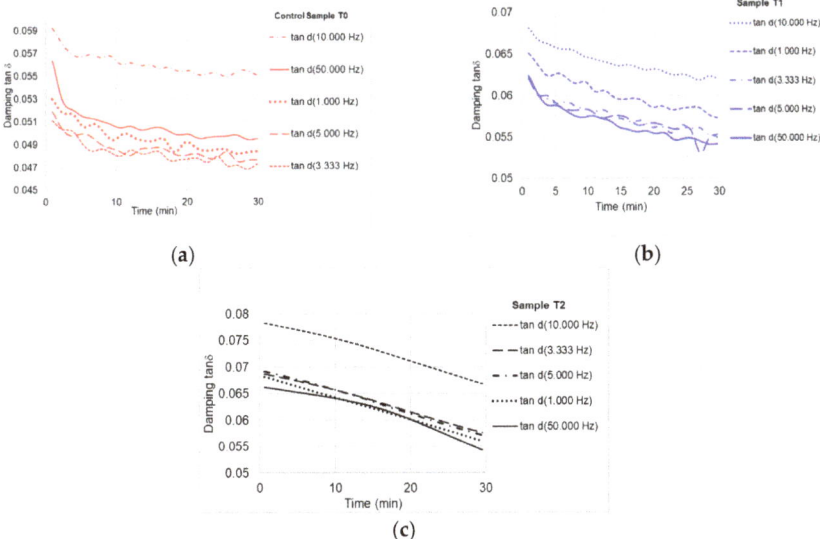

Figure 12. Variation of damping tanδ with loading frequencies: (**a**) control sample; (**b**) sample T1 (first treatment); (**c**) sample T2 (second treatment).

(2) Temperature Scanning

Temperature affects the stiffness and resilience of wood characterized by thermoset behaviour. In the case of the control samples, the storage modulus increases slightly up to 75–80 °C, then the recording of a decreased trend (Figure 13, a—red lines) is noticed; in the case of treated samples (T1—blue lines and T2—black lines), the storage modulus has the trend to remain constant between 30–65 °C, after that, a trend to increase between 65 and 100 °C can be seen (Figure 13).

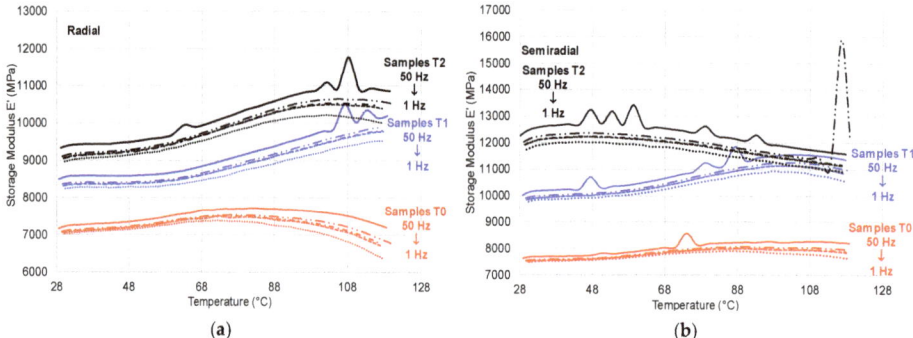

Figure 13. Variation of storage modulus E' versus temperature: (**a**) samples cut in radial–longitudinal direction; (**b**) samples cut in semi-radial–longitudinal direction. Legend: red line—untreated samples; blue line—samples exposed to first ammonia treatment; black line—samples exposed to the second ammonia treatment; solid line (50 Hz); long dash dot dot line (10 Hz); dash dot line (5 Hz); square dot line (3.3 Hz); round dot line (1 Hz).

The cross-section of wood and the relative position of loading with respect to grains (radial and semi-radial) influence the mechanical response of samples: for untreated samples and samples T1, the trend is similar with that of the samples cut in radial direction, but the samples T2 have shown the tendency of decreasing the storage modulus (E') starting with the temperature of 50 °C. The overshoot (peak) recorded for both treated samples, at the loading frequency of 50 Hz, is caused by molecular rearrangements that occur due to the increased free volume at the transition [22]. With increasing the temperature, the loss modulus E' increases too for all types of samples; at temperatures higher than 100 °C, the values are double (Figure 14). An interesting behaviour regarding the variation of the loss modulus with temperature is recorded by semi-radial ammonia treated samples using the second treatment T2, which tends to decrease starting with 48 °C (Figure 14b). The viscous-elastic behavior is influenced by temperatures higher than 48 °C, also revealed in damping variation (Figure 15).

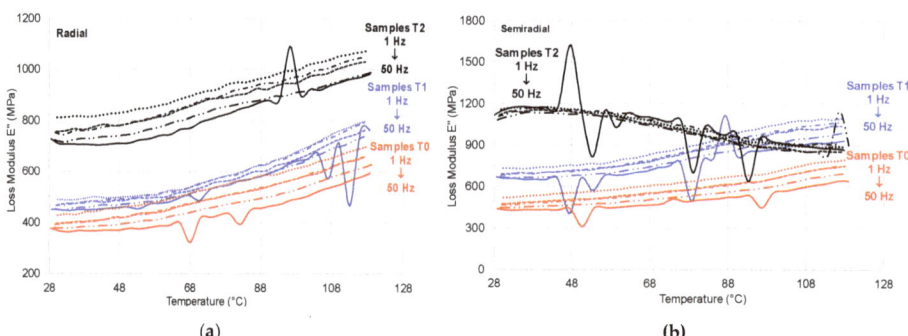

Figure 14. Variation of loss modulus E'' with temperature: (**a**) samples cut in radial–longitudinal direction; (**b**) samples cut in semi-radial–longitudinal direction. Legend: red line—untreated samples; blue line—samples exposed to first ammonia treatment; black line - samples exposed to the second ammonia treatment; solid line (50 Hz); long dash dot dot line (10 Hz); dash dot line (5 Hz); square dot line (3.3 Hz); round dot line (1 Hz).

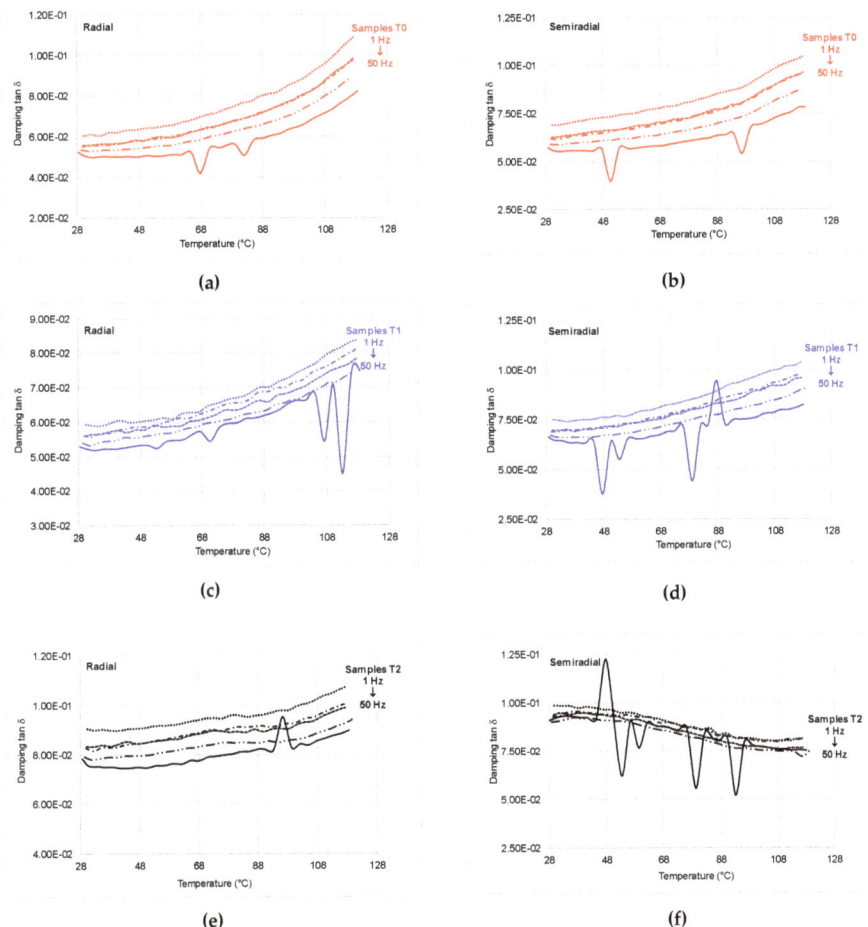

Figure 15. tanδ with temperature: (**a**) control sample—radial cross-section; (**b**) control sample—semi-radial cross-section; (**c**) sample T1 (first treatment)—radial cross-section; (**d**)—sample T1 (first treatment)—semi-radial cross; (**e**) sample T2 (second treatment)—radial cross-section; (**f**) sample T2 (second treatment)—semi-radial cross-section; solid line (50 Hz); long dash dot dot line (10 Hz); dash dot line (5 Hz); square dot line (3.3 Hz); round dot line (1 Hz).

3.4. Scanning Electron Microscope Hitachi S3400N

In Figures 16 and 17, the SEM capture of untreated and ammonia treated black locust wood samples is presented. At microscopic level, no differences between the three types of samples can be observed. The control samples cross-section shows that the wide early wood vessels (150–220 μm) are arranged in a 2–3 vessels-thick ring and are heavily clogged by tyloses. Latewood vessels have smaller diameters (70–140 μm). The longitudinal and the ray parenchyma often contain crystalline deposits. Molnar et al. and Nemeth et al. presented numerous studies on microstructure of black locust wood, highlighting the microscopic characteristics [20,21].

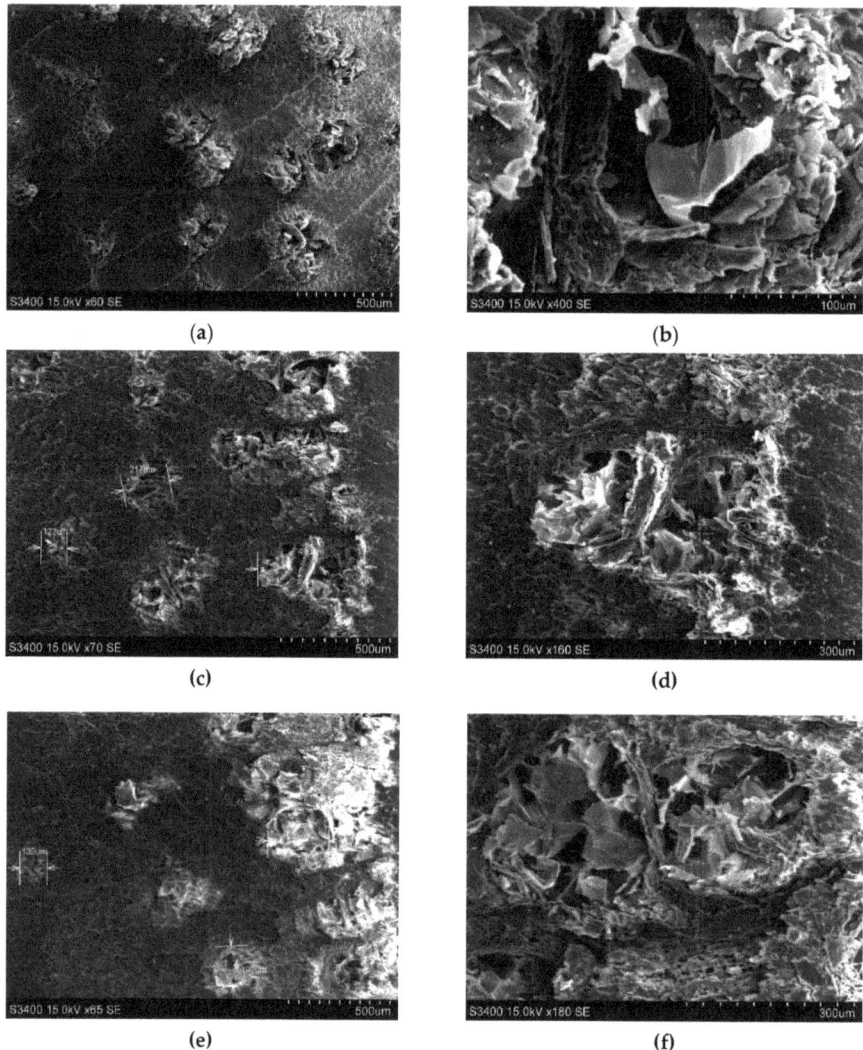

Figure 16. SEM view on cross-section of wood: (**a**) control sample magnification ×60 SE; (**b**) control sample magnification ×400 SE; (**c**) sample T1 (first treatment) magnification ×70 SE; (**d**) sample T1 (first treatment) magnification ×160 SE; (**e**) sample T2 (second treatment) magnification ×65 SE; (**f**) sample T2 (second treatment) magnification ×180 SE.

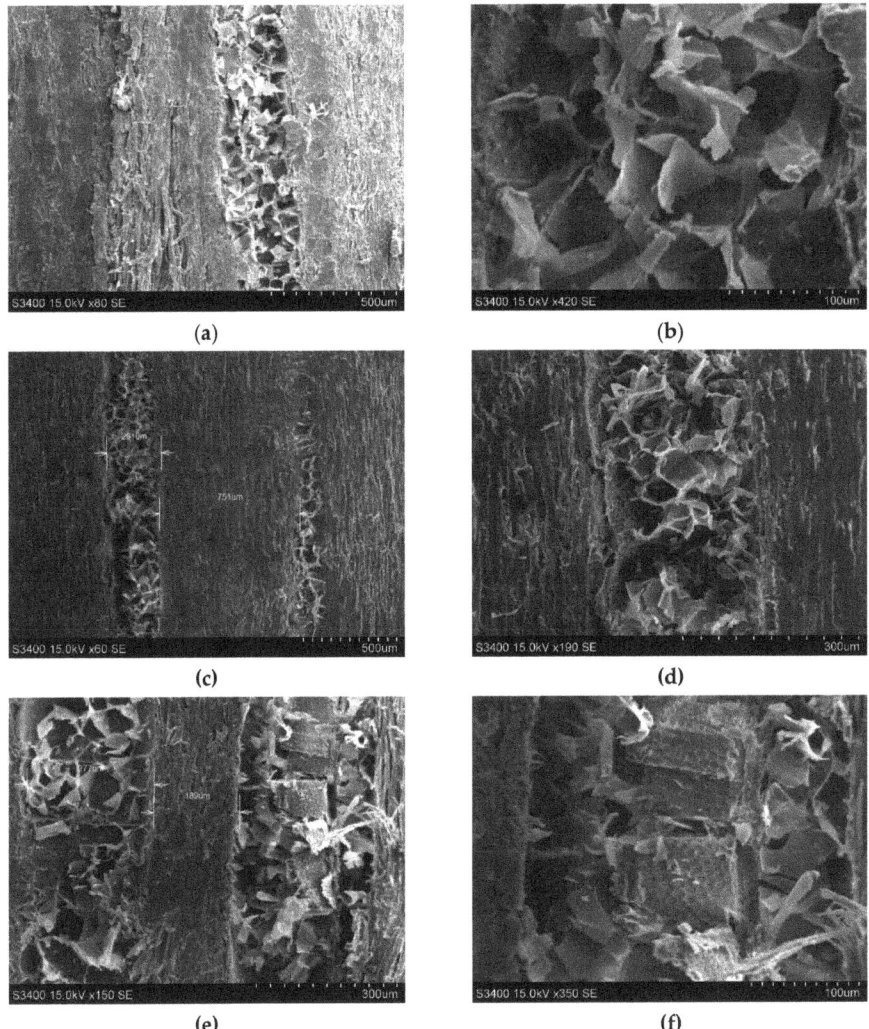

Figure 17. SEM view in longitudinal–radial section of wood: (**a**) control sample magnification ×80 SE; (**b**) control sample magnification ×420 SE; (**c**) sample T1 (first treatment) magnification ×60 SE; (**d**) sample T1 (first treatment) magnification ×190 SE; (**e**) sample T2 (second treatment) magnification ×150 SE; (**f**) sample T2 (second treatment) magnification ×350 SE.

4. Conclusions

The paper presents the experimental results of numerous studies on black locust wood samples, uncoloured and coloured, as a result of exposure to ammonium hydroxide.

- The overall colour change in the case of the second treatment is 27% in comparison with 7% recorded for the control samples. The lightness and yellowness are the most affected colour values after the second treatment of black locust wood with ammonia hydroxide (the lightness decreased with almost 40% after ammonia treatment and the yellowness—with 50%).
- The density increased with almost 20% due to ammonium fuming (10% concentration/60 days).

- At the static three-point bending, MOE recorded a decrease with almost 11% and MOR increased for ammonia treated samples with 15% to 17% in comparison with control samples.
- In the DMA test, the storage modulus (E') is higher with almost 60% in the case of the second treatment of samples in comparison with control samples.
- The viscous behaviour is more evident when the temperature increases above 40–60 °C for all types of samples.
- The exposure of treated samples (T2) to temperature led to different behaviours according to the direction of wood grain related to the orthogonal directions of the samples (radial versus semiradial).
- The microscopic views captured with SEM did not highlight a specific surface modification of wood.

Author Contributions: Conceptualization, M.D.S. and D.S.; methodology, S.M.D, A.S., G.A.G. and N.I.; software, G.G.A.; validation, N.I., A.S. and G.A.G.; formal analysis, S.D and N.I.; investigation, S.M.D and A.S.; writing—original draft preparation, M.D.S., A.S.; writing—review and editing, D.S.; visualization, G.A.G.; supervision, N.I. All authors have read and agreed to the published version of the manuscript.

Funding: This paper was supported by Program partnership in priority domains -PNIII under the aegis of MECS -UEFISCDI, project no. PN-III-P2-2.1-BG-2016-0017/85 SINOPTIC, project manager Stanciu Mariana Domnica.

Acknowledgments: We are grateful to thetechnical staff of S.C. Hora S.A Reghin, Romanian manufacturer of musical string instrumentsfor supplying the specimens for the experimental research of the present article. We are grateful to Cosnita Mihaela from R&D Institute of Transilvania University of Brasov for her help in taking SEM views.

Conflicts of Interest: The authors declare no conflict of interest.

References

1. Valkenburgh Van, M.; Noone, S.; Zoli, T.; Lavender, D. Black Locust Lumber: A Sustainable Alternative. San Diego, 2011. Available online: https://www.asla.org/search.aspx?q=Valkenburgh (accessed on 28 December 2019).
2. Papadopoulos, A.N.; Mantanis, G.I. Surface treatment technologies applied to wood surfaces. *FDM Asia-Solid Wood and Panel Technology* **2011**, *5*, 36–39. Available online: users.teilar.gr/~{}mantanis/Papadopoulos-Mantanis-FDM-Asia.pdf (accessed on 28 December 2019).
3. Côte, W. Chemical composition of wood. In *Principles of Wood Science and Technology*; Springer: Berlin, Germany, 1968; pp. 55–78.
4. Kapitovic, S.; Klasnja, B.; Guzina, V. Importance of structural, physical and chemical properties of robinia wood for its mechanical characteristics. *Drevarski Vyskum.* **1989**, *122*, 13–30.
5. Kollmann, F.F.P. *Principles of Wood Science and Technology: Solid Wood*; Springer: Berlin, Germany, 1968.
6. Weigl, M.; Pockl, J.; Grabner, M. Selected properties of gas phase ammonia treated wood. *Eur. J. Wood Prod.* **2009**, *6*, 103–109. [CrossRef]
7. Miklečić, J.; Španić, N.; Vlatka, J.R. Wood Colour Changes by Ammonia Fuming. *BioResources* **2012**, *7*, 3767–3778.
8. Pawlak, Z.; Pawlak, A.S. A review of infrared spectra from wood and wood components following treatment with liquid ammonia and solvated electrons in liquid ammonia. *Appl. Spectrosc. Rev.* **1997**, *32*, 349–383. [CrossRef]
9. Troppova, E.; Tippner, J.; Hrčka, R.; Halacan, P. Quasi-stationary measurements of Lignamon thermal properties. *BioResources* **2013**, *8*, 6288–6296. [CrossRef]
10. Čermák, P.; Dejmal, A. The effect of heat and ammonia treatment on colour response of oak wood *(Quercus Robur)* and comparison of some physical and mechanical properties. *Maderas. Ciencia y Tecnología* **2013**, *15*, 375–389.
11. Borysiuk, P.; Jablonski, M.; Policimska-Serwa, A.; Ružinská, E. Mechanical properties of glue bonds in black locust wood treated with ammonia. *Ann. WULS-SGGW For. Wood Technol.* **2011**, *73*, 162–166.
12. Kačík, F.; Ďurkovič, J.; Kačíková, D.; Zenková, E. Changes in the chemical composition of black locust wood after hot-water pretreatment before bioethanol production. *Acta Facultatis Xylologiae Zvolen.* **2016**, *58*, 15–23.

13. Adamopoulos, S. Flexural properties of black locust (Robinia pseudoblack locust L.) small clear wood specimens in relation to the direction of load application. *Holz als Roh-und Werkstoff.* **2002**, *60*, 325–327. [CrossRef]
14. Niklas, K.J. Mechanical Properties of Black Locust (Robinia pseudoacacia L.) Wood. Size and Age-dependent Variations in Sap- and Heartwood. *Ann. Bot.* **1997**, *79*, 265–272. [CrossRef]
15. Pollet, C.; Verheyen, C.; Hebert, J.; Jourez, B. Physical and mechanical properties of black locust (Robinia pseudoacacia) wood grown in Belgium. *Can. J. For. Res.* **2012**, *42*, 831–840. [CrossRef]
16. Weigl, M.; Müller, U.; Wimmer, R.; Hansmann, C. Ammonia vs. thermally modified timber—Comparison of physical and mechanical properties. *Eur. J. Wood Prod.* **2012**, *70*, 233–239. [CrossRef]
17. Rousek, R.; Rademacher, P.; Brabec, M.; Dejmal, A.; Horníček, S.; Baar, J.; Šprdlík, V. Beech wood modification with ammonia gas improved properties. *Pro Ligno.* **2015**, *11*, 230–238.
18. Šprdlík, V.; Brabec, M.; Mihailović, S.; Rademacher, P. Plasticity increase of beech veneer by steaming and gaseous ammonia treatement. *Maderas. Ciencia y Tecnología* **2016**, *18*, 91–98.
19. Kamperidou, V.; Barboutis, I.; Vassiliou, V. Prospects for the utilization of black locust wood (*Robinia pseudoacacia* L.) coming from plantations in Furniture Manufacturing. In Proceedings of the 27th International Conference on Wood Modification and Technology, Zagreb, Croatia, 13–14 October 2016; pp. 123–128.
20. Molnar, S.; Peszlen, I.; Richter, H.G.; Tolvaj, L.; Varga, F. *Influence of Steaming on Selected Wood Properties of Black Locust.* In Proceedings "Environment and Wood Science"; Acta Facultatis Ligniensis, University of Sopron: Sopron, Hungary, 1998; pp. 38–45.
21. Nemeth, R.; Molnar, S. Utilisation of Walnut (Juglans), Robinia (Robinia pseudoacacia) and Ash (Fraxinus) on the Basis of Hungarian Experiences. Available online: http://www.valbro.uni-freiburg.de/pdf/pres_thes_util_walnut.pdf (accessed on 20 November 2019).
22. Menard, K.P. *Dynamic Mechanical Analysis–A Practical Introduction*; CRC Press LLC: Boca Raton, FL, USA, 1999; p. 117.

© 2020 by the authors. Licensee MDPI, Basel, Switzerland. This article is an open access article distributed under the terms and conditions of the Creative Commons Attribution (CC BY) license (http://creativecommons.org/licenses/by/4.0/).

Article

Improving Fire Retardancy of Beech Wood by Graphene

Ayoub Esmailpour [1], Roya Majidi [1], Hamid R. Taghiyari [2,*], Mehdi Ganjkhani [2], Seyed Majid Mohseni Armaki [3] and Antonios N. Papadopoulos [4,*]

1. Department of Physics, Faculty of Sciences, Shahid Rajaee Teacher Training University, Tehran 22970021, Iran; esmailpour@sru.ac.ir (A.E.); royamajidi@gmail.com (R.M.)
2. Wood Science and Technology Department, Faculty of Materials Engineering & New Technologies, Shahid Rajaee Teacher Training University, Tehran 22970021, Iran; mahdi.ganjkhani1@gmail.com
3. Faculty of Physics, Shahid Beheshti University, Evin, Tehran 22970021, Iran; majid.mohseni19@yahoo.com
4. Laboratory of Wood Chemistry and Technology, Department of Forestry and Natural Environment, International Hellenic University, GR-661 00 Drama, Greece
* Correspondence: htaghiyari@sru.ac.ir (H.R.T.); antpap@for.ihu.gr (A.N.P.)

Received: 8 January 2020; Accepted: 21 January 2020; Published: 3 February 2020

Abstract: The aim of this paper was to improve the fire retardancy of beech wood by graphene. Six fire properties, namely time to onset of ignition, time to onset of glowing, back-darkening time, back-holing time, burnt area and weight loss were measured using a newly developed apparatus with piloted ignition. A set of specimens was treated with nano-wollastonite (NW) for comparison with the results of graphene-treated specimens. Graphene and NW were mixed in a water-based paint and brushed on the front and back surface of specimens. Results demonstrated significant improving effects of graphene on times to onset of ignition and glowing. Moreover, graphene drastically decreased the burnt area. Comparison between graphene- and NW-treated specimens demonstrated the superiority of graphene in all six fire properties measured here. Fire retardancy impact of graphene was attributed to its very low reaction ability with oxygen, as well as its high and low thermal conductivity in in-plane and cross-section directions, respectively. The improved fire-retardancy properties by the addition of graphene in paint implied its effectiveness in hindering the spread of fire in buildings and structures, providing a longer timespan to extinguish a fire, and ultimately reducing the loss of life and property. Based on the improvements in fire properties achieved in graphene-treated specimens, it was concluded that graphene has a great potential to be used as a fire retardant in solid wood species.

Keywords: fire retardants; fire retardancy; graphene; nano-materials; wollastonite

1. Introduction

Wood is a versatile material with a myriad of applications and therefore, its plantation and harvesting are vastly studied all over the world [1,2]. The idea of protecting solid wood, and wood and cellulose-based materials against different physical and chemical damages, and against the attacks of living micro-organisms, and fire as well, is as old as human civilizations [3–7]. Over time, numerous methods and a variety of materials have been examined and developed. Some methods changed the pathway of pyrolysis in wood cell-wall polymers [8]. This is considered one of the easiest and inexpensive ways for wood. In another method, the surface of the wood is improved, acting as an isolating layer. Intumescent coatings are also categorized in the surface protection method. Alteration in the thermal properties through changing its density, specific and heat thermal conductivity is another one that can also be used to improve fire retardancy in wood and wood-based composites. Other techniques involve decreasing wood ignitability by diluting pyrolysis gases or inhibiting the chain reactions of burning. Though several practical and effective methods and techniques have so far been developed, research for more effective and non-toxic materials to improve fire retardancy is still in progress [9–14].

Improving the effects of nanotechnology on different materials has been vastly and intensively elaborated [15–21]. Wood-based materials and composites are no exception [22–24]. Different nano-metals and nano-minerals were utilized to improve heat-transfer property in solid wood species and wood composite mats; they were also used to improve biological resistance against different deteriorating fungi to decrease hot-press time as a costly bottle-neck in nearly all wood-composite manufacturing factories and to increase thermal conductivity in solid wood and composite mats [3,15,18,25–33].

In this connection, graphene is a one-atom-thick planar sheet of a hexagonally arranged carbon atoms and therefore, it is considered a nano-material. These sheets are densely packed in a honeycomb crystal lattice. Graphene has got appreciable attention over the last two decades due to its special structure and exceptional properties [34,35]. These properties have made graphene ideal to be used in electronics, sensors, energy-saving devices, different composites, and emerging modern materials, and many other new applications to be investigated in the future. However, the authors came across little or no research projects studying the outcome of graphene as a fire retardant in wood. Therefore, the present research project was carried out to primarily find out if graphene may improve fire retardancy in beech, as an important industrial solid wood species. However, a parallel study was conducted with nano-wollastonite (NW), as a nano-material that has been reported to improve fire retardancy in both solid wood species and wood-composite panels [36–38], for comparison purposes. As the surface of wooden bodies and parts in buildings and structures are painted, graphene and NW were mixed with a popular water-based paint as a carrier of the nano-materials. Separate sets of specimens without any paint (plain wood), and with paint that contained no nano-materials, were prepared and tested for comparison purposes with the graphene- and NW-treated specimens.

2. Materials and Methods

2.1. Specimen Preparation

For the present research project, beech boards (*Fagus orientalis*) were purchased from the Khavaran Wood Bazar (Tehran, Iran). The density of beech specimens was measured to be 0.62 g·cm^{-3}. Boards were kept in the wood workshop of Shahid Rajaee Teacher Training University (Tehran, Iran) for four months before cutting (35–40 °C; relative humidity 26%–30%). They were then cut to size, 220 mm in length, 140 mm in width (plain sawn), and 5 mm in thickness. Specimens were free from any fungal or insect attack, checks or cracks, and knots. Twenty specimens were selected and divided into four treatments; for each treatment, five replicate specimens were prepared. Treatments included: control (without any paint and nano-material), painted, NW+painted (nano-wollastonite was mixed in paint and applied on the surface of specimens), and NG+painted (graphene was mixed with paint and applied on the surface of specimens. Acrylic paints are popular in wood products [39] and therefore, a water-based acrylic-latex paint was used in this experiment (code number ALCO-6510), purchased from Alvan Paint and Resin Production Co. (Tehran, Iran). The solid content of the paint was 37% ± 1%. Two coats were brushed on each specimen to achieve a 190–200 μm dry paint film on the front and back surfaces of specimens. A 24-h time was given between the two coats to let the first coat being dried out. Once the two coats were applied, all specimens were kept in a conditioning chamber (25 ± 1 °C, and 40% ± 2% relative humidity) for two weeks. For the nanomaterial-treated specimens, 15% of nano-material was mixed with the paint, based on the wet weight of the paint, before being brushed on specimens. NW gel was produced by Vard Manufacturing Company of Mineral and Industrial Products (Iran). At least 70% of NW particles ranged from 30 to 110 nm. The formulation of NW gel has been reported by Taghiyari et al. [29–31]. Nanomaterials were mixed with the paint for 20 min, using a magnetic stirrer. While mixing the nanomaterials, solid content of the final paint was kept constant by the addition of a calculated amount of distilled water. The moisture content of specimens at the time of fire tests was 8% ± 0.5%.

2.2. Graphene Production Technique

Electrochemical exfoliation of graphite was performed in a system with a cathode electrode of platinum (Pt 0.5 × 10 cm^2) and the anode electrode of graphite foil (2 × 10 cm^2). Two electrodes were placed at a distance of 2.7 cm from each other. NiCl$_2$.6H$_2$O powder (98.0% Merck) was dissolved in water (concentration of 0.05 M, and pH 6.5–7.0) to prepare electrolyte. To provide expansion, exfoliation of graphite, and deposition of Ni, a voltage of 10 V was applied. The Pt electrode was washed every 20 min with HCl and water to avoid the accumulation of the product on the cathode. Finally, the products were collected using vacuum filtration, and then they were washed with water. The end product was dispersible for sonication in water. In the electrochemical exfoliation mechanism, hydroxyl ions (OH) were first produced in the cathode region by applying the voltage between the electrodes, and then the ions accelerate towards the anode and hit the graphite surface. The collision of OH ions with graphite and oxidation at the edge sides and grain boundaries lead to expansion of the graphite layers, penetration of Cl ions through the graphite layers, and reduction of Cl ions to finally produce Cl gas. An excessive force exerted to graphite layers upon Cl gas caused the separation of the graphite layers [40,41]. Then, graphene sheets distributed in the solution trapped Ni^{2+} ions and proceeded towards the negative electrode under an electric field. Finally, black composite on the Pt electrode was created. Generation of hydroxyl at cathode via ionization of water is as follows [40]:

$$2H_2O + 2e \rightarrow H_2 + 2OH.$$

The OH generation together with other electrons and ions lead to formation of Ni and Ni(OH)$_2$ on graphene sheets. Finally, Ni and Ni(OH)$_2$ as crystalline layers were deposited on graphene flakes (G-flakes). The final product in form of powders was pressed (under a pressure of 5 MPa for 30 min) into pellets. The average dimensions were 6.5 mm (radius) and 1 mm (thickness), approximately.

The characterization of Ni-graphene composite was done by X-ray diffraction (XRD, Cu Kα λ = 0.154 nm) radiation, X-ray photoelectron spectroscopy (XPS, ESCA/AES, CHA, Specs model EA10 plus), tunneling electron microscopy (TEM-Philips model CM120), and dynamic light scattering (DLS, Malvern Instruments Ltd., Worcestershire). TEM and XRD patterns of Ni-graphene represented that the nano-crystals (Ni, Ni(OH)$_2$, Ni-oxides) were randomly distributed on graphene sheets. The size of Ni and Ni(OH)$_2$ nanocrystals obtained by TEM and XRD patterns were found to be 30–40 nm. The brightness of graphene sheets appearing in TEM images showed a small thickness (1.2 µm) of graphene sheets [42].

2.3. Fire Test Apparatus

Fixed Fire Test Apparatus (FFTA) was designed and built, using piloted ignition, as depicted in Figure 1 [39,43]. Natural gas was used as the fuel; it mainly comprised of methane CH$_4$ (90–98%). The producer reported that other hydrocarbons accompanied methane (C$_2$H$_6$: 1–8%; C$_3$H$_8$: 2%; H$_4$H$_{10}$+C$_5$H$_{12}$: less than 1%; and also N$_2$ + H$_2$S + H$_2$O: less than 1.5%). The gas flew steadily at the rate of 0.097 lt/s through a Bunsen type burner hold vertically and the specimen is mounted at a 45° angle to it. The internal diameter of the burner was 11 mm. The Bunsen-type burner provides a fairly mild and localized fire exposure to the testing specimens. The time from the point of burner application that it takes for the specimen to develop a visible flame that is sustained for more than one second will be registered as the "time to onset of ignition"; and the time from the point of burner application that it takes for the specimen to sustain glowing for more than one second will be registered as the "time to onset of glowing". The higher the amount of time of these two properties, the better from the viewpoint of fire-safety. As the burning continues, the back face of the specimen nearest to the flame of the burner starts blackening and after some more time, a small hole or split appears. These times are also registered as back-blackening and back-holing times. The test is terminated once the back-holing occurs. Then, the burnt area, as well as weight loss, is measured. The whole apparatus is put in a three-wall-compartment in order to protect the burning flame from wind and air movements.

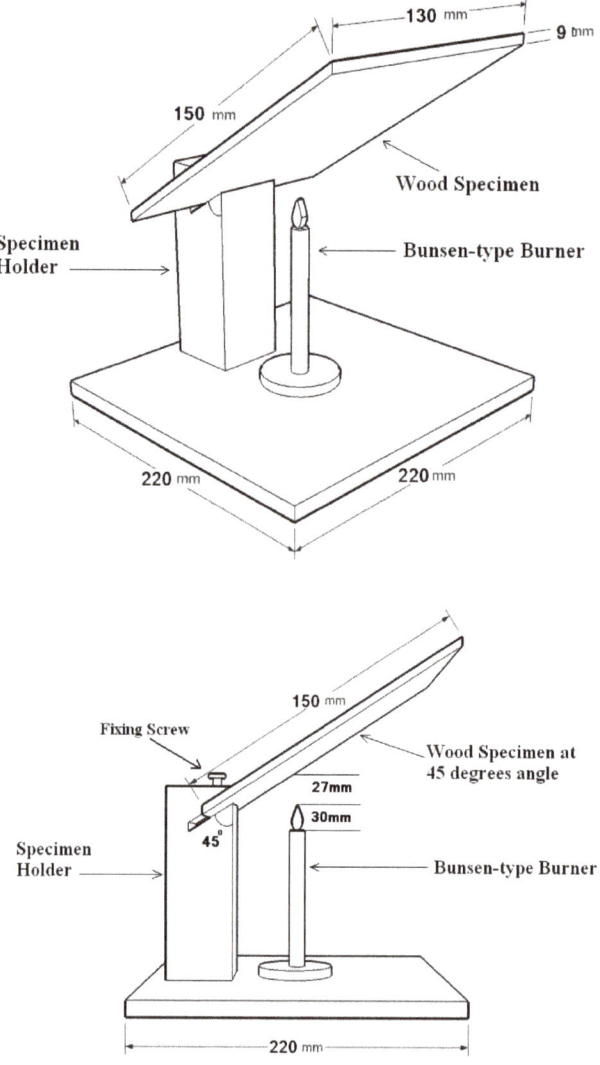

Figure 1. Schematic picture of fixed fire testing apparatus (Iranian Patent No. 67232; approved by Iranian Research Organization for Science and Technology under license No. 3407; USPTO Pub. No.: US 2019/0212283 A1) [38,43].

2.4. Computational Modeling and Simulation

Density functional theory (DFT) calculations were carried out for graphene and oxygen molecule with OpenMX3.8 package to calculate the adsorption energy of oxygen molecule on graphene [44–46]. The generalized gradient approximation (GGA) function with the Perdew-Burke-Ernzerhof (PBE) correction was employed to analyze the exchange-correlation energy functional between graphene and oxygen molecule [47]. The energy cutoff was set 100 Ry. The van der Waals interactions between adsorbed molecules and sheets were described with Grimme's method [48].

In the present simulations, one model for pure graphene flake (G-flake) was investigated. The fully optimized atomic structure of pure G-flake is shown in Figure 2a. This G-flake consists of 24 C atoms. The edge C atoms of the G-flake were terminated by H atoms. The second model represents Ni-doped G-flake in which one of C atom in the pure G-flake was replaced by a Ni atom to model Ni-doped G-flake. The optimized structure of G-flake doped with Ni atom is shown in Figure 2b.

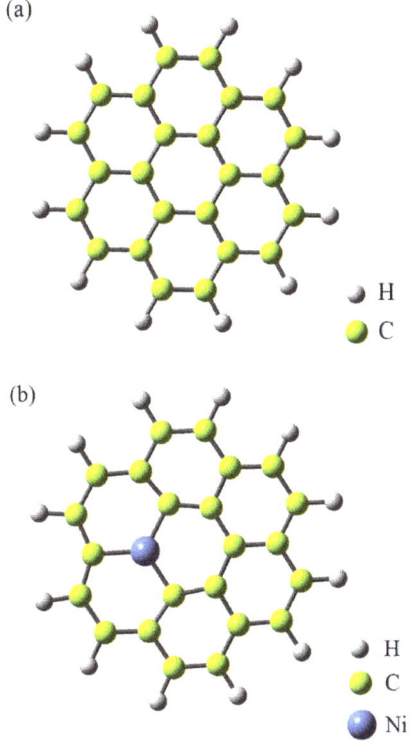

Figure 2. Atomic structure of pure (**a**) and Ni-doped (**b**) graphene flakes.

The adsorption energy, E_{Ads}, was calculated to evaluate stability of structures by the Equation (1).

$$E_{Ads} = E_{G-flake+O_2} - (E_{G-flake} + E_{O_2}), \qquad (1)$$

where $E_{G-flake+O_2}$ is the total energy of G-flake with adsorbed oxygen molecule, E_{O_2} and $E_{G-flake}$ are the total energies of oxygen molecule and G-flake, respectively.

2.5. Statistical Analysis

Statistical analysis was carried out by a SAS software, version 9.2 (Cary, NC, USA). One-way analysis of variance (ANOVA) was performed on the average values to ascertain significant differences at the 95% level of confidence. Hierarchical cluster analysis, including dendrograms and Ward methods (using squared Euclidean distance intervals), was carried out using SPSS/18, version 18 (IBM, Armonk, NY, USA). The scaled indicator on top of cluster analysis shows similarities and differences between treatments; lower scale numbers show more similarities while higher ones show dissimilarities. Contour and surface plots were designed in Minitab software, version 16.2.2 (Minitab Inc., Philadelphia, PA, USA).

3. Results and Discussion

Results demonstrated outstanding improving effects of graphene (NG) on both times to onset of ignition and to the onset of glowing (Figure 3). NG resulted in 184% and 162% increases in times to onset of ignition and glowing, respectively, in comparison to un-painted specimens. The favorable increases are primarily attributed to the very low reaction ability of graphene with oxygen [49]. A weak bond with basically no charge transfer was reported to form between oxygen molecules and graphene sheets or tubes with very low adsorption energy [49,50]. In fact, graphene here acted as an impermeable insulating layer towards penetration of fire into its substrate, significantly delaying in its ignition and glowing. DFT analysis in the present study also revealed that energy of oxygen molecule on Pure and Ni-doped G-flakes were −1.07 and −1.20 eV, respectively (Figures 4 and 5). The shortest distances from O atom in oxygen molecule to C atom in pure and Ni-doped G-flakes were calculated to be 3.40 and 2.37 Å, respectively. These small adsorption energies and large adsorption distances clearly demonstrated that oxygen molecules were weakly physisorbed by pure and Ni-doped G-flake, ultimately illustrating very low reaction ability between them.

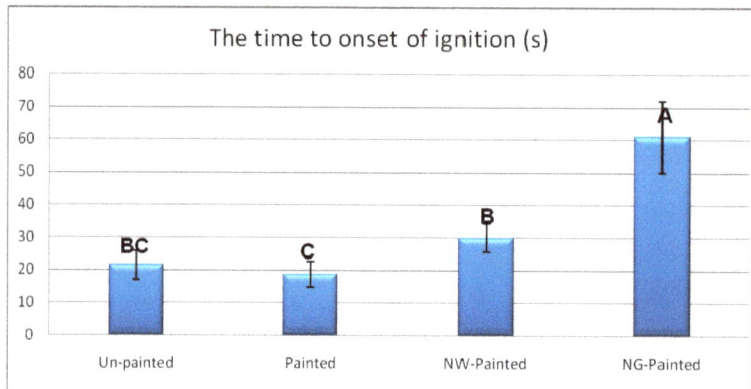

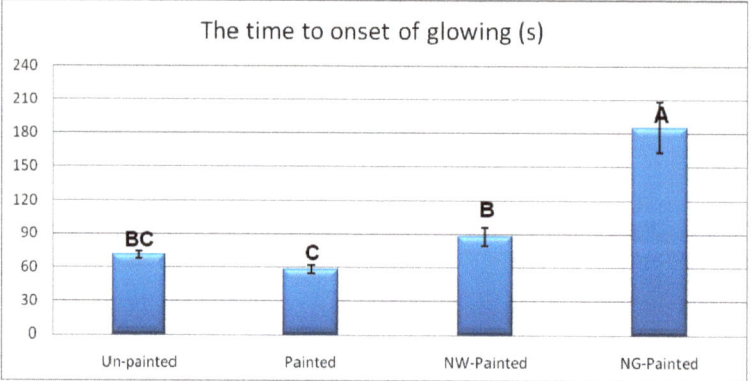

Figure 3. Time to onset of ignition and glowing, in the four treatments of beech specimens (NW = nanowollastonite; NG = nano-graphene) (Letters on each column represent Duncan groupings at 95% level of confidence).

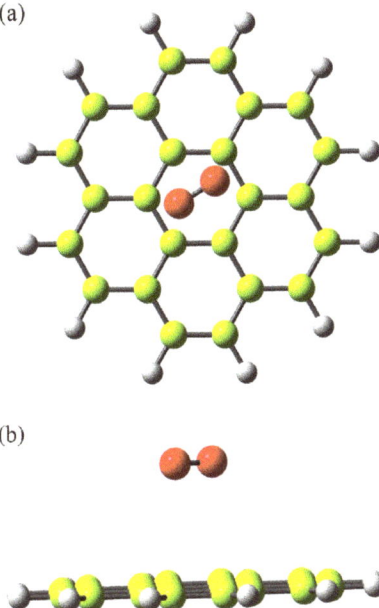

Figure 4. Side (a) and top (b) views of oxygen molecule adsorbed on pure graphene flake.

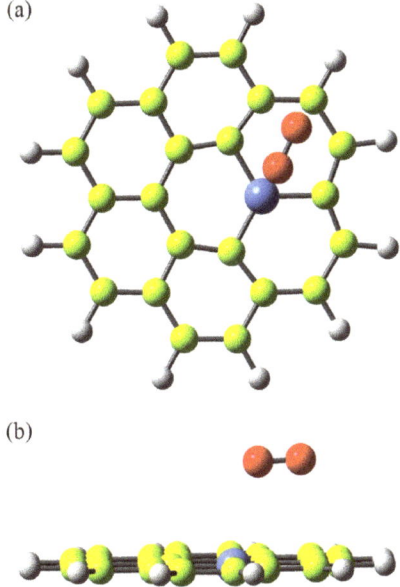

Figure 5. Side (a) and top (b) views of oxygen molecule adsorbed on Ni-doped graphene flake.

The improvement in fire properties by graphene can be elaborated from its thermal conductivity as well. With regard to its two-dimensional structure, the thermal conductivity of graphene can also be studied in two directions. The in-plane thermal conductivity of graphene is reported to be one among materials the highest thermal conductivity, about 2000–4000 $W \cdot m^{-1} \cdot K^{-1}$, at room temperature [51]. For comparison purposes, the thermal conductivity of natural diamond is about 2200 W·m^{-1}·K^{-1}, and that of a purified diamond can be up to 50% higher. However, the thermal conductivity of graphene in cross-section direction (that is, along the nano-size direction of z-axis) is as low as only 6 W·m^{-1}·K^{-1} [51]. This extreme difference in thermal conductivity of in-plane and cross-section directions played a vital role in improving fire properties in beech specimens. From one perspective, very low thermal conductivity in cross-section direction prevented heat flow from the piloted ignition to pass through and reach the beech wood substrate, delaying the substrate in catching fire, ultimately improving fire properties. The increases in both times to onset of ignition and glowing partially resulted from this low thermal conductivity in cross-section direction. From another perspective, the high thermal conductivity in the in-plane direction prevented accumulation of heat at the point nearest to the piloted fire, eventually postponing its ignition. Similar transfer of heat to the surrounding area of woody specimens by nano-wollastonite was previously reported to improve fire properties as well [36,38]. Overall, the thermal conductivity of graphene in both directions helped its improving fire properties in beech specimens.

Painted specimens showed a little bit of decrease in time values, though the decreases were not statistically significant. The decrease, which is not favorable as far as fire retardancy is concerned, was attributed to the easier ignitability of the chemical ingredients of the paint applied on the surface of specimens. The addition of NW to the paint resulted in an increase in both times to onset of ignition and glowing in comparison to the unpainted specimens, though the increase was not statistically significant in time to onset of glowing. The increase was partially attributed to the mineral and unignitable nature of wollastonite. In fact, NW acted as an insulating layer, too, towards the penetration of the piloted fire from the Bunsen burner, protecting the wood substrate beneath the paint.

Results of assessing the back of specimens showed that both NG and NW increased the times to back-darkening and to back-holing (Figure 6), showing they are both reliable to hinder the passing of fire through specimens. NG performed a bit more efficiently with regard to these two fire-retarding properties in relation to the backside properties of specimens. This may be attributed to the unignitable nature of graphene, acting an insulating layer towards penetration and passing of flame.

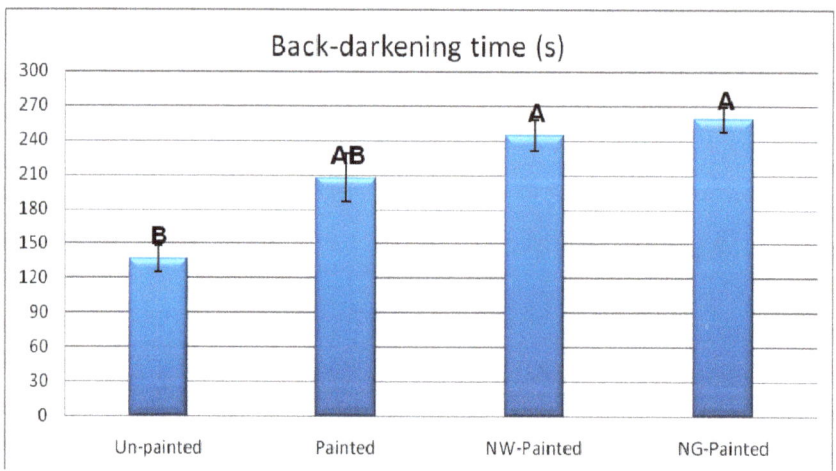

Figure 6. *Cont.*

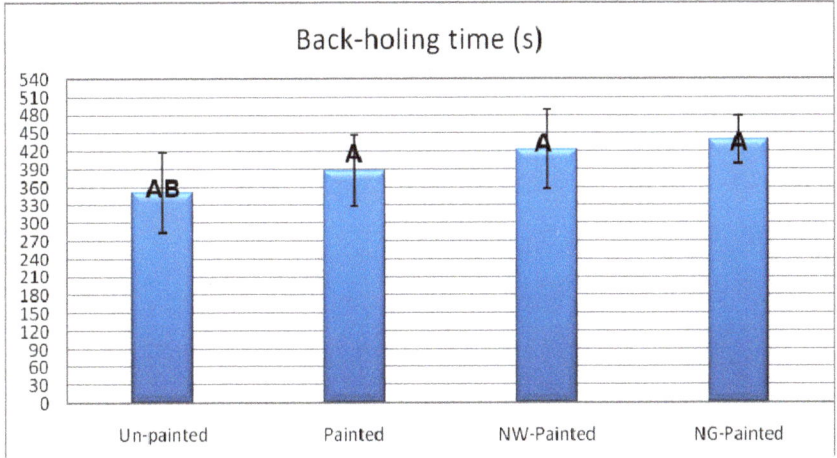

Figure 6. Back-darkening and back-holing times, in the four treatments of beech specimens (NW = nanowollastonite; NG = nano-graphene) (Letters on each column represent Duncan groupings at 95% level of confidence.).

Measurement of the burnt area illustrated that NG was very effective in hindering the spread of fire throughout specimens (Figure 7). The spread of fire in materials made of wood is of vital importance as its decrease would limit the potentiality of other surrounding parts to catch on fire. A decrease in the burnt area would eventually provide fire-fighters with a longer time-span to extinguish the fire. This would ultimately result in a decrease in losses, both to life and property. As to NW, thought it decreased the burnt area in comparison to un-painted specimens, the area was bigger than that of NG. This was attributed to an increased thermal conductivity that caused heat to be more easily transferred to the surrounding areas, ultimately increasing the burnt area. Similar increase in thermal conductivity was previously reported [22,30].

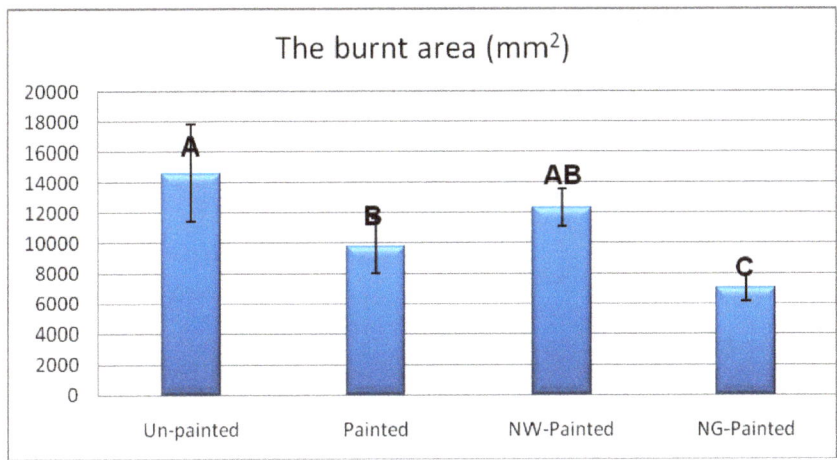

Figure 7. *Cont.*

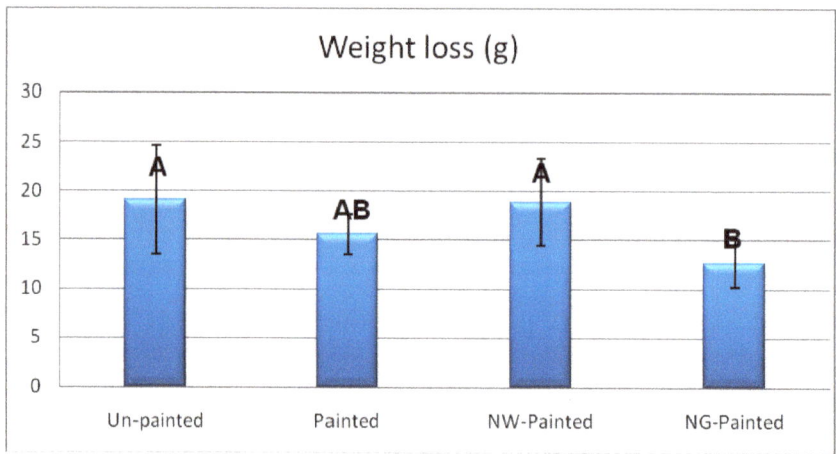

Figure 7. Burnt area and weight loss in the four treatments of beech specimens (NW = nanowollastonite; NG = nano-graphene) (Letters on each column represent Duncan groupings at 95% level of confidence.).

The weight measurement of specimens before and after being exposed to fire showed that the lowest weight losses occurred in specimens treated with NG (Figure 7). The weight loss of NG-painted specimens was 50% lower than that of the control of unpainted specimens. This clearly indicated that NG can significantly decrease the volume of burnt materials.

A significant direct relationship was calculated between most of the fire properties. The highest R-square (99%) was found between the two times to onset of ignition and glowing. Contour and surface plots also illustrated a direct trend among different properties; for instance, weight loss increased as the two times to onset of glowing and ignition increased (Figure 8A,B).

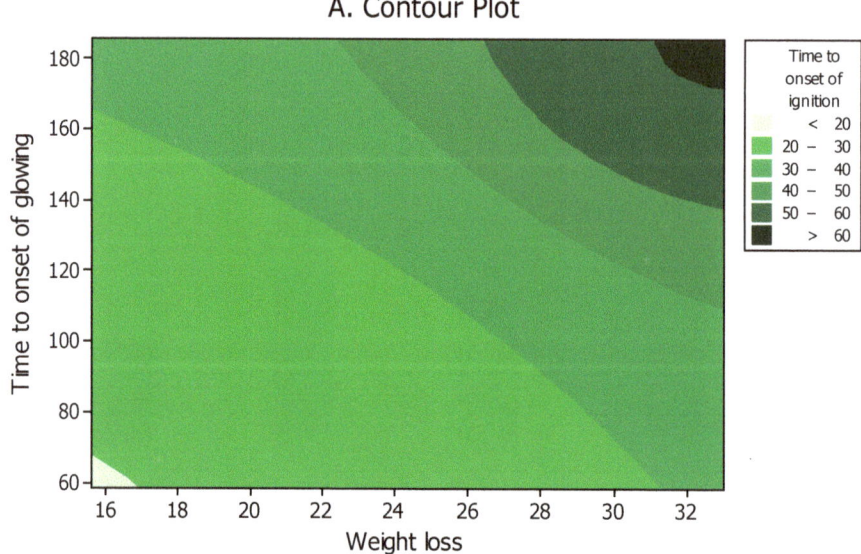

Figure 8. *Cont.*

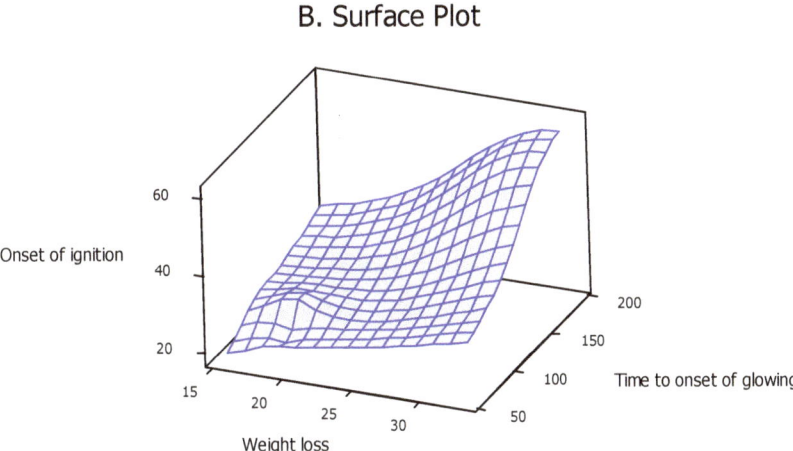

Figure 8. Contour (**A**) and surface plots (**B**) among fire properties of weight loss versus times to onset of glowing and ignition.

The cluster analysis based on all six fire properties studied in the present project demonstrated the distinct difference of specimens treated with graphene in comparison to the other three treatments (control, painted, and NW-painted) (Figure 9). This indicated the great improving impact of graphene on fire properties in beech specimens. Painted specimens were closely clustered with the control specimens, indicating no significant difference between these two treatments. NW-painted specimens were in-between position of control and NG-painted; this indicated significant improving effects of NW on fire properties, similar to previous studies on solid wood and wood-composite panels [36,38]. However, the difference in clustering with NG-painted specimens showed higher effectiveness of graphene in improving fire properties in comparison to NW. Based on the results discussed above, it was concluded that graphene has great potential in improving fire properties in beech. Easy application of graphene on the surface of materials makes it a good prospect for the industry sector to develop an effective fire-retardant. However, as the paint used in the present project can even be more elaborated to find a better carrier liquid.

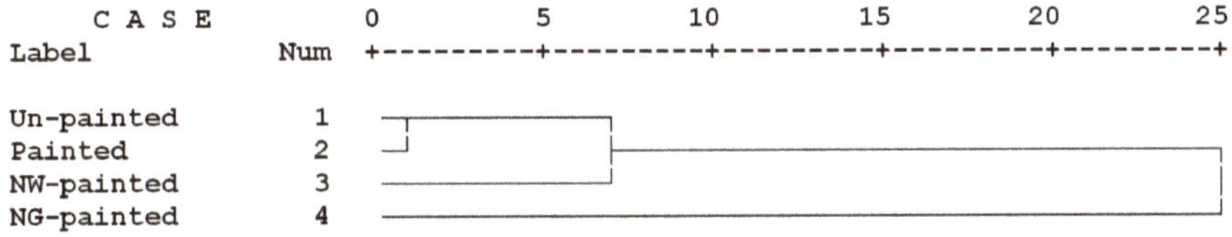

```
          C A S E            0         5        10        15        20        25
        Label          Num   +---------+---------+---------+---------+---------+

        Un-painted      1    ┐
        Painted         2    ┘
        NW-painted      3    ─────────────────────────────────────────────────┐
        NG-painted      4    ──────────────────────────────────────────────────┘
```

Figure 9. Cluster analysis among the four treatments of beech specimens based on all the fire properties studied in the present project (NW = nano-wollastonite; NG = nano-graphene).

4. Conclusions

Fire properties of beech specimens were tested by a newly developed apparatus using piloted fire. Specimens were prepared to be surface-treated by graphene to investigate its improving effects on fire properties. Graphene was mixed at 5% with water-based paint and applied on the front and back surfaces of specimens. As a basis for comparison, three separate groups of specimens were prepared, including specimens without any paint (the control group), painted with no nano-material mixed with it, and painted with 5% nano-wollastonite. Results showed significant improving effects of graphene on all six fire properties studied in the present project. Graphene demonstrated outstanding improvement in times to onset of ignition and glowing, as two very important properties that determine the ignitability of materials. Moreover, graphene illustrated a high impact on decreasing the burnt area. The decrease was also important from the viewpoint of the spread of fire to other surrounding parts in an area. Overall, the improved fire properties provide fire-fighters with a larger timespan to extinguish the fire, ultimately saving both life and property. It was concluded that graphene has a great potential to be used as an effective fire-retardant in wood and wood-composite materials for surface protection against fire. Therefore, further studies should be carried out to investigate graphene from other perspectives, including FT-IR analysis.

Author Contributions: Methodology, H.R.T. and A.E.; Validation, H.R.T., R.M., and S.M.M.A.; Investigation, H.R.T., and M.G.; Writing-Original Draft Preparation, H.R.T., and A.N.P.; Writing-Review & Editing, H.R.T., and A.N.P.; Visualization, H.R.T.; Supervision, H.R.T. and A.N.P. All authors have read and agreed to the published version of the manuscript.

Funding: This research received no external funding.

Acknowledgments: The authors appreciate constant scientific support of Jack Norton (Retired, Horticulture & Forestry Science, Queensland Department of Agriculture, Forestry and Fisheries, Australia).

Conflicts of Interest: The authors declare no conflict of interest.

References

1. Behling, M.; Piketty, M.G.; Morello, T.F.; Bouillet, J.P.; Mesquita Neto, F.; Laclau, J.P. Eucalyptus plantations and the steel industry in Amazonia—A contribution from the 3-PG model. *Bois et Forets Des Tropiques* **2011**, *309*, 37–49. [CrossRef]
2. Cherelli, S.G.; Sartori, M.M.P.; Prospero, A.G.; Ballarin, A.W. Heartwood and sapwood in eucalyptus trees: Non-conventional approach to wood quality. *Anais da Academia Brasileira de Ciencias* **2018**, *90*, 425–438. [CrossRef] [PubMed]
3. Mantanis, G.; Terzi, E.; Kartal, S.N.; Papadopoulos, A. Evaluation of mold, decay and termite resistance of pine wood treated with zinc- and copper- based nanocompounds. *Int. Biodeterior. Biodegrad.* **2014**, *90*, 140–144. [CrossRef]
4. Papadopoulos, A.N.; Duquesnoy, P.; Cragg, S.M.; Pitman, A.J. The resistance of wood modified with linear chain carboxylic acid anhydrides to attack by the marine wood borer Limnoria quadripunctata Hothius. *Int. Biodegrad. Biodeterior.* **2008**, *61*, 199–202. [CrossRef]
5. Papadopoulos, A.N.; Avtzis, D.; Avtzis, N. The biological effectiveness of wood modified with linear chain carboxylic acid anhydrides against the subterranean termites Reticulitermes flavipes. *Holz als Roh und Werkstoff* **2003**, *66*, 249–252. [CrossRef]
6. Papadopoulos, A.N. Chemical modification of solid wood and wood raw materials for composites production with linear chain carboxylic acid anhydrides: A brief Review. *BioResources* **2010**, *5*, 499–506.
7. Lin, C.F.; Karlsson, O.; Mantanis, G.I.; Sandberg, D. Fire performance and leach resistance of pine wood impregnated with guayl-urea phosphate/boric acid and a melamin-formaldehyde resin. *Eur. J. Wood Wood Prod.* **2020**. [CrossRef]
8. Östman, B.A.L.; Tsantaridis, L.D. Heat release and classification of fire retardant wood products. *Fire Mater.* **1995**, *19*, 253–258. [CrossRef]
9. LeVan, S.; Winandy, J.E. Effect of fire retardant treatments on wood strength: A review. *Wood Fiber Sci.* **1990**, *22*, 113–131.
10. White, R.H.; Sweet, M.S. Flame retardancy of wood: Present status, recent problems, and future fields. In Proceedings of the 3rd Annual BCC Conference on Flame Retardance, Stamford, CT, USA, 19–21 May 1992; pp. 250–257.
11. Winandy, J.E. Thermal degradation of the fire-retardant treated wood. *For. Prod. J.* **2001**, *51*, 47–54.
12. Winandy, J.E.; Wang, Q.; White, R.H. Fire-retardant-treated strandboard: Properties and fire performance. *Wood Fiber Sci.* **2008**, *40*, 62–71.
13. Mantanis, G. *Aqueous Fire Retardant*; WO 02/102926 A1; World Intellectual Property Organisation: Geneva, Switzerland, 2002.

14. Mantanis, G.; Martinka, J.; Lykidis, C.; Ševčík, L. Technological properties and fire performance of medium density fibreboard (MDF) treated with selected polyphosphate-based fire retardants. *Wood Mater. Sci. Eng.* **2019**. [CrossRef]
15. Papadopoulos, A.N.; Bikiaris, D.N.; Mitropoulos, A.C.; Kyzas, G.Z. Nanomaterials and chemical modification technologies for enhanced wood properties: A review. *Nanomaterials* **2019**, *9*, 607. [CrossRef] [PubMed]
16. Bayani, S.; Taghiyari, H.R.; Papadopoulos, A.N. Physical and mechanical properties of thermally-modified beech wood impregnated with silver nano-suspension and their relationship with the crystallinity of cellulose. *Polymers* **2019**, *11*, 1535. [CrossRef] [PubMed]
17. Taghiyari, H.; Esmailpour, A.; Papadopoulos, A. Paint Pull-Off Strength and Permeability in Nanosilver-Impregnated and Heat-Treated Beech Wood. *Coatings* **2019**, *9*, 723. [CrossRef]
18. Papadopoulos, A.N.; Taghiyari, H.R. Innovative wood surface treatments based on nanotechnology. *Coatings* **2019**, *9*, 866. [CrossRef]
19. Mantanis, G.; Papadopoulos, A.N. The sorption of water vapour of wood treated with a nanotechnology compound. *Wood Sci. Technol.* **2010**, *44*, 515–522. [CrossRef]
20. Sandeep, N.; Sulochana, C.; Kumar, B.R. Flow and heat transfer in MHD dusty nanofluid past a stretching/shrinking surface with non-uniform heat source/sink. *Walailak J. Sci. Technol.* **2017**, *14*, 117–140.
21. Suganya, S.; Kumar, P.S.; Saravanan, A. Construction of active bio-nanocomposite by inseminated metal nanoparticles onto activated carbon: Probing to antimicrobial activity. *IET Nanobiotechnol.* **2017**, *11*, 746–753. [CrossRef]
22. Hassani, V.; Taghiyari, H.R.; Schmidt, O.; Maleki, S.; Papadopoulos, A.N. Mechanical and physical properties of Oriented Strand Lumber (OSL): The effect of fortification level of nanowollastonite on UF resin. *Polymerss* **2019**, *11*, 1884. [CrossRef]
23. Mantanis, G.I.; Papadopoulos, A.N. Reducing the thickness swelling of wood based panels by applying a nanotechnology compound. *Eur. J. Wood Wood Prod.* **2010**, *68*, 237–239. [CrossRef]
24. Tajvidi, M.; Gardner, D.J.; Bousfield, D.W. Cellulose Nanomaterials as Binders: Laminate and Particulate Systems. *J. Renew. Mater.* **2016**, *4*, 365–376. [CrossRef]
25. Ayata, U.; Akcay, C.; Esteves, B. Determination of decay resistance against *Pleurotus ostreatus* and *Coniophora puteana* fungus of heat-treated Scots pine, oak and beech wood species. *Maderas Ciencia y Tecnologia* **2017**, *19*, 309–316.
26. Da Silveira, A.G.; Santini, E.J.; Kulczynski, S.M.; Trevisan, R.; Wastowski, A.D.; Gatto, D.A. Tannic extract potential as natural wood preservative of Acacia mearnsii. *Anais da Academia Brasileira de Ciencias* **2017**, *89*, 3031–3038. [CrossRef] [PubMed]
27. Humar, M.; Lesar, B.; Thaler, N.; Krzisnik, D.; Kregar, N.; Drnovsek, S. Quality of copper impregnated wood in Slovenian hardware stores. *Drvna Industrija* **2018**, *69*, 121–129. [CrossRef]
28. Karim, M.; Ghodskhah Daryaei, M.; Torkaman, J.; Oladi, R.; Tajick Ghanbary, M.A.; Bari, E.; Yilgor, N. Natural decomposition of hornbeam wood decayed by the white rot fungus Trametes versicolor. *Anais da Academia Brasileira de Ciências* **2017**, *89*, 2647–2655. [CrossRef]
29. Taghiyari, H.R.; Bari, E.; Schmidt, O.; Tajick Ghanbary, M.A.; Karimi, A.; Tahir, P.M.D. Effects of nanowollastonite on biological resistance of particleboard made from wood chips and chicken feather against Antrodia vaillantii. *Int. Biodeterior. Biodegrad.* **2014**, *90*, 93–98. [CrossRef]
30. Taghiyari, H.R.; Ghorbanali, M.; Tahir, P.M.D. Effects of improvement in thermal conductivity coefficient by nano-wollastonite on physical and mechanical properties in medium-density fiberboard (MDF). *BioResources* **2014**, *9*, 4138–4149. [CrossRef]
31. Taghiyari, H.R.; Majidinajafabadi, R.; Vahidzadeh, R. Wollastonite to hinder growth of Aspergillus niger fungus on cotton textile. *Anais da Academia Brasileira de Ciencias* **2018**, *90*, 2797–2804. [CrossRef]
32. Taghiyari, H.R.; Tajvidi, M.; Taghiyari, R.; Mantanis, G.I.; Esmailpour, A.; Hosseinpourpia, R. Nanotechnology for wood quality improvement and protection. *Nanomater. Agric. For. Appl.* **2019**. [CrossRef]
33. Taghiyari, H.R.; Karimi, A.; Tahir, P.M.D.; Choo, A.C.Y. Effects of nanotechnology on fluid flow in agricultural and wood-based composite materials. In *Agricultural Biomass Based Potential Materials*; Springer: Cham, Switzerland, 2015; pp. 73–89.
34. Geim, A.K.; Novoselov, K.S. The rise of grapheme. *Nat. Mater.* **2007**, *6*, 183–191. [CrossRef] [PubMed]
35. Katsnelson, M.; Iosifovich, M. *Graphene: Carbon in Two Dimensions*, 1st ed.; Cambridge University Press: Cambridge, NY, USA, 2012.

36. Haghighi Poshtiri, A.; Taghiyari, H.R.; Karimi, A.N. Fire-retarding properties of nano-wollastonite in solid wood. *Philipp. Agric. Sci.* **2014**, *97*, 45–52.
37. Soltani, A.; Hosseinpourpia, R.; Adamopoulos, S.; Taghiyari, H.R.; Ghaffari, E. Effects of heat-treatment and nano-wollastonite impregnation on fire properties of solid wood. *BioResources* **2016**, *11*, 8953–8967. [CrossRef]
38. Esmailpour, A.; Taghiyari, H.R.; Nouri, P.; Jahangiri, A. Fire-retarding properties of nanowollastonite in particleboard. *Fire Mater.* **2018**, *42*, 306–315. [CrossRef]
39. Yuningsih, I.; Sekartining Rahayu, I.; Lumongga, D.; Darmawan, W. Wettability and adherence of acrylic paints on long and short rotation teaks. *Wood Mater. Sci. Eng.* **2019**. [CrossRef]
40. Aghazadeh, M.; Golikand, A.N.; Ghaemi, M. Synthesis, characterization, and electrochemical properties of ultrafine β-Ni(OH)$_2$ nanoparticles. *Int. J. Hydrog. Energy* **2011**, *36*, 8674–8679. [CrossRef]
41. Parvez, K.; Wu, Z.S.; Li, R.; Liu, X.; Graf, R.; Feng, X.; Müllen, K. Exfoliation of Graphite into Graphene in Aqueous Solutions of Inorganic Salts. *J. Am. Chem. Soc.* **2014**, *136*, 6083–6091. [CrossRef]
42. Sheykhifard, Z.; Majid Mohseni, S.M.; Tork, B.; Hajiali, M.R.; Jamilpanah, L.; Rahmati, B.; Haddadi, F.; Hamdi, M.; Morteza Mohseni, S.; Mohammadbeigi, M.; et al. Magnetic graphene/Ni-nano-crystal hybrid for small field magnetoresistive effect synthesized via electrochemical exfoliation/deposition technique. *J. Mater. Sci. Mater. Electron.* **2018**, *29*, 4171–4178. [CrossRef]
43. Taghiyari, H.R. Fire-retarding properties of nano-silver in solid woods. *Wood Sci. Technol.* **2012**, *45*, 939–952. [CrossRef]
44. Ozaki, T.; Kino, H.; Yu, J.; Han, M.J. User's Manual of OpenMX, Version 3.8. 2018. Available online: http://www.openmx-square.org/openmx_man3.8/openmx.html (accessed on 1 January 2018).
45. Majidi, R. A biosensor for hydrogen peroxide detection based on electronic properties of carbon nanotubes. *Mol. Phys.* **2012**, *111*, 89–93. [CrossRef]
46. Majidi, R. Electronic properties of graphyne nanotubes filled with small fullerenes: A density functional theory study. *J. Comput. Electron.* **2016**, *15*, 1263–1268. [CrossRef]
47. Perdew, J.P.; Burke, K.; Ernzerhof, M. Generalized Gradient Approximation Made Simple. *Phys. Rev. Lett.* **1996**, *77*, 3865. [CrossRef] [PubMed]
48. Grimme, S. Semiempirical GGA-type density functional constructed with a long-range dispersion correction. *J. Comput. Chem.* **2006**, *27*, 1787–1799. [CrossRef]
49. Giannozzi, P.; Car, R.; Scoles, G. Oxygen adsorption on graphite and nanotubes. *J. Chem. Phys.* **2003**, *118*, 1003–1006. [CrossRef]
50. Sorescu, D.C.; Jordan, K.D.; Avouris, P. Theoretical study of oxygen adsorption on graphite and the (8,0) single-walled carbon nanotube. *J. Phys. Chem. B* **2001**, *105*, 11227–11232. [CrossRef]
51. Pop, E.; Varshney, V.; Roy, A.K. Thermal properties of graphene: Fundamentals and applications. *MRS Bull.* **2012**, *37*, 1273–1281. [CrossRef]

© 2020 by the authors. Licensee MDPI, Basel, Switzerland. This article is an open access article distributed under the terms and conditions of the Creative Commons Attribution (CC BY) license (http://creativecommons.org/licenses/by/4.0/).

Article

Flexural Creep Behavior of High-Density Polyethylene Lumber and Wood Plastic Composite Lumber Made from Thermally Modified Wood

Murtada Abass A. Alrubaie [1,*], Roberto A. Lopez-Anido [1] and Douglas J. Gardner [2]

1. Department of Civil and Environmental Engineering, Advanced Structures and Composites Center, University of Maine, Orono, ME 04469, USA; rla@maine.edu
2. School of Forest Resources, Advanced Structures and Composites Center, University of Maine, Orono, ME 04469, USA; douglasg@maine.edu
* Correspondence: murtada.alrubaie1@maine.edu

Received: 26 November 2019; Accepted: 9 January 2020; Published: 24 January 2020

Abstract: The use of wood plastic composite lumber as a structural member material in marine applications is challenging due to the tendency of wood plastic composites (WPCs) to creep and absorb water. A novel patent-pending WPC formulation that combines a thermally modified wood flour (as a cellulosic material) and a high strength styrenic copolymer (high impact polystyrene and styrene maleic anhydride) have been developed with advantageous viscoelastic properties (low initial creep compliance and creep rate) compared with the conventional WPCs. In this study, the creep behavior of the WPC and high-density polyethylene (HDPE) lumber in flexure was characterized and compared. Three sample groupings of WPC and HDPE lumber were subjected to three levels of creep stress; 7.5, 15, and 30% of the ultimate flexural strength (Fb) for a duration of 180 days. Because of the relatively low initial creep compliance of the WPC specimens (five times less) compared with the initial creep compliance of HDPE specimens, the creep deformation of HDPE specimens was six times higher than the creep deformation of WPC specimens at the 30% creep stress level. A Power Law model predicted that the strain (3%) to failure in the HDPE lumber would occur in 1.5 years at 30% Fb flexural stress while the predicted strain (1%) failure for the WPC lumber would occur in 150 years. The findings of this study suggest using the WPC lumber in structural application to replace the HDPE lumber in flexure attributable to the low time-dependent deformation when the applied stress value is withing the linear region of the stress-strain relationship.

Keywords: viscoelasticity; WPC; HDPE; composite; wood; creep; thermoplastic; flexure; power law; modeling

1. Introduction

Wood plastic composites (WPCs) are commonly used as deck boards and railings thanks to their low maintenance and high durability compared with conventional pressure-treated lumber [1]. However, extensive efforts have been made to expand the use of WPCs to include structural applications [2–8] because of their mechanical properties, longer lifetime, and their competing commercial prices with conventional types of lumber [2,3,5,9,10]. Furthermore, WPCs made from thermally modified wood have shown potential to be used in structural applications, since they have been shown to exhibit relatively low time-dependent deformation under sustained flexural loads [11,12]. Likewise, plastic lumber is also used in low-cost structural applications. One type of plastic lumber, high-density polyethylene (HDPE) lumber, is used in the construction of aquaculture-offshore fish cages (a.k.a. Aquapod Net Pen cages) [13,14], however, the HDPE lumber experienced damage during its service life attributable to exposure to severe ocean conditions (wave action and high temperatures during the summer, ca. 48 °C in the Gulf of Mexico [14]) when these cages are partially exposed to air [14], and lounging sea lions causing damage to the exposed struts of the cage structure (in the partially exposed cages) [15–17], as shown in Figure 1.

Figure 1. Buckled Aquapod cage made from HDPE lumber and netting (covered with biofouling) with two lounging sea lions on the exposed struts [1].

The need to have a material that has a reasonable cost for the construction of aquaculture cages that also exhibits satisfactory structural performance during the service life of these cages [11,12] suggests that WPC lumber can be considered a potential alternative to HDPE lumber [11,12]. Although WPCs have been explored for use in structural applications, the material's long-term behavior is still a subject of concern among researchers and end-users, especially in marine applications. WPC lumber exhibits viscoelastic behavior. When a constant stress is applied to a viscoelastic material, the sum of the elastic strain (instantaneous strain) and the time-dependent strain will represent the total strain (creep strain) of the viscoelastic composite material [18,19]. One dimensional (1D) viscoelastic models [power law, Maxwell, Kelvin, Prony series, and four element viscoelastic models] have been used in previous studies to describe both the short, and long-term creep-behavior of viscoelastic materials [2,3,20,21]. Alrubaie et al. [12] implemented a 1D power law viscoelastic model to describe the 180-day creep behavior of WPC lumber made from thermally modified wood with a span L = 853 mm in 4-point bending (flatwise). The power law model among other models were investigated in a preliminary study that has shown a good agreement with the short and long-term creep behavior of WPC and the HDPE lumber. Alvarez-Valencia [3] conducted a full-scale 90-day creep rupture in 4-point bending of a Z-shape WPC sheet piling with 4.70 m in length, to evaluate the time-dependent structural behavior of the WPC sheet piling, and the 1D Findlay's power law model was used to predict the creep behavior of the WPC sheet piling that has shown good agreement with experimental data. Dura [7] conducted one, seven, and 15-day creep experiments on WPC dumbbell-shaped tensile specimens at 15, 30, and 45% of the average tensile strength, to evaluate the time-dependent behavior of the WPCs. In addition to the creep in tension, Dura [7] also conducted creep tests in compression at the same stress levels used for the tensile creep experiments, but with respect to the average maximum compression stress and to the same creep duration. Many researchers [3,6,7] have studied the large-scale flexural creep behavior of WPC specimens (i.e., when the WPC specimens have length to span ratios (L/h) that exceed the ratio recommended by the Standards [22]). Dura [7] conducted a 90-day flexural creep experiments (edgewise) on WPC specimens with a span length of 2515 mm with and without a layer of fiber reinforced polymer layer (FRP) and their creep behaviors were reported. [7]. Dura used the experimental response to verify a nonlinear 1D long-term viscoelastic model [7]. Alvarez-Valencia [3] conducted a flexural creep rupture experiment on Z-shape WPC sheet pile with a span length of 4700 mm subjected to 55% of the flexural load at failure (11.7 kN). Hamel [6] performed a three-year tensile creep test experiment on WPC dumbbell shaped specimens subjected to two different levels of stress, 20% and 50% of the average maximum stress at failure, to predict the creep behavior of 2.13 m WPC boards in flexure. Hamel [23] developed a 2D finite element (FE) model that predicted the flexural creep behavior (edgewise) based on the uniaxial quasi-static testing using the Abaqus [24] software.

The two objectives of the research presented here were: (1) to experimentally characterize the long-term (180 days) flexural creep behavior (flatwise) of WPC lumber made from thermally modified wood and compare it with the flexural creep behavior of HDPE lumber currently used in the construction of aquaculture fish cages (Aquapod Net Pen cages), and (2) to implement a power law model to describe the long-term viscoelastic creep behavior of WPC and HDPE lumber in flexure (flatwise) for a duration of 180 days, respectively. Furthermore, the model was implemented to predict the failure occurrence at the outer fiber of the WPC and HDPE lumber for a duration longer than the 180 days.

In this study, thirty 4-point bending creep frames (flat wise) located in a climate control creep room in the Advanced Structures and Composite Center (ASCC) at the University of Maine (Orono, ME, USA) were utilized to conduct 180-day creep experiments in 4-point bending (flatwise) of the WPC and HDPE lumber subjected to three different levels of stresses and each level of stress was applied to five specimens (i.e., the total number of WPC and HDPE specimens is 30).

2. Experimental

2.1. Material

The WPC lumber with cross section dimensions [width (w), thickness (h)], (139 mm, 33.5 mm) was produced using a twin-screw Woodtruder™ (Davis-Standard, Orono, Maine, USA) in the ASCC at the University of Maine (Orono, ME) [20]. The WPC lumber cross section has two grooves along the longitudinal direction (extrusion direction) of the lumber at the top layer with 3 mm width and 1.8 mm depth, and these grooves are located at 21.9 mm from the short edges of the WPC lumber, as shown in the cross-section A-A in Figure 2A. The WPC examined here is based on a patent-pending formulation, in accordance with the International Publication Number WO 2018/142314 A1 dated in 09 August, 2018, combining thermally modified wood flour (as a cellulosic material) that has been produced at Uimaharju sawmill in Finland and a high strength styrenic copolymer system (high impact polystyrene (HIPS) and styrene maleic anhydride (SMA)) in an equivalent weight ratio to each of the two constituents. Section A-A in Figure 2A shows the cross section of WPC and HDPE lumber. However, a simplifying assumption was made to consider the WPC cross-section is a rectangular cross-section and eliminate the grooved areas at the top layer in the computations. The commercially available HDPE lumber has a rectangular cross section with the width of 140 mm and the thickness of 38 mm is used in the construction of the Aquapod Net Pen cages and was provided by InnovaSea [11], to conduct this study.

2.2. WPC and HDPE Sample Preparation

WPC and HDPE lumber specimens with cross section dimensions (width, thickness), (139 mm, 33.5 mm) and (140 mm, 38 mm), respectively, were cut to an adequate length to fit the span of the creep test rig, L = 853 mm with an appropriate overhang at each support of the test rig [51 mm at each overhang (a) in Figure 2A], as shown in Figure 2B. To achieve the magnetic mounting of the string potentiometer that measures the creep deflection to the mid-span of the specimens, a 3-min flame treatment to each specimen followed by application of a 5-min epoxy to adhere a square metal piece (19 × 19 mm) to the mid-span of each specimen (flatwise). Thereafter, a magnetic hook was mounted on the square metal and the string potentiometer was attached to the hook during the creep loading, and hence, the creep mid-span deflection was acquired, accordingly.

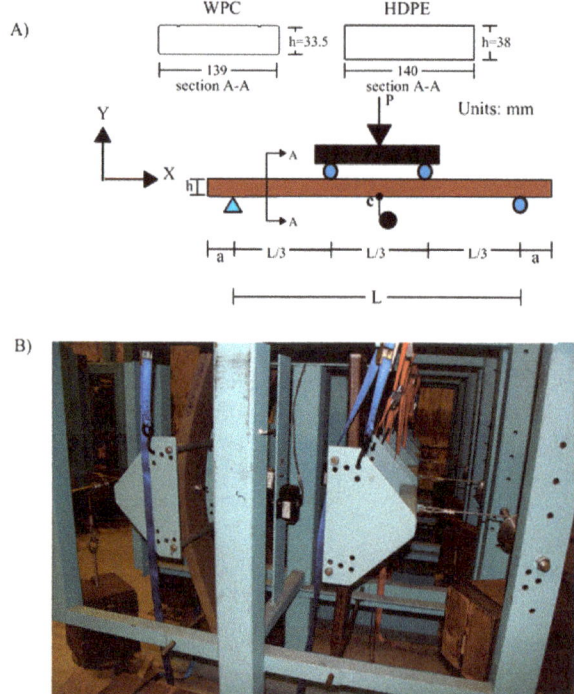

Figure 2. (**A**) Four-point bending test configuration used for both quasi-static tests and creep tests, (**B**) Creep frames experimental setup.

2.3. 180-Day Creep Experimental Setup

Prior to the creep loading and in accordance with ASTM D618 [23], WPC and HDPE specimens were preconditioned in the climate control creep room at the ASCC for one week. Thereafter, and according to ASTM D6109 and ASTM D6815 [25], the long-term WPC and HDPE specimens were loaded in 4-point bending (flat wise) with values of L/h 22 and 20, respectively. The relative humidity (RH) and temperature were controlled during the 180 days of the creep experiment to be 50 ± 5% and 21 ± 2 °C. The crosshead speed used to load the WPC and HDPE specimens for creep was the same crosshead speed used in the quasi-static testing to obtain the mean ultimate flexural stress (i.e., to ensure the initial applied loading will be applied to the specimens not less than one minute and not greater than 10 min). The measurements and the recordings of the; applied flexural level, creep displacements, and the relative humidity and the temperature of the climate control creep room, are managed by a data acquisition system (DAQ) located at the climate control room at the ASCC at the University of Maine.

Based on the applied flexural stress level relative to the flexural strength (Fb), the WPC and HDPE specimens have been divided into three groups: 7.5% of Fb, 15% of Fb, and 30% of Fb, respectively. The selection of the stress levels was made based on the level of the linear region which is below 40% of the ultimate flexural strength in the stress strain relationship in flexural tests specified in ASTM

D6109, to avoid the failure occurrence during the creep duration if the selected levels of stress were higher than 40% of the ultimate flexural strength.

2.4. Quasi-Static Tests

To obtain the apparent elastic modulus (E) and the mean of the flexural strength, five specimens of each of the WPC and HDPE lumber were cut with a span to depth ratio 16:1 with an adequate overhang length over the supports of the fixture, and were tested in 4-point bending in accordance with ASTM D6109 [21], as shown in Figure 2A. The support spans of the WPC and HDPE specimens were L = 545 mm and L = 620 mm, respectively. The crosshead rate used on the WPC and the HDPE specimens during the 4-point bending test were selected in accordance with ASTM D 6109 [22], to be 15.9 and 18 mm/min, respectively. For the 180-day creep experiments, three levels of flexural creep stress were applied to the WPC and HDPE specimens (five specimens in each level). These three levels were: 7.5%, 15%, and 30% of the mean of the flexural strength obtained from the quasi-static tests. The flexural test was conducted in accordance with ASTM D6109. The flexural stress versus strain relationships of the WPC and HDPE lumber used in this study were reported elsewhere [26,27] The selection of the stress levels was made based on; (1) the use of the WPC and HDPE lumber in submerged Aquapod Net Pen cages is expected to be under low stresses (the structural members of the cage does not carry the weight of the cage, except to withstand the mooring and the buoyancy system [14,15,28], (2) researchers in previous studies [6,7,20,29–31] have studied the creep behavior of WPCs under stress levels that were greater than or equal to 30% and recommended further studies using low stress levels [6,32], thus, it is important to investigate the creep behavior of WPCs under low stress levels. Table 1 shows the values of the apparent elastic modulus of the WPC specimens and the HDPE with their standard deviation values and the selected levels of the creep flexural stress. The determination of the apparent elastic modulus of WPC and HDPE specimens was performed in accordance with ASTM D6109 [22], by computing the slope of the line obtained from the linear regression to the linear portion in the load-midspan deflection curve. Since the span to depth ratio (L/h) of the tested WPC and HDPE specimens was 16 which met the recommended L/h in the ASTM standards, the shear deformation was ignored in the computation of the apparent elastic modulus (further discussion on shear deformation in the computation of the elastic modulus of the WPCs with similar formulation was described elsewhere [11,12]). Then, the flexural strength (Fb) was determined: (1) for WPC, as the ultimate flexural stress at midspan at failure, (2) for HDPE, as the flexural stress at midspan corresponding to 3% of outer fiber strain. The results are reported in Table 1. The mechanical properties of the HDPE lumber tested in this study agreed with the mechanical properties reported in the data sheet of the manufacturer [33]. In accordance with ASTM D6109, the flexural strength is determined as the maximum stress in the outer fibers at failure or when the strain in the outer fibers equals 3%, whichever occurs first.

Table 1. Values of elastic modulus (E), flexural strength, and the applied creep stress level of WPC and HDPE lumber obtained from 4-point quasi-static testing.

Material	Name of the Group	Applied Stress Level	E (GPa)	Mean Fb (MPa)	Applied Flexural Creep Stress Level (MPa)
WPC	group 7.5% Fb	7% Fb	4.34 ± 0.26	41.2 ± 4.53	3.0 ± 0.08
	group 15% Fb	14% Fb			5.9 ± 0.04
	group 30% Fb	29% Fb			11.8 ± 0.09
HDPE	group 7.5% Fb	8% Fb	0.93 ± 0.03	14.1 ± 0.70	1.1 ± 0.05
	group 15% Fb	16% Fb			2.2 ± 0.04
	group 30% Fb	31% Fb			4.4 ± 0.09

3. Results and Discussion

3.1. Determination of the Creep Stress Levels

The applied flexural stress levels for WPC and HDPE lumber were selected to be as percentages of the mean of the flexural strength obtained from the quasi-static tests, Fb = 41.2 MPa, and Fb = 14.1 MPa, respectively. Thus, the flexural creep stress levels applied on the three groups of each of WPC and HDPE lumber were approximately 7.5%, 15%, and 30% of the ultimate flexural strength, as shown in Table 1. Since the cross section of the WPC lumber has a depth (d) which is 88% of the depth of the HDPE lumber and according to ASTM D 6109 the expected mid-span creep displacement of the WPC lumber is expected to be 14% higher than the mid-span creep displacement of the HDPE lumber under the same applied stress with the assumption that the both materials have the same strength and elastic modulus. Thus, to ignore this difference in the cross section of each materials, the applied creep stresses were selected to be approximately at the same level to each group of WPC and HDPE lumber, as percentages of the flexural strength of each material (Table 1). The applied stresses to each group of HDPE lumber is approximately 14% higher than the applied stresses of each group of WPC lumber. This difference was applied to overcome the difference between the cross section (depth) of the WPC lumber and the cross section (depth) of the HDPE lumber. However, each group of HDPE and WPC lumber was given a name based on the applied stress to be; group 7.5%, group 15%, and group 30%.

3.2. Experimental Comparison Between the Long-Term Creep of WPC and HDPE Lumber

Three levels of stress were applied on each group of five specimens of WPC and HDPE lumber. The mean of the mid-span creep deflection of each group of WPC and HDPE lumber was reported, as shown in the log-log space axes in Figure 3.

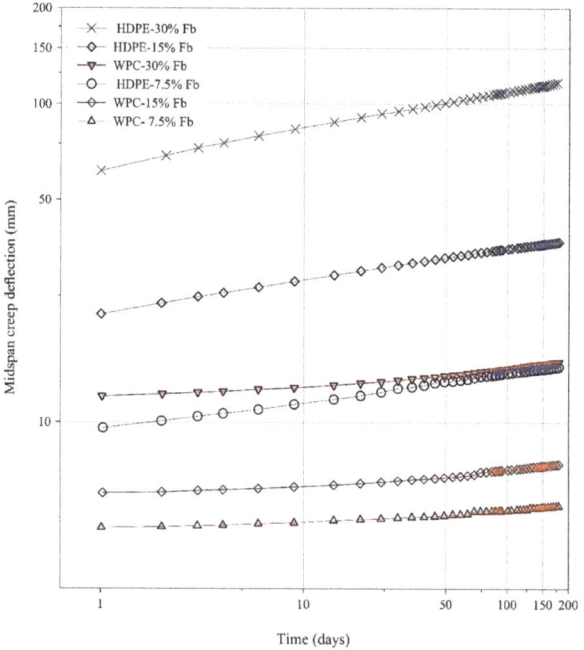

Figure 3. Time-dependent mid-span creep displacement for WPC and HDPE specimens at different stress levels as percentages from the flexural strength Fb.

In accordance with ASTM D 6815 [25], the acceptance criteria of the creep behavior of the specimen is evaluated via: (1) the decrement in the creep rate (all the subsequent creep rate data should be decreasing during the duration of the creep test), (2) the fractional deflection (FD) should not exceed 2, which is obtained from dividing the mid-span creep deflection at the end of the creep experiment by the initial mid-span deflection (D0) [25]. The values of initial midspan displacement measured during the first four minutes of the creep test and were reported in Table 2. In addition to D0, the initial strain (ε_0) was reported in Table 2. The computation of the initial strain was made in accordance with ASTM D 6109. The creep rate in this study was measured at each 30 days as reported in Table 3. Table 3 shows the 30-day creep rate of the three groups of each of WPC and HDPE specimens during the 180-day creep experiment. It can be seen that the values of the WPC fractional deflection under the three different flexural stress levels were within the acceptable limit recommended by ASTM D 6815, whereas, the values of the HDPE fractional deflection failed to meet the recommended fractional deflection limit. However, all the WPC and HDPE groups exhibited a decreasing creep rate during the 180-day creep experiment as reported in Table 3, except a noticeable increase in the creep rate of the HDPE group-15% Fb for the time between the 150 and 180 days. This increase can be attributable to the assumption that the creep of HDPE specimens entered the steady-state of creep in the secondary region [34].

Table 2. Initial midspan deflection (D_0) and strain (ε_0) of WPC and HDPE lumber at three different stress levels.

Material % of Fb	D_0 (mm)	ε_0 (%)
WPC-7.5%	2.96	0.1
WPC-15%	5.8	0.2
WPC-30%	11.3	0.3
HDPE-7.5%	5.5	0.2
HDPE-15%	8.74	0.3
HDPE-30%	18.71	0.6

Table 3. Values of creep rate deflection (D) (mm) of all the groups of WPC and HDPE specimens at 30th, 60th, 90th, 120th, 150 and 180th day respectively and the fractional deflection (FD) at the 180th day with respect to the initial deflection D0.

Creep Rate and FD	Material-% of Fb					
	WPC-7.5%	WPC-15%	WPC-30%	HDPE-7.5%	HDPE-15%	HDPE-30%
$D_{30}-D_0$	0.54	0.99	2.35	7.31	15.57	72.54
$D_{60}-D_{30}$	0.13	0.21	0.57	0.77	1.72	7.80
$D_{90}-D_{60}$	0.12	0.28	0.45	0.51	1.04	4.62
$D_{120}-D_{90}$	0.05	0.12	0.35	0.36	0.65	3.45
$D_{150}-D_{120}$	0.08	0.12	0.29	0.25	0.5	2.87
$D_{180}-D_{150}$	0.06	0.09	0.23	0.22	0.62	2.47
FD_{180}	1.22	1.33	1.28	2.71	3.88	5.11

For further comparison between the creep behavior of WPC and HDPE specimens, a statistical analysis of variance (ANOVA) study of the mid-span creep deflection of each specimen at each group of the WPC and HDPE was conducted and the results are shown in Figure 4. At the applied flexural stress level of 7.5% of the flexural strength, HDPE specimens showed a mid-span creep deflection exceeding two times the mid-span creep deflection of the WPC specimens. As the levels of applied flexural stress increased from 7.5% to 15% and 30%, the HDPE specimens showed mid-span creep deflections exceeding five times and seven times the mid-span creep deflection of the WPC at the same applied flexural levels of stress, respectively. The rate of increase in the mid-span creep deflection between the HDPE specimens subjected to 7.5 and 15% (i.e., HDPE specimens for-7.5% Fb, and 15% Fb) of the flexural strength was below 150%, whereas it was below 35% for the WPC specimens (WPC specimens in group-7.5% and 15% of Fb). When the applied flexural stress levels increased from 15% to 30% of the flexural strength, the creep rate between groups-15% and 30% of Fb was below 215% for the HDPE specimens, and below 110% for WPC specimens. This low time-dependent mid-span deflection creep behavior of the WPC specimens compared with the behavior of HDPE specimens can be anticipated based on their initial compliances (the reciprocal of the elastic modulus); 0.232 GPa-1 and 1.11 GPa-1, respectively. In regards to the comparison of the time-dependent viscoelastic behavior of the WPC with the WPC in previous studies; a short-term time-dependent behavior comparison of the WPC with the same formulation of WPC in this study was presented elsewhere [11], and Alrubaie et al. [12] have presented a comparison between the creep behavior of the group-30% of Fb of WPC presented in this study and the creep behavior of WPC from previous studies. Thus, a comparison to the creep behavior of the WPC used in this study with WPC material from previous studies is not discussed here.

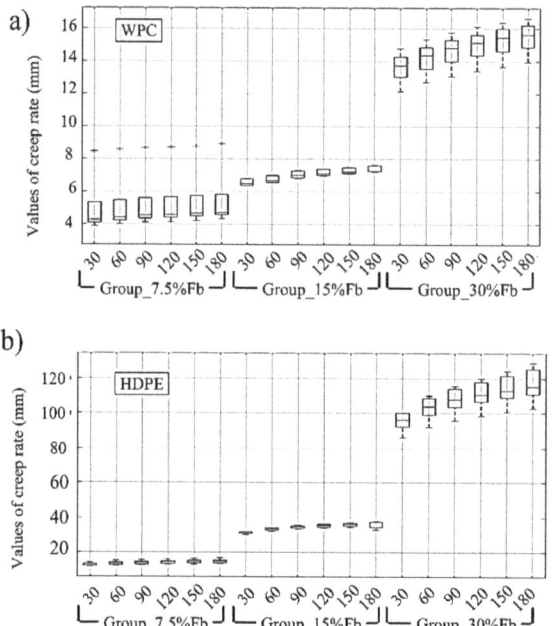

Figure 4. (a) Statistical analysis of variance (ANOVA) that investigates the reduction in creep rate of the WPC specimens subjected to three applied flexural creep stress levels. (b) ANOVA that investigates the reduction in creep rate of the HDPE specimens subjected to three applied flexural creep stress levels as percentages from the flexural strength Fb.

3.3. Time-Dependent Creep Modeling

An empirical power law model was used to describe the 180-day mid-span flexural creep displacement. The model showed a good degree of agreement with the experimental data of the WPC and HDPE lumber in 4-point bending creep test (flatwise). Based on the assumption that the WPC should fail at a flexural strain in outer fiber of 1%, and the HDPE lumber should fail at a flexural strain in outer fiber of 3% (similar to the failure strain value mentioned in ASTM D 6109), the computed mid-span creep the predicted failure occurrence for WPC and HDPE in flexure and under a flexural stress of 30% of Fb will occur after 150 years and 1.5 years, respectively, as shown in Figure 5. To investigate the stress-independency behavior (viscoelastic behavior) of the WPC and HDPE lumber with regards the three applied stress levels (7.5%, 15%, and 30% of Fb), a power law model was implemented to describe the normalized mid-span creep displacement behavior ($d(t)$). Equation (1) describes the normalized midspan creep displacement behavior:

$$d(t) = \frac{D(t)}{D_0} \quad (1)$$

where $d(t)$ is the time dependent midspan deflection. For a 4-point bending test configuration, the initial mid-span creep displacement (D_0) is related to the applied flexural stress, as shown in Equation (2):

$$D_0 = \frac{23}{108} \frac{F_b}{E} \frac{L^2}{h} \qquad (2)$$

where F_b and E are the flexural stress and elastic modulus, respectively, L is the support span, and h is the depth of the WPC and HDPE specimen. The normalized mid-span creep displacement is predicted, as shown in Equation (3):

$$d(t) = 1 + d_1 t^m \qquad (3)$$

where d_1 and m are the stress-independent power law parameters. These parameters (d_1 and m) were computed from the experimental least square error data fitting using a Matlab code. The creep behavior of HDPE lumber and WPC lumber has been predicted for ten years using the power law model, as was reported in Table 4. According to InnovaSea Systems Inc. (Morril, Maine, USA), the estimated service life of aquaculture cages is ten years. The prediction showed the failure occurrence (maximum strain at outer fiber layer) will not occur for both WPC and HDPE specimens for the stress levels 7.5% and 15% of Fb. Whereas, the failure occurrence was predicted in 1.5 years for the HDPE lumbers subjected to 30% of Fb. For this reason, WPCs are considered in the construction of aquaculture cage structures subjected to stress levels 30% below Fb.

Values of the normalized mid-span creep displacement are reported in Table 5. The normalized power law model showed the stress-independency [18] of the WPC and HDPE lumber by having similar values of the normalized power law model (d_1 and m) at different flexural stress levels, respectively. Figures 6 and 7 illustrate the stress-independency behavior of each group of the WPC and HDPE lumber via describing the normalized mid-span creep displacement by the normalized creep behavior.

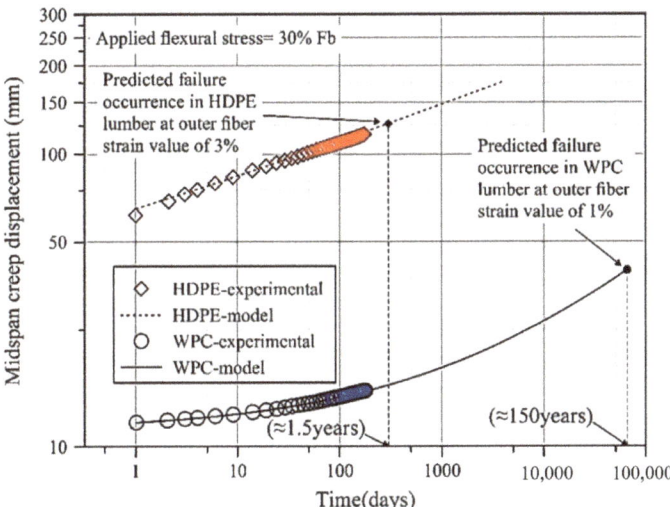

Figure 5. Predicted failure occurrence in the outer fiber strain of WPC and HDPE lumber for the specimens subjected to 30% Fb flexural stress using the power law model.

Table 4. 10-year prediction of the creep displacement of the WPC and HDPE lumber (in accordance with ASTM D6109).

Material Name-% of Fb	Outer Fiber Strain at Failure %	Mid-Span Displacement at Failure (mm)	Predicted Mid-Span Creep Displacement in 10 Years (mm)
WPC-7.5%			6
WPC-15%	1.040	46	11
WPC-30%			22
HDPE-7.5%			21
HDPE-15%	3.004	120	50
HDPE-30%			165

Table 5. Power law model parameters.

Material Type	Model Parameters	
	d_1	m
WPC	0.011	0.596
HDPE	0.018	0.494

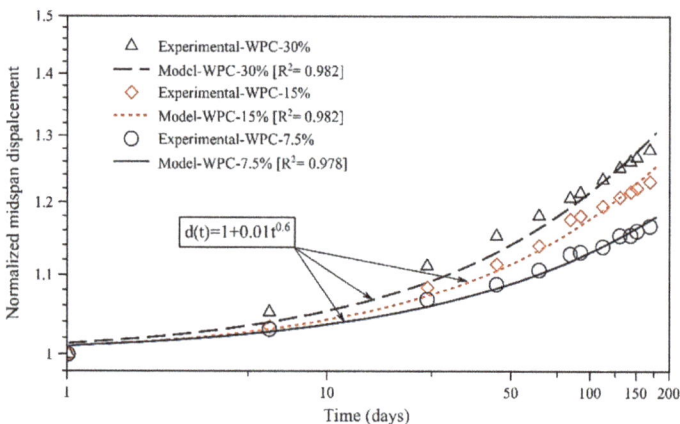

Figure 6. Comparison of power law model and experimental creep result for WPC lumber.

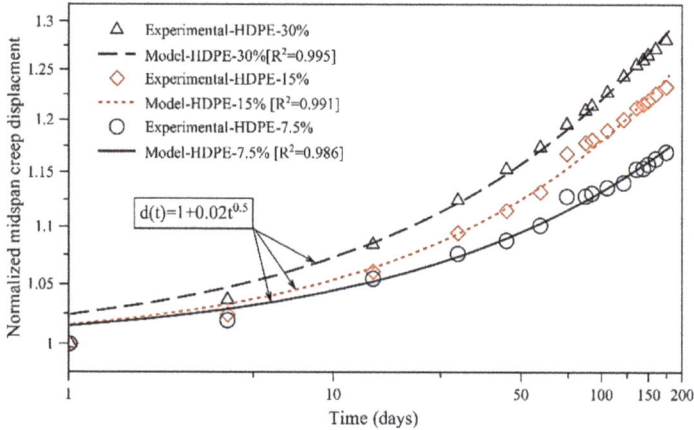

Figure 7. Comparison of power lay model and experimental creep results for HDPE lumber.

4. Conclusions

The WPC in this study showed a reduced time-dependent creep behavior compared to HDPE. WPCs thus show potential to replace HDPE lumber in the construction of aquaculture cage structures. While previous studies have studied the creep behavior of WPC at relatively high stress levels, this study conducted the creep experiments using levels of stresses that were below 30% of the ultimate flexural strength, which are typical for the intended design application. During the comparison between the creep behavior of WPC and HDPE specimens at the low stress levels (7.5% and 15% of Fb), the fractional deflections (FD) of HDPE were 122% and 192% higher than the FD of the WPC specimens, respectively. Whereas, the FD of HDPE specimens at 30% stress level was 300% higher than the FD of the WPC specimens. This can be advantageous for using WPC lumber as a replacement of the HDPE lumber in the construction of aquaculture cages.

The power law model was a useful tool to describe and predict the creep behavior of both WPC and HDPE lumber for all the stress levels (7.5%, 15%, and 30% of Fb). This model predicted that both HDPE lumber and WPC lumber show low creep rate during ten years at stress levels below 15% of Fb. Whereas, at stress level 30% of Fb, failure occurrence at outer fiber is predicted to happen at 1.5 years for HDPE lumber and at 150 years for WPC lumber.

Author Contributions: Conceptualization, M.A.A.A. and D.J.G.; methodology, M.A.A.A.; software, M.A.A.A.; validation, M.A.A.A., D.J.G. and R.A.L.-A.; formal analysis, M.A.A.A.; investigation, M.A.A.A. and D.J.G.; resources, D.J.G.; data curation, M.A.A.A.; writing—original draft preparation, M.A.A.A.; writing—review and editing, M.A.A.A., D.J.G., and R.A.L.-A.; visualization, M.A.A.A., D.J.G., and R.A.L.-A.; supervision, D.J.G.; project administration, D.J.G.; funding acquisition, D.J.G. All authors have read and agreed to the published version of the manuscript.

Funding: This research received no external funding.

Acknowledgments: The work described in this document was conducted at the Advanced Structures and Composites Center at the University of Maine, Orono, Maine (USA). The first author was supported through a scholarship provided by the Higher Committee for Education Development (HCED) in Iraq. The University of Maine research reinvestment funds (RRF) Seed Grant entitled (Development of structural wood plastic composite timber for innovative marine application), the United Stated Department of Agriculture (USDA)-the agricultural research service (ARS) Funding Grant Number (58-0204-6-003), and This project was supported by the USDA National Institute of Food and Agriculture, McIntire-Stennis, Project Number ME0-41809.through the Maine Agricultural & Forest Experiment Station. Maine Agricultural and Forest Experiment Station Publication Number 3715 have provided the financial support for this project. The wood plastic composite is based on a patent-pending formulation that has the publication number (WO2018/142314). The thermally modified wood fiber used in this research is supplied by Stora Enso (Finland).

Conflicts of Interest: The authors declare no conflicts of interest.

References

1. Klyosov, A.A. *Wood-Plastic Composites*; Wiley-Interscience: Hoboken, NJ, USA, 2007.
2. Slaughter, A.E. Design and fatigue of a structural wood-plastic composite. Master of Science thesis, Washington State University, Pullman, WA, USA, 2004.
3. Daniel, A.-V. Structural Performance of Wood Plastic Composite Sheet Piling. *J. Mater. Civil Eng.* **2010**, *12*, 1235–1243.
4. Gardner, D.; Han, Y. Towards Structural Wood-Plastic Composites: Technical Innovations. In Proceedings of the 6th Meeting of the Nordic-Baltic Network in Wood Material Science and Engineering (WSE), Tallinn, Estonia, 21–22 October 2010.
5. Haiar, K.J. *Performance and Design of Prototype Wood-Plastic Composite Sections*; Washington State University: Pullman, WA, USA, 2000.
6. Hamel, S.E. Modeling the Time-Dependent Flexural Response of Wood-Plastic Composite Materials. Ph.D. Thesis, University of Wisconsin-Madison, Madison, WI, USA, January 2011.
7. Dura, M.J. Behavior of Hybrid Wood Plastic Composite-Fiber Reinforced Polymer Structural Members for Use in Sustained Loading Applications. Master's Thesis, The University of Maine, Orono, ME, USA, May 2005.

8. Melissa, K. Structural Design of Hollow Extruded WPC Sheet Piling. Master's Thesis, January 2006. Available online: http://digitalcommons.library.umaine.edu/etd/117 (accessed on 12 December 2018).
9. Bright, K.D.; Smith, P.M. Perceptions of New and Established Waterfront Materials by US Marine Decision Makers. *Wood Fiber Sci.* **2007**, *34*, 186–204.
10. Tamrakar, S.; Lopez-Anido, R.A. Water Absorption of Wood Polypropylene Composite Sheet Piles and Its Influence on Mechanical Properties. *Constr. Build. Mater.* **2011**, *25*, 3977–3988. [CrossRef]
11. Alrubaie, M.A.; Lopez-Anido, R.A.; Gardner, D.J.; Tajvidi, M.; Han, Y. Experimental investigation of the hygrothermal creep strain of wood–plastic composite lumber made from thermally modified wood. *J. Thermoplast. Compos. Mater.* **2019**. [CrossRef]
12. Alrubaie, M.A.; Lopez-Anido, R.A.; Gardner, D.J.; Tajvidi, M.; Han, Y. Modeling the hygrothermal creep behavior of wood plastic composite (WPC) lumber made from thermally modified wood. *J. Thermoplast. Compos. Mater.* **2019**. [CrossRef]
13. InnovaSea Systems, Inc. A4700 BRIDLE SYSTEM IN GRID MOORING CELL. 2016 [cited September 2017]. Available online: www.innovasea.com (accessed on 17 September 2017).
14. Vandenbroucke, K.; Metzlaff, M. Abiotic stress tolerant crops: Genes, pathways and bottlenecks. In *Sustainable Food Production*; Springer: Berlin/Heidelberg, Germany, 2013; pp. 1–17.
15. InnovaSea Systems, Inc. *Report on Structural Damage to A4800 AquaPod*; InnovaSea Systems, Inc.: Orono, ME, USA, 2015.
16. Gardner, D.J. Development of Structural Wood Plastic Composite Timber for Innovative Marine Application. In *Research Reinvestment Funds (RRF) Seed Grant Program*; The University of Maine: Orono, ME, USA, 2015; p. 23.
17. Commerce, D.O. Water Temperature Table of All Coastal Regions. Available online: https://www.nodc.noaa.gov/dsdt/cwtg/all.html (accessed on 23 June 2018).
18. Gibson, R.F. *Principles of Composite Material Mechanics*; CRC Press: Boca Raton, FL, USA, 2016.
19. Barbero, E.J. *Finite Element Analysis of Composite Materials Using AbaqusTM*; CRC Press: Boca Raton, FL, USA, 2013.
20. Tamrakar Sandeep, R.A.L.-A.; Kiziltas, A.; Gardner, D.J. Time and temperature dependent response of a wood–polypropylene composite. *Compos. Part A* **2011**, *42*, 834–842. [CrossRef]
21. Pooler, D.J. *The Temperature Dependent Non-Linear Response of a Wood Plastic Composite*; Washington State University: Pullman, WA, USA, 2001.
22. ASTM International. *Standard Test Methods for Flexural Properties of Unreinforced and Reinforced Plastic Lumber and Related Products, D6109-13*; ASTM International: West Conshohocken, PA, USA, 2013.
23. Hamel, S.E.; Hermanson, J.C.; Cramer, S.M. Predicting the flexure response of wood-plastic composites from uni-axial and shear data using a finite-element model. *J. Mater. Civ. Eng.* **2014**, *26*, 04014098. [CrossRef]
24. Abaqus/CAE. *Computer Software*; SIMULIA Inc.: Providence, RI, USA, 2017.
25. ASTM International. *Standard Specification for Evaluation of Duration of Load and Creep Effects of Wood and Wood-Based Products, D6815-09 (Reapproved 2015)*; ASTM International: West Conshohocken, PA, USA, 2015.
26. Alrubaie, M.A. Investigating the Time-dependent and the Mechanical Behavior of Wood Plastic Composite Lumber Made from Thermally Modified Wood in the Use of Marine Aquacultural Structures. Ph.D. Thesis, The University of Maine, Orono, ME, USA, 8 April 2019.
27. Alrubaie, M.A.A.; Gardner, D.J.; Lopez-Anido, R.A. Structural Performance of HDPE and WPC Lumber Components Used in Aquacultural Geodesic Spherical Cages. *Polymers* **2020**, *12*, 26. [CrossRef] [PubMed]
28. Page, S.H. Aquapod Systems aquaculture Aquapod systems for Sustainable Ocean Aquaculture Aquaculture. In *Sustainable Food Production*; Springer: Berlin/Heidelberg, Germany, 2013; pp. 223–235.
29. King, D.; Hamel, S. The Tensile Creep Response of a Wood-Plastic Composite in Cold Regions. In Proceedings of the 10th International Symposium on Cold Regions Development, Anchorage, AK, USA, 2–5 June 2013; pp. 771–778.
30. Chassagne, P.; Saïd, E.B.; Jullien, J.F.; Galimard, P. Three dimensional creep model for wood under variable humidity-numerical analyses at different material scales. *Mech. Time-Depend. Mater.* **2005**, *9*, 1–21. [CrossRef]
31. Hamel, S.E.; Hermanson, J.C.; Cramer, S.M. Mechanical and time-dependent behavior of wood–plastic composites subjected to tension and compression. *J. Thermoplast. Compos. Mater.* **2013**, *26*, 968–987. [CrossRef]
32. Chang, F.-C. *Creep Behaviour of Wood-Plastic Composites*; University of British Columbia: Vancouver, BC, Canada, 2011.

33. Tangent Technologies, L. Polyforce Sturctural Recycled Plastic Lumber. 2015 [cited November 2018]. Mechanical propertied of the HDPE Polyforce Lumber. Available online: http://tangentusa.com/wp-content/uploads/2016/01/PolyForce_DataSheet_01_20_16.pdf (accessed on 12 December 2018).
34. ASTM International. *Standard Test Methods for Compressive and Flexural Creep and Creep-Rupture of Plastic Lumber and Shapes, D6112-13*; ASTM International: West Conshohocken, PA, USA, 2013.

© 2020 by the authors. Licensee MDPI, Basel, Switzerland. This article is an open access article distributed under the terms and conditions of the Creative Commons Attribution (CC BY) license (http://creativecommons.org/licenses/by/4.0/).

Article

Properties of Injection Molded Biocomposites Reinforced with Wood Particles of Short-Rotation Aspen and Willow

Anuj Kumar [1,*], Tuula Jyske [1] and Veikko Möttönen [2]

1. Natural Resources Institute Finland (Luke), Production Systems, Tietotie 2, FI-02150 Espoo, Finland; tuula.jyske@luke.fi
2. Natural Resources Institute Finland (Luke), Production Systems, Yliopistokatu 6, FI-80100 Joensuu, Finland; veikko.mottonen@luke.fi
* Correspondence: anuj.kumar@luke.fi; Tel.: +358-295322088

Received: 12 November 2019; Accepted: 11 January 2020; Published: 22 January 2020

Abstract: Injection molded biocomposite specimens were prepared by using four different weight percentages, i.e., 10%, 20%, 30%, and 40% of aspen (*Populus tremula* L.) and willow (*Salix caprea* L.) wood particles in a biopolymeric matrix. Dog-bone test specimens were used for testing the physical, mechanical, and thermal properties, and microstructure of biocomposites. The tensile and bending strength changed with the change in weight percentages of wood particles and the bending stiffness increased with the increasing weight percentage of wood. In Brinell hardness, similar changes as a function of wood particle weight percentage were shown, and a relationship between hardness and tensile strength with wood content was also investigated. The prepared biocomposites could be an alternative for plastic-based materials and encourage the use of fast growing (aspen and willow) wood from short-rotation forests in biocomposites.

Keywords: short-rotation; aspen; willow; injection molding; biocomposite; tensile strength; bending strength; microstructure behavior

---

1. Introduction

Future social and economic development globally depends on our success in mitigating climate change by transforming our dependence on finite fossils fuels into use of sustainable resources. The key concept of the circular economy is to reduce waste levels and increase the utilization of side-streams and low-value wood that can be transformed into biocomposites and other value added products from the view point of wood products cluster [1–3]. Biomass materials, such as wood, represent environmentally friendly alternatives for fossil resources that play a key role while also turning societies towards sustainable and circular bioeconomy [4,5]. Various types of wood products, such as engineered wood, and wood-based panels incorporate wood, as raw material in varied forms, into industrial applications that are manufactured by using effective processing methods [6,7]. Such manufacturing methods are able to utilize wood with inconvenient shapes, such as branches and side-streams, or fast-growing, small-diameters species (i.e., aspen or willow), being otherwise difficult to convert into valuable products. Currently, the fast-growing coppice species aspen and willow are either used for energy generation or particleboard production [8]. Fast-growing species could also be used for producing higher added value design biocomposites. Biocomposites are defined as materials where the polymeric matrix or resin and the reinforcement (fibers, particles, powder, etc.) are entirely made from renewable resources. In recent years, they have attracted considerable interest due to their sustainability with great potential to become eco-friendly, biodegradable substitutes for petroleum-based polymeric matrices [4,9].

Many types of natural origin fibers are predominately consumed in biocomposites production. Flax, hemp, jute, coir, cotton, sisal, kenaf, silk, and bamboo fibers are the most explored cellulosic fibers [10–12]. Migneault et al. [13] studied the effects of wood fiber origin, proportion, and chemical composition on the properties of wood-plastic composites (WPC). Interestingly, pulp and paper sludge-based WPC showed better overall properties when compared with the other raw materials. Csikós et al. [14] fabricated the poly(lactic) acid (PLA) and Filtracel EFC 1000 (Rettenmaier and Söhne GmbH) wood fiber composite and studied the surface of wood fibers on the interfacial bonding between wood fibers and polymer matrix. Porebska et al. [15] found that the polymer matrix influenced the properties of wood polymer composites when they prepared cellulose fiber reinforced composites with polypropylene, polystyrene, polyoxymethylene, acrylonitrile butadiene styrene, polyester resin, and PLA with different contents of cellulose fibers, by using injection molding process. The size of wood fibers could influence the processing and properties of wood polymer composites [16]. The saw dust of spruce wood could be potential filler for high-density polyethylene (HDPE)-based composite [17]. The properties of wood fibers are dependent on the species, contents, defects, physical, and mechanical properties, as well as the interaction of a fiber with the polymer in the wood polymer composites [18]. Hardwood and softwood fibers could both be suitable for wood polymer composites while using injection molding technique with different particle size and dimension [19]. Surface treatment of wood fiber with alkali improved the compatibility with polymer matrices by creating rough surface, cavities, and much interspace between smaller fibrils [20]. Effah et al. [21] fabricated wood polymer composites while using different wood species, such as pine, eucalyptus, black wattle, long-leaved wattle, port jackson and beefwood, with a low density polyethylene (LDPE) matrix. The different wood fibers interacted differently with polymer matrix due to the differences in chemical and physical properties. Recently, fast-growing willow (*Salix viminialis*) and high-density polyethylene (PEHD)-based injection molded composites were compared with the properties other commercial Lignocel C-120 fibers-based composite [22].

Short-rotation forest (SRF) plantations are gaining attention in many countries, especially when grown for energy production [23]. Among the different fast-growing hardwoods that were proposed for energy uses, willow (*Salix*) is one of the few that has been planted commercially to a significant extent in the European Union (EU). In Northern Europe, it presents the advantages of high productivity in Nordic conditions [23,24]. European aspen (*Populus tremula* L.) and hybrid aspen (*Populus tremula* L. x *P. tremuloides* Michx.) have proved to be one of the fastest growing deciduous tree species in Nordic countries, with successful breeding and cultivation of hybrid aspen since the early twentieth century [25]. In Finland, aspen is mainly used for paper and energy production [26,27].

The aim of this study was to investigate the effect of using the short-rotation wood particles as filler on the physical and mechanical properties and microstructure of injection molded biocomposite. The physical (density, color change, and water uptake), chemical (Fourier transform infrared spectrophotometer (FTIR)), and mechanical properties (tensile strength, bending strength, and stiffness), as well as microstructure of biocomposites were evaluated. Further, the relationship between Brinell hardness and tensile strength of biocomposites as a function of wood particles content was also investigated.

2. Materials and Methods

2.1. Preparations of Wood Particles

Short-rotation European aspen (*Populus tremula* L.) and willow (*Salix caprea* L.) trees (two aspen and 19 willow stems) were harvested in Tuusula (60°33′09″N, 24°58′06″E; 43 m a.s.l.) in southern Finland. The tree height and stem diameter at butt (0 m), breast height (1.3 m), and 6 m (aspen only) were measured for each tree (Table 1).

Table 1. Characteristics of the sample trees in Tuusula, Finland (mean ± standard deviation).

Wood Species	Height (m)	Stem Diameter (cm) *		
		0 m	1.3 m	6 m
Aspen	19.6 ± 0.7	26.1 ± 1.4	20.3 ± 0.6	13.0 ± 0.6
Willow	7.0 ± 2.9	4.7 ± 2.2	1.7 ± 0.5	not measured

* Diameters were measured from two compass directions across the stem and arithmetic mean values calculated.

The stems were then sawn into smaller blocks, from which bark, branches, larger knots, and defects and whorls were removed. The blocks were converted into wood chips while using a lab-based chipping machine, as described in Figure 1. The chips were dried with warm air at a room temperature for two to three weeks to stabilize the moisture content. The corn starch-based polylactic acid (PLA) was purchased from Sigma Aldrich (Helsinki, Finland) as a biopolymer matrix for biocomposites fabrication.

Figure 1. Harvesting and chipping process of short-rotation aspen and willow.

Aspen and willow chips were milled with Fritsch Pulverisette mill (Helsinki, Finland). A two-step milling process was applied, where the second, fast rotating cutting blade followed the first slowly rotating crushing blade. Against the second blade, a sieve with 2.0 mm openings was used. The particle size varied from 0.2 mm × 1.0 mm × 2.0 mm to 0.5 mm × 2.0 mm × 6.0 mm. The milled powder was dried for at least 4 h at 105 °C, and the mixtures with natural binder were made at 80 °C to avoid moisture absorption to the raw materials. The wood particles of both species were mixed to natural binder in 10%, 20%, 30%, and 40% based on dry weight percentage of PLA before molding process.

2.2. Injection Molding Process

The test specimens (Figure 2) of aspen and willow wood with PLA matrix were molded at industrial scale with a twin-extruder (Engel ES 200/50 HL, Eschweiler, Germany), according to ASTM D638 standards. Twenty tensile bars of each weight percentages of wood in biocomposites were prepared for tests of tensile, bending strength, and Brinell hardness. All of the specimens were conditioned at 20 °C and 65% relative humidity for 48 h prior to testing. The nominal dimensions of specimens were, as follows: gauge length 80 mm, width 10 mm, and thickness 4 mm. The following temperature settings were used for injection molding process: Feed zone: 185 °C, compression zone: 190 °C, homogenizing zone: 190 °C, machine nozzle: 195 °C, and mold temperature: 30 °C. The gravimetric method measured the density of all the prepared biocomposites.

Figure 2. Dog-bone shaped biocomposite test samples of aspen and willow.

2.3. Mechanical Properties

The mechanical properties were tested while using Zwick Z050 (Kennesaw, GA, USA) material testing machine. Tensile strength was measured according to EN ISO 527, bending strength (σ_w, MPa), and modulus of elasticity (E_w, GPa) were measured according to EN ISO 178, and Brinell hardness (HB, MPa) was measured according to EN 1534 [28].

2.4. Color Measurement

The surface reflectance spectrum was measured from the range of 8 mm in diameter of each specimen in the visible light wavelength range 360–740 nm while using a Konica Minolta CM-2600d spectrophotometer (New Jersey, NY, USA) Spectral data was converted to CIEL*a*b* color coordinates using 2° standard observer and D65 light source for lightness (L*), redness (a*), and yellowness (b*), according to CIEL*a*b* color space (ISO 11664-4:2008). For each sample group, the mean and standard deviations of the color coordinates were calculated.

2.5. Microstructures of Biocomposites

Small specimens (ca. 20 mm × 5 mm × 4 mm) having a trapezoid-shaped head were cut from the biocomposite samples by using a saw and a razor blade. Semi-thin sections (ca. 5–6 µm thick) were cut from trapezoids by using glass knives in a rotary microtome (Leica RM2265, Leica Microsystems, Wetzlar, Germany). The sections were stained with an aqueous solution of 0.1% toluidine blue, air-dried, and then mounted in Ultrakitt M540 mountant (TAAB, Reading, UK). Optical microscope images of sections were taken by using a digital camera (MicroPublisher 3.3 RTV, QImaging, Surrey, BC, Canada; 6.6 PL-B686CF-KIT, PixeLINK, Ottawa, ON, Canada) that was attached to a light microscope (Olympus BX60 or Olympus BX50) at 10X -magnification and with a resolution of 0.343 µm/pixel.

2.6. FTIR Measurement

Fourier transform infrared spectrophotometer (Shimadzu Cooperation, Kyoto, Japan IRPrestige-21/IRAffinity-1/FTIR-8000 series) coupled with IRsolution software to control them and data processing used in this work. Semi-thin layers were cut from biocomposites, a razor blade, and then dried at 60 °C for two hours. The prepared samples were scanned while using Attenuated Total Reflection (ATR) setup in the absorbance range of 400–4000 cm^{-1} with a scanning rate of 2 cm^{-1} and 50 scans per run.

2.7. Water Absorption

Three specimens of each biocomposite samples with nominal dimensions of 32 mm × 20 mm × 4 mm were immersed into water for 24 h and for four days and weight change was measured to calculate the water absorption percentage.

2.8. Statistical Analysis

One-way analysis of variance (ANOVA) at P < 0.05 level was performed to identify the statistical difference between the control PLA sample and wood particles reinforced biocomposite samples by using Microcal Origin statistical software.

3. Results and Discussion

3.1. Density and Water Absorption

The density of pure PLA injection molded composite varied between 1.22 to 1.26 g/cm^3. The density of biocomposite increased with an increasing weight percentage of wood particles (Table 2). The density of aspen-based biocomposite was increased from 1.22 g/cm^3 (wood content 10%) to 1.32 g/cm^3 (wood content 40%), while the pure biopolymer possesses density between. The willow-based biocomposite showed higher density in comparison to aspen-based biocomposite.

Table 2. Density and water absorption of biocomposites for different weight percentage of wood particles.

Species and Wood Weight Percentage	Density (g/cm^3)	Water Absorption (%)	
		24 h	4 days
PLA	1.22 ± 0.05	0.20 ± 0.04 ns	0.59 ± 0.07 ns
Aspen 10%	1.23 ± 0.05 ns	0.33 ± 0.04 ns	0.98 ± 0.06 *
Aspen 20%	1.26 ± 0.06 ns	0.53 ± 0.07 ns	1.5 ± 0.15 *
Aspen 30%	1.29 ± 0.05 *	0.75 ± 0.05 ns	2.27 ± 0.2 *
Aspen 40%	1.32 ± 0.04 *	0.90 ± 0.08 ns	2.79 ± 0.18 *
Willow 10%	1.30 ± 0.05 *	0.40 ± 0.04 ns	1.41 ± 0.21 *
Willow 20%	1.31 ± 0.03 *	0.51 ± 0.03 ns	2.01 ± 0.10 *
Willow 30%	1.32 ± 0.04 *	1.25 ± 0.1 *	5.30 ± 0.30 *
Willow 40%	1.33 ± 0.05 *	1.56 ± 0.03 *	6.20 ± 0.27 *

(ns—no signifiecant different). (*—Significant difference at $p < 0.05$).

The stability of biocomposite in humid condition or water contact is a very essential characteristic that is required for several applications. Table 2 shows the water absorption of biocomposites after being immersed into water for 24 h and four days. The water absorption capacity of biocomposite increased with increase in weight percentage of wood particles. After 24 h, the 10% aspen sample showed 0.33% water absorption, while the 40% aspen sample reached 0.90% water absorption. Similar trend was observed for willow-based biocomposites, but with higher average water absorption than that of aspen-based biocomposites. Increasing trend of water uptake with wood particle content was also observed after four days of immersion. However, the water absorption was significantly higher, i.e., 2.79% and 6.20% for 40% aspen and willow samples, respectively. These results are in accordance to studies that were reported in literature, where wood-based material was used in biocomposite preparation [29,30].

3.2. Tensile Testing

To evaluate the effect of reinforcement by wood particles on the PLA matrix, the mechanical properties of biocomposites were determined. The tensile strength of PLA was 64 ± 3 MPa. The addition of 10% of wood particles significantly reduced the tensile strength by 26% to 49 ± 2 MPa, as shown in Figure 3. This might be due to the random distribution of wood particles in biopolymer matrix, which has created different load transfer points within biocomposite due to packing frication. Tensile strength increased with further increase in wood particle content up to 30. However, increasing the wood particles content beyond 30% was found to have a negative impact on bond formation between polymer matrix and wood particles, thus creating weak interfacial regions [31]. Earlier studies also showed similar behavior in the reduction of tensile strength of wood particles (flour and fibers) reinforced PLA-based biocomposites with increasing wood content [32–34].

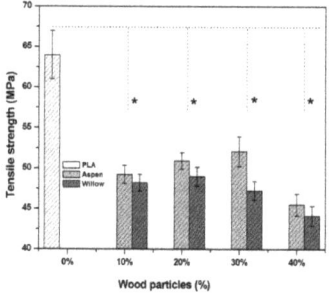

Figure 3. Tensile strength comparison between different species. (* Significant difference at $p < 0.05$)

3.3. Bending Strength and Stiffness

The pure biopolymer of this study showed a strong bending strength of 100 MPa, which is comparable to the bending strengths that were reported for other biopolymers, such as PLA and polyester resin [15,21]. Figure 4a shows the bending strength of the prepared biocomposites. The bending strengths of wood particle-reinforced biocomposites were significantly decreased as compared to the pure PLA. The bending strength reduced to 88–90 MPa from 100 MPa with different weight percentages of wood particles of both species and the lowest values were shown for the samples with 40% of weight percentage. This reduction in bending strength is attributed to the redistribution of binding forces between wood particles and biopolymer. At higher wood filler contents, the bending stiffness of biocomposites significantly increased (50%–90%) with the increasing wood particle content of both species in comparison with the bending stiffness of pure biopolymer. The bending stiffness of pure biopolymer was 3.4 GPa and the stiffness of biocomposites increased to 3.7, 4.7, 5.7, and 6.8 GPa with 10%, 20%, 30%, and 40% of aspen wood weight percentages, respectively, as shown in Figure 4b. Similar, trend was shown by willow wood particles reinforcement, with the highest stiffness of 5.6 GPa with 40% of wood particle content.

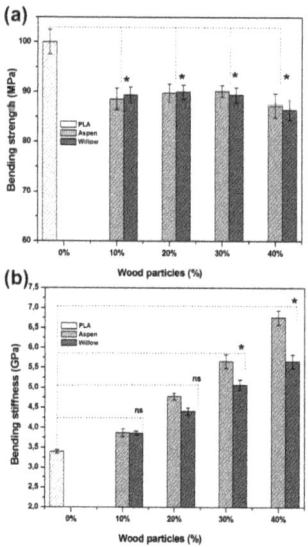

Figure 4. Comparison of the bending properties between the PLA and biocomposite samples: (**a**) Bending strength and (**b**) Bending stiffness. (* significant difference at $p < 0.05$ and ns not significant).

The stiffness and brittleness of the biocomposites increases with increasing wood particle content (Figures 4b and 5). For all cases, there is a linear increase in the bending stiffness with wood particle content. The load-displacement curves (Figure 5a,b) show that as the content of wood particles increases, the biocomposites show the brittle-behavior with reduced displacement towards bending force.

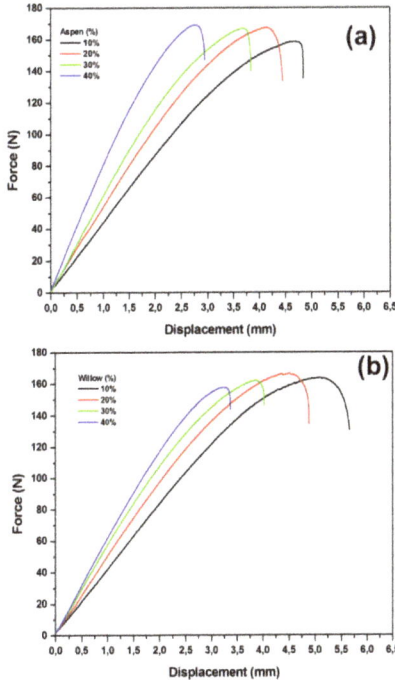

Figure 5. Load-displacement curves of biocomposites for bending strength for (**a**) aspen wood particles and (**b**) willow wood particles.

3.4. Brinell Hardness

The technical hardness is the resistance that a body opposes the penetration of another. The hardness mainly characterizes plastic or mainly elastic deformation, depending on the type of deformation of the materials to be tested [34]. Figure 6 demonstrates the change in the Brinell hardness of biocomposites with wood particles content. The hardness significantly increases with the wood particles content for both wood species. With 10% wood particles, the hardness of biocomposite was 89.27 and 91.80 MPa for aspen and willow, respectively. It was increased to 102.79 and 98.80 MPa for aspen and willow-based biocomposites, respectively, with 40% wood particles content. The average mean Brinell hardness for pure biopolymer was 75 ± 5 MPa, so Brinell hardness of biocomposites significantly increased with increasing wood particles content. Similar results have been reported in literature, where wood particles of other species were used [31,35]. The natural aspen and willow wood have a significantly lower, i.e., 18 ± 3 MPa Brinell hardness when compared to that of biopolymer (75 ± 5 MPa). The hardness of the produced biocomposites increased with increasing wood contents when wood particles were used as fillers in biopolymer.

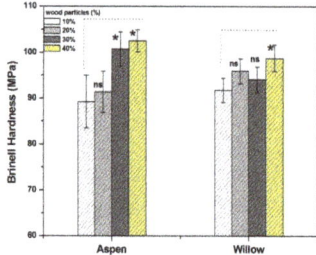

Figure 6. Brinell hardness of biocomposite with different weight percentages of wood particles. (* significant difference at $p < 0.05$ and ns not significant)

3.5. Correlation between Brinell hardness and Tensile Strength

Hardness is of persistent interest to understand the relationships between hardness and other fundamental properties of material [36]. In the present work, the ultimate tensile strength of biocomposite was estimated with different wood particle content of two different wood species. The relationship between the Brinell hardness and the tensile strength of the biocomposite was determined by linear regression analysis and the coefficient of correlation was established between them, as shown in Figure 7. The linear correlation showed a decreasing trend with increasing wood particles content. The highest coefficient correlation R^2 found was 0.987 for 10% aspen wood particles biocomposite, whereas 10% willow particles filled biocomposite showed R^2 of 0.972. With 40% wood particles content, the R^2 was reduced to 0.958 and 0.954 for aspen- and willow-based biocomposites, respectively. A similar value of correlation coefficient was reported [30] for wood plastic composite (polypropylene beech and pine wood mixed wood flour) between the Brinell hardness and tensile strength, where R^2 significantly reduced when wood flour content of ≥40% was used for polypropylene-based injection molded composite. The interfacial bonding between biopolymer and wood particles might be reduced with increasing wood content, also possibly resulting in lower R^2. According to literature, this is a typical behavior of thermoplastic composites that are filled with lignocellulosic material [16,18,31,35,37].

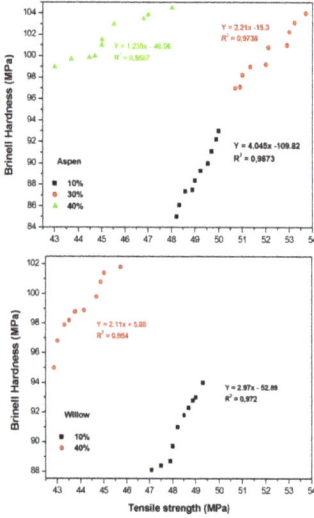

Figure 7. The correlation between Brinell hardness and tensile strength of the biocomposite reinforced with aspen and willow wood particles.

3.6. FTIR Analysis

Figure 8 shows the typical hardwood FTIR spectra that represent aspen and willow wood with characteristic peaks at 3350 cm^{-1} for O–H stretch (hydrogen-bonded), C–H stretching at 2926 and 2854 cm^{-1}; and 1739 cm^{-1} for C=O stretch; 1593 cm^{-1} and 1502 cm^{-1} for aromatic skeletal vibration of lignin; 1234 cm^{-1} for C–O of guaiacyl ring; and, at 1031 cm^{-1} for C–O of primary alcohol and guaiacyl C–H, respectively [38]. The stretching frequencies for C=O and C–O, –CH$_3$ asymmetric, –CH$_3$ symmetric at 1746 cm^{-1}, 2995 cm^{-1}, 2946 cm^{-1}, and 1080 cm^{-1}, respectively, as shown in Figure 9 for biopolymer. The –CH$_3$ asymmetric and –CH$_3$ symmetric frequencies at 1452 and 1361 cm^{-1}, respectively, are the identification of PLA [39].

The biocomposite that was reinforced with aspen and willow wood particles showed the characteristic FTIR peaks (see Figure 9) of biopolymer and aspen wood, as discussed above. The O–H stretch (hydrogen-bonded) at 3350 cm^{-1} of wood was not presented in biocomposite, due to bonding between biopolymer and reactive hydroxyl group of wood. The C–O at 2995 cm^{-1}, –CH$_3$ asymmetric at 2946 cm^{-1} has shifted to 2926 cm^{-1}, 2854 cm^{-1}, respectively, when 40% of aspen wood particles reinforced the biopolymer matrix. The peak stretch intensity at 1746 cm^{-1} that represented the C=O stretch of wood and –CH$_3$ symmetric of biopolymer significantly increased for biocomposites with increasing wood content.

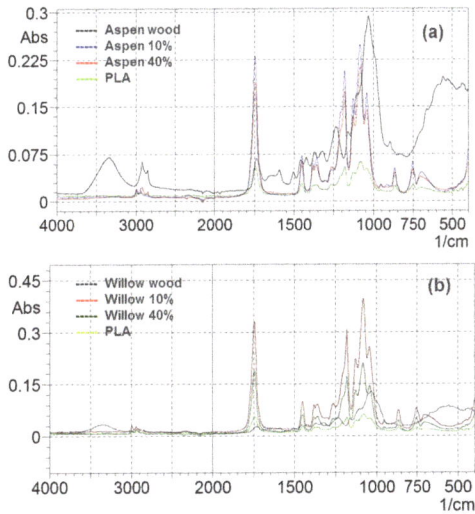

Figure 8. Fourier transform infrared spectrophotometer (FTIR) of wood particles, biopolymer and the biocomposites.

3.7. Color Difference

The lightness values were at the lowest in the groups of 10% of wood particles content, as shown in Figure 9. The low level of the lightness in those biocomposites were probably due to the fact that the glasslike binder of the composite material passed through most of the light and the light was not reflected back to the detector from the sample surface or the surface behind it. The redness of biocomposites increased with increase in wood particles content. Aspen-based biocomposites showed lower redness values as compared to willow-based biocomposites. While yellowness was consistent with different wood particles contents, the difference in yellowness between the wood species was apparent: aspen-based biocomposites showed yellowness values of 30 or more, while willow-based biocomposites had values that were lower than 30.

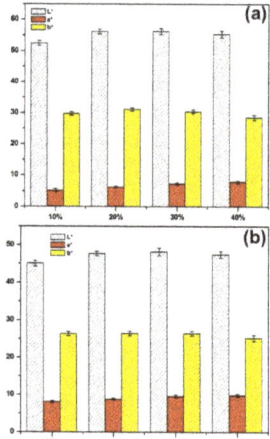

Figure 9. Color measurement of biocomposites with different weight percentage of aspen (**a**) and willow (**b**) wood particles.

3.8. Microstructure of Biocomposites

Figure 10 shows the visual appearance of microstructure of biocomposites with different wood particles content of aspen and willow mixed with the PLA matrix. The wood particles were uniformly mixed with the binder and cellular characteristics of wood were well recognizable. The interfacial bonding of wood particles with the biopolymer was homogenous, which resulted in even and high quality of the composite material. It was clear from microstructures that the wood particles were covered by the polymer matrix that could account for good strength performance of biocomposites. Any types of bubbles or voids were not observed between the biopolymer matrix and the wood particles; however, with higher wood particles content, the homogeneity decreased, possibly causing negative impacts on mechanical properties. The arrows in Figure 10 represent the interfacial zone formation between wood particles and PLA materials. At low wood content, the wood particles formed a bigger interface and random distribution with PLA matrix (Figure 10a,c), and wood particles seem to appear ruptured. On the other hand, the high wood content loading into PLA matrix formed smaller interface zone and wood particles appeared to be less ruptured. The higher wood content loading showed lower tensile strength due to the formation of smaller interface zone.

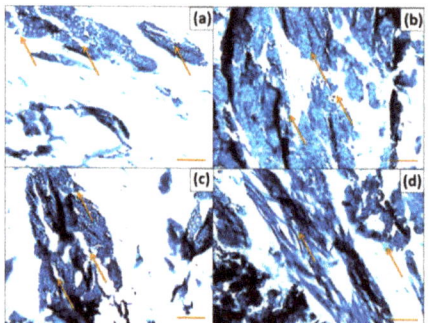

Figure 10. Microstructure of biocomposites with different weight percentage of wood particles mixed to natural binder before the molding process (white color for PLA matrix and blue color for wood particles in PLA matrix): (**a**) 10% of aspen wood, (**b**) 40% of aspen wood, (**c**) 20% of willow wood, and (**d**) 40% of willow wood. Scale bars: 100 μm.

4. Conclusions

The present study was focused on exploring the potential utilization of wood raw materials from short-rotation forests, namely aspen and willow, in the production of injection molded biocomposites. The aspen and willow wood particles were mixed as filler in different weight percentages (10%, 20%, 30%, and 40%) into the PLA matrix to produce biocomposites at the industrial scale setup. Biocomposites were analyzed to evaluate their physical (color, density and water absorption, microstructure), chemical (FTIR), and mechanical (tensile strength, bending behavior, and Brinell hardness) properties. The results revealed that the tensile and bending strength initially decreased with 10% weight percentage of wood particles when compared to pure biopolymer, but showed increasing trend with higher wood particles contents. However, the bending stiffness was higher than that of pure PLA already at the lowest wood particles content and increased with the increase in weight percentages of wood particles. The linear correlation between tensile strength and Brinell hardness varies with wood particles content, as the values of the linear coefficient of regression (R^2) was decreased with increasing wood particles percentage. The microstructure analysis revealed the formation of good interfacial bonding between wood particles and biopolymer, but also variations in the homogeneity with different weight percentages of wood particles. As a conclusion, wood of short-rotation tree species has excellent potential to be used for production of biocomposites and contribute to the sustainable bioeconomy.

Author Contributions: Conceptualization, T.J. and V.M.; formal analysis, A.K., T.J. and V.M.; resources, T.J.; writing—original draft preparation, A.K.; writing—review and editing, A.K., T.J., and V.M.; visualization, T.J.; project administration, T.J.; funding acquisition, T.J. All authors have read and agreed to the published version of the manuscript.

Funding: Authors would like to acknowledge the Luke's strategic financing for project entitled as "More, faster, higher quality: potential of short-rotation aspen and willow biomass for novel products in bioeconomy" (AspenWill).

Acknowledgments: Authors would like to acknowledge the support of Juha Metros, Tapio Nevalainen, Kalle Kaipanen, and Tapio Laakso for skillful assistance with the field and laboratory work.

Conflicts of Interest: The authors declare no conflict of interest.

References

1. Jiang, W.; Kumar, A.; Adamopoulos, S. Liquefaction of lignocellulosic materials and its applications in wood adhesives—A review. *Ind. Crop. Prod.* **2018**, *124*, 325–342. [CrossRef]
2. Kumar, A.; Staněk, K.; Ryparová, P.; Hajek, P.; Tywoniak, J. Hydrophobic treatment of wood fibrous thermal insulator by octadecyltrichlorosilane and its influence on hygric properties and resistance against moulds. *Compos. Part B: Eng.* **2016**, *106*, 285–293. [CrossRef]
3. Kumar, A.; Petrič, M.; Kričej, B.; Žigon, J.; Tywoniak, J.; Hajek, P.; Škapin, A.S.; Pavlič, M. Liquefied-wood-based polyurethane–nanosilica hybrid coatings and hydrophobization by self-assembled monolayers of orthotrichlorosilane (OTS). *ACS Sustain. Chem. Eng.* **2015**, *3*, 2533–2541. [CrossRef]
4. D'Amato, D.; Droste, N.; Allen, B.; Kettunen, M.; Lähtinen, K.; Korhonen, J.; Leskinen, P.; Matthies, B.; Toppinen, A. Green, circular, bio economy: A comparative analysis of sustainability avenues. *J. Clean. Prod.* **2017**, *168*, 716–734. [CrossRef]
5. Husgafvel, R.; Linkosalmi, L.; Hughes, M.; Kanerva, J.; Dahl, O. Forest sector circular economy development in Finland: A regional study on sustainability driven competitive advantage and an assessment of the potential for cascading recovered solid wood. *J. Clean. Prod.* **2018**, *181*, 483–497. [CrossRef]
6. Ramage, M.H.; Burridge, H.; Busse-Wicher, M.; Fereday, G.; Reynolds, T.; Shah, D.U.; Wu, G.; Yu, L.; Fleming, P.; Densley-Tingley, D.; et al. The wood from the trees: The use of timber in construction. *Renew. Sustain. Energy Rev.* **2017**, *68*, 333–359. [CrossRef]
7. Kumar, A.; Verkasalo, E. Wood-based panel industries in Finland–current status and development potential. In Proceedings of the 12th Edition of the International Conference "Wood Science and Engineering in the Third Millennium"—ICWSE 2019, Brasov, Romania, 7–9 November 2019.
8. Hemmilä, V.; Adamopoulos, S.; Karlsson, O.; Kumar, A. Development of sustainable bio-adhesives for engineered wood panels—A Review. *RSC Adv.* **2017**, *7*, 38604–38630. [CrossRef]

9. Kumar, A.; Vlach, T.; Ryparovà, P.; Škapin, A.S.; Kovač, J.; Adamopoulos, S.; Hajek, P.; Petrič, M. Influence of liquefied wood polyol on the physical-mechanical and thermal properties of epoxy based polymer. *Polym. Test.* **2017**, *64*, 207–216. [CrossRef]
10. Väisänen, T.; Das, O.; Tomppo, L. A review on new bio-based constituents for natural fiber-polymer composites. *J. Clean. Prod.* **2017**, *149*, 582–596. [CrossRef]
11. Ramamoorthy, S.K.; Skrifvars, M.; Persson, A. A review of natural fibers used in biocomposites: Plant, animal and regenerated cellulose fibers. *Polym. Rev.* **2015**, *55*, 107–162. [CrossRef]
12. Gurunathan, T.; Mohanty, S.; Nayak, S.K. A review of the recent developments in biocomposites based on natural fibres and their application perspectives. *Compos. Part A: Appl. Sci. Manuf.* **2015**, *77*, 1–25.
13. Migneault, S.; Koubaa, A.; Perré, P. Effect of fiber origin, proportion, and chemical composition on the mechanical and physical properties of wood-plastic composites. *J. Wood Chem. Technol.* **2014**, *34*, 241–261. [CrossRef]
14. Csikós, Á.; Faludi, G.; Domján, A.; Renner, K.; Móczó, J.; Pukánszky, B. Modification of interfacial adhesion with a functionalized polymer in PLA/wood composites. *Eur. Polym. J.* **2015**, *68*, 592–600. [CrossRef]
15. Porebska, R.; Rybak, A.; Kozub, B.; Sekula, R. Polymer matrix influence on stability of wood polymer composites. *Polym. Adv. Technol.* **2015**, *26*, 1076–1082. [CrossRef]
16. Akpan, E.I.; Wetzel, B.; Friedrich, K. Processing and properties of short wood fiber/acrylate resin composites. *Polym. Compos.* **2019**, *40*, 91–98. [CrossRef]
17. Tazi, M.; Sukiman, M.S.; Erchiqui, F.; Imad, A.; Kanit, T. Effect of wood fillers on the viscoelastic and thermophysical properties of HDPE-wood composite. *Int. J. Polym. Sci.* **2016**, *2016*, 9032525. [CrossRef]
18. Dányádi, L.; Janecska, T.; Szabo, Z.; Nagy, G.; Moczo, J.; Pukánszky, B. Wood flour filled PP composites: Compatibilization and adhesion. *Compos. Sci. Technol.* **2007**, *67*, 2838–2846. [CrossRef]
19. Bledzki, A.K.; Faruk, O. Injection moulded microcellular wood fibre–polypropylene composites. *Compos. Part A: Appl. Sci. Manuf.* **2006**, *37*, 1358–1367. [CrossRef]
20. Jiang, X.; Wang, J.; Wu, G.; Peng, X.; Ma, X. Significant reinforcement of polypropylene/wood flour composites by high extent of interfacial interaction. *J. Thermoplast. Compos. Mater.* **2018**, *32*, 577–592. [CrossRef]
21. Effah, B.; Van Reenen, A.; Meincken, M. Mechanical properties of wood-plastic composites made from various wood species with different compatibilisers. *Eur. J. Wood Wood Prod.* **2018**, *76*, 57–68. [CrossRef]
22. Barton-Pudlik, J.; Czaja, K. Fast-growing willow (*Salix viminalis*) as a filler in polyethylene composites. *Compos. Part B: Eng.* **2018**, *143*, 68–74. [CrossRef]
23. Weih, M. Intensive short rotation forestry in boreal climates: Present and future perspectives. *Can. J. For. Res.* **2004**, *34*, 1369–1378. [CrossRef]
24. Mola-Yudego, B. Regional potential yields of short rotation willow plantations on agricultural land in Northern Europe. *Silva Fenn.* **2010**, *44*, 63–76. [CrossRef]
25. Tullus, A.; Rytter, L.; Tullus, T.; Weih, M.; Tullus, H. Short-rotation forestry with hybrid aspen (*Populus tremula* L.× *P. tremuloides* Michx.) in Northern Europe. *Scand. J. For. Res.* **2012**, *27*, 10–29. [CrossRef]
26. Heräjärvi, H.; Junkkonen, R. Wood density and growth rate of European and hybrid aspen in southern Finland. *Balt. For.* **2006**, *12*, 2–8.
27. Latva-Karjanmaa, T.; Penttilä, R.; Siitonen, J. The demographic structure of European aspen (*Populus tremula*) populations in managed and old-growth boreal forests in eastern Finland. *Can. J. For. Res.* **2007**, *37*, 1070–1081. [CrossRef]
28. Herrmann, K. *Hardness Testing: Principles and Applications*; ASM International: Cleveland, OH, USA, 2011.
29. Sinha, P.; Mathur, S.; Sharma, P.; Kumar, V. Potential of pine needles for PLA-based composites. *Polym. Compos.* **2018**, *39*, 1339–1349. [CrossRef]
30. Averous, L.; Le Digabel, F. Properties of biocomposites based on lignocellulosic fillers. *Carbohydr. Polym.* **2006**, *66*, 480–493. [CrossRef]
31. Kaymakci, A.; Ayrilmis, N. Investigation of correlation between Brinell hardness and tensile strength of wood plastic composites. *Compos. Part B: Eng.* **2014**, *58*, 582–585. [CrossRef]
32. Csizmadia, R.; Faludi, G.; Renner, K.; Móczó, J.; Pukánszky, B. PLA/wood biocomposites: Improving composite strength by chemical treatment of the fibers. *Compos. Part A: Appl. Sci. Manuf.* **2013**, *53*, 46–53. [CrossRef]

33. Peltola, H.; Pääkkönen, E.; Jetsu, P.; Heinemann, S. Wood based PLA and PP composites: Effect of fibre type and matrix polymer on fibre morphology, dispersion and composite properties. *Compos. Part A: Appl. Sci. Manuf.* **2014**, *61*, 13–22. [CrossRef]
34. Petinakis, E.; Yu, L.; Edward, G.; Dean, K.; Liu, H.; Scully, A.D. Effect of matrix–particle interfacial adhesion on the mechanical properties of poly (lactic acid)/wood-flour micro-composites. *J. Polym. Environ.* **2009**, *17*, 83–94. [CrossRef]
35. Kord, B. Investigation of reinforcing filler loading on the mechanical properties of wood plastic composites. *World Appl. Sci. J.* **2011**, *13*, 171–174.
36. Zhang, P.; Li, S.; Zhang, Z. General relationship between strength and hardness. *Mater. Sci. Eng. A* **2011**, *529*, 62–73. [CrossRef]
37. Yu, Y.; Yang, Y.; Murakami, M.; Nomura, M.; Hamada, H. Physical and mechanical properties of injection-molded wood powder thermoplastic composites. *Adv. Compos. Mater.* **2013**, *22*, 425–435. [CrossRef]
38. Pandey, K. A study of chemical structure of soft and hardwood and wood polymers by FTIR spectroscopy. *J. Appl. Polym. Sci.* **1999**, *71*, 1969–1975. [CrossRef]
39. Chieng, B.W.; Azowa, I.N.; Zin, W.Y.W.M.; Hussein, M.Z. Effects of graphene nanopletelets on poly (lactic acid)/poly (ethylene glycol) polymer nanocomposites. *Adv. Mater. Res.* **2014**, *1024*, 136–139. [CrossRef]

 © 2020 by the authors. Licensee MDPI, Basel, Switzerland. This article is an open access article distributed under the terms and conditions of the Creative Commons Attribution (CC BY) license (http://creativecommons.org/licenses/by/4.0/).

Article

Multiple Analysis and Characterization of Novel and Environmentally Friendly Feather Protein-Based Wood Preservatives

Yan Xia [1,2], Chengye Ma [1], Hanmin Wang [1], Shaoni Sun [1], Jialong Wen [1,*] and Runcang Sun [1,3,*]

1. Beijing Key Laboratory of Lignocellulosic Chemistry, Beijing Forestry University, Beijing 100083, China; xiayan@swfu.edu.cn (Y.X.); chengye.ma@foxmail.com (C.M.); wanghanmin798@163.com (H.W.); sunshaoni@bjfu.edu.cn (S.S.)
2. College of Material Science and Engineering, South-West Forestry University, Kunming 650224, China
3. Center for Lignocellulose Science and Engineering, Dalian Polytechnic University, Dalian 116034, China
* Correspondence: wenjialong@bjfu.edu.cn (J.W.); rcsun3@bjfu.edu.cn (R.S.); Tel./Fax: +86-10-62336903 (J.W. & R.S.)

Received: 20 December 2019; Accepted: 16 January 2020; Published: 19 January 2020

Abstract: In this study, feather was used as the source of protein and combined with copper and boron salts to prepare wood preservatives with nano-hydroxyapatite or nano-graphene oxide as nano-carriers. The treatability of preservative formulations, the changes of chemical structure, micromorphology, crystallinity, thermal properties and chemical composition of wood cell walls during the impregnation and decay experiment were investigated by retention rate of the preservative, Fourier transform infrared spectroscopy (FT-IR), scanning electronic microscopy-energy dispersive spectrometer (SEM-EDS), X-ray diffraction (XRD), thermoanalysis (TG), and confocal Raman microscopy (CRM) techniques. Results revealed that the preservatives (particularly with nano-carrier) successfully penetrated wood blocks, verifying the enhanced effectiveness of protein-based preservative with nano-carrier formulations. Decay experiment demonstrated that the protein-based wood preservative can remarkably improve the decay resistance of the treated wood samples, and it is an effective, environmentally friendly wood preservative. Further analysis of these three preservative groups confirmed the excellent function of nano-hydroxyapatite as a nano-carrier, which can promote the chelation of preservatives with higher content of effective preservatives.

Keywords: feather protein; wood preservatives; nano-carrier; treatability; decay resistance

1. Introduction

Wood is a conventional construction material, but wood products that have direct contact with outdoor soil without protection easily become less stable and present serious deterioration through the decay and degradation in the ambient environment, which may result huge economic losses and resource waste because wood materials are susceptible to being damaged and destroyed by microorganism such as fungi, bacteria and insects. Based on the above reasons, wooden constructions and architectures are mostly protected and chemically-modified to obtain significant improvements in their stability and durability [1]. In most cases, preservative treatment should be performed on wood products, and the durability and resistance of treated wood products against biological attacks during their service period can be improved [2,3]. Chemical preservatives are common in wood preserving treatments, in which water-soluble preservatives are mostly used. Chromated Copper Arsenate (CCA) preservatives have been the most extensively used in the past decades, especially for wood-framed building timbers. Nevertheless, CCA has been prohibited for residential purposes by the U.S. Environmental Protection Agency since 2004 due to its toxic effect on the environment during manufacture, treatment, and disposal [4,5]. In recent years, some wood preservatives have been subjected to restrictions on its application considering public concern regarding their high toxicity [6–9]. Based on the above considerations, low toxicity and environmental benign wood preservatives is the research focus in this field, and the development of effective and environment-benign wood preservatives for new preservative systems without chromium and arsenic is necessary [10,11].

In this perspective, copper and boron salts are attracting more and more attention in recent years since they are poisonous to microorganisms and insects but with low toxicity [12–16]. In current circumstances, copper-based preservatives, such as Ammonium Copper Quaternary (ACQ) and Chromated Copper Arsenate (CCA) are the most common preservatives. Since some species of fungi can develop resistance to copper salts, copper-based preservatives are usually applied with other active ingredients to achieve a better preservative effect. Boron-based preservatives are also commonly used because of their low toxicity, resistance against fungi and insects, and low-cost characteristics. In contrast, borates are not suitable for outdoor application because they are easily to leach out due to borates' preferable water-soluble property. To avoid the leaching of active ingredients from preservative caused by their water solubility, proteins can be used as fixative agents, such as soy isolates and egg albumin, to chelate boron and copper in the preservative by chelation, coagulation, and/or chemical reactions to form insoluble complexes thus increase the fixation and durability of preservatives during the wood treating process [17–23]. Nevertheless, since protein is a kind of nutritious matter, excessive amounts of protein may precipitate in wood blocks. This suggests that the excessive protein can serve as a nutrient source consumed by fungi, which might cause a loss of Cu and B in the decay process. Therefore, the ideal protein-based wood preservative should chelate more preservative components and contain less protein.

Based on this, many researchers seek alternative solutions to solve the problem about preservative fixation. Furthermore, to promote the penetration depth and uniformity of the active preservative components, nano-carriers are frequently used in the preparation of wood preservative to enhance the content of effective preservative ingredients [24–28]. In this study, feather protein-based preservatives with different nano-carriers were firstly developed. The treatability of preservative formulations, the changes of chemical structure, micromorphology, crystallinity, thermal properties, and chemical composition of wood cell walls during the modification and decay test were comprehensively investigated. Results show that the feather protein-based preservative studied in this paper can significantly improve the resistance performance against decay fungi of the treated wood, and it is believed that the protein-based preservatives with different nano-carriers have great potential in the fabrication of eco-friendly wood products.

2. Materials and Methods

2.1. Preparation of Feather Protein-Based Wood Preservatives

The preservative formulations were made from hydrolyzed feather protein, copper sulfate ($CuSO_4 \cdot 5H_2O$) and sodium borate ($Na_2B_4O_7 \cdot 10H_2O$). Protein hydrolysate was obtained by hydrolyzing chicken feather powder at 140 °C for 4 h after immersed in 6 wt% aqueous sodium hydroxide at room temperature for 24 h. The concentration of hydrolyzate was condensed to 50% and it was added into the suspension of copper sulfate and sodium borate with the ratio of protein to total amounts of Cu and B in the formulations of 1:1, w/w. Then, a few drops of glacial acetic acid were added into the mixture. Commercial ammonium hydroxide (NH_4OH) with a one-tenth volume of the suspension was added to dissolve the water-insoluble mixture and obtain preservative solution, which was named feather protein-based wood preservative in this study.

To further increase the performance of the preservative, nano-hydroxyapatite or nano-graphene oxide was added into feather protein-based preservative as nano-carriers and blended by the ultrasonic vibration, which facilitates the uniform distribution of these nano-particles in the newly developed preservative. In this study, there are three preservative formulations (P_1, P_2 and P_3). P_1 was prepared by feather protein combined with copper sulfate and sodium borate, accordingly named the feather protein-based preservative (Cu-B-Pr). Preservative P_2 and P_3 were prepared by the feather protein combined with copper sulfate and sodium borate and nano-hydroxyapatite, nano-graphene oxide, respectively. Based on these formulations, P_2 and P_3 were correspondingly named nano-hydroxyapatite protein-based preservative, and the nano-graphene oxide protein-based preservative (Cu-B-Pr-HA, Cu-B-Pr-Go). The preparation procedure for the protein-based preservative was shown below (Scheme 1).

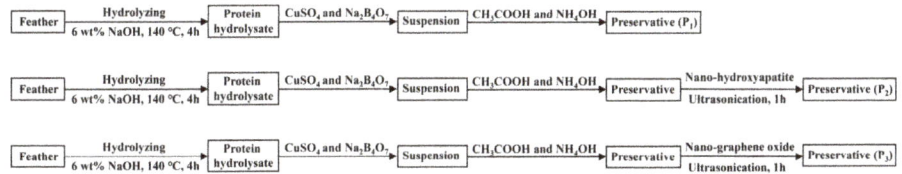

Scheme 1. The preparation process for the feather protein-based preservative.

2.2. Preservative Treatment of the Wood Blocks

Wood blocks sawed from *Pinus yunnanensis* sapwood with a dimension of 150.0 × 20.0 × 20.0 mm (size in axial, radial, and tangential) were each immersed in preservative formulation for 24 h at normal temperature and pressure conditions, and then oven-dried at 60 °C for 24 h, followed by air-dried over 24 h. Each preservative formulation with size of 10.0 × 20.0 × 20.0 mm (size in axial, radial, and tangential) was used for determining decay experiments.

2.3. Treatability of the Preservatives

To measure the solution uptake of treated samples in preservative formulations, treated wood blocks were air conditioned for 24 h and then oven-dried at 60 °C for 24 h. Treatability, representing actual percent retention of the preservatives in the treated samples, was calculated through the ratio of measured retention and target retention for the preservative.

2.4. Decay Resistance of Treated Wood Samples

Decay resistance of control and the treated wood blocks exposed to decay fungi was evaluated according to the method described in ASTM Standard D 1413-07. Brown-rot fungi *Gloeophyllum trabeum* (GT) was used as the test fungi in decay experiments [29].

Fungus cultured on potato dextrose agar was inoculated on the feeder strips on the surface of a mixture composed of river sand, sawdust, corn flour and brown sugar. After the fungal mycelia covered the surface, sterilized wood blocks were placed onto the feed strips, two blocks per bottle. Culture bottles were sterilized for 1 h before being inserted into the decay chamber. The soil-block culture was incubated at 26 ± 1 °C and 75% relative humidity for 12 weeks. After the incubation, wood blocks were moved out from the culture bottles in the decay chambers, and the fungal mycelia were cleaned, then dried overnight in an oven at 80 °C and weighed to determine weight loss of the wood samples. The decay rate of the wood block was represented by the percentage weight loss during exposure to the decay fungus. The treated wood blocks (24 pieces of wood) with different preservatives formulations were evaluated by the decay test.

2.5. Multi-Analysis of Control and Treated Samples

To better investigate the effect of preservative treatment on the treatability of protein-based preservatives, multi-analysis of control and treated samples (12 weeks decay test) was conducted to investigate the decay resistance performances against fungi of the three preservative formulations.

Fourier transform infrared (FT-IR) spectra analysis was used on Thermo Scientific Nicolet iN10 FT-IR microscope (Thermo Nicolet Corporation, Waltham, MA, USA), which was conducted in the range from 4000 to 400 cm^{-1} with 64 scansions per sample at a resolution of 4 cm^{-1} [30,31]. Control and wood samples were milled into powders (40–60 mesh) and then analyzed to elucidate the changes before and after the treatment with preservatives.

Crystallinity index was measured by X-ray diffraction (XRD) using Ni-filtered CuKa radiation at 40 kV and 30 mA. The crystallinities of the specimens were calculated by the ratio of areas under crystalline peaks and amorphous curve according to previous publications [32,33]. Thermogravimetric analysis (TGA) was performed on a simultaneous thermal analyzer DTG-60 (Shimadzu, Kyoto, Japan). 3-5 mg samples were heated in an alumina crucible at a heating rate of 10 °C·min^{-1} from room temperature to 600 °C under nitrogen atmosphere [34].

Scanning Electron Microscope (SEM) can clearly observe the micromorphology and microstructure change of plant cell walls. Energy Dispersive Spectrometer (EDS) can analyze the chemical composition of the cell walls of controlled and treated samples. SEM images were executed with a Hitachi S-3400 N II (Hitachi, Tokyo, Japan) instrument at 10 kV and 81 mA [35].

Raman spectra were acquired with a confocal Raman microscope (CRM, LabRam Xplora, HORIBA, Kyoto, Japan), which was equipped with a piezo scanner and a high numerical aperture (NA) microscope objective from Olympus (100oil NA = 1.40). The Labspect5 software (HORIBA) was used for measurement setup and image processing to remove spikes, smooth the spectra by the Savitsky-Golay algorithm at a moderate level, correct baselines, and the data was further smoothed by Fourier transformation coupled with cosine apodization function [34–36]. Cross sections of 10 μm thickness were cut from wood sample using a rotary microtome RM 2255 (Leica, Wetzlar, Germany) to obtain a full wafer and then covered with glass cover slips. The chemical images allowed us to separate cell wall layers into secondary wall (S) and the cellular corner middle layer (CCML) with different chemical compositions, and to mark distinct cell wall regions for constructing average spectra.

3. Results and Discussion

3.1. Treatability of Protein-Based Preservatives

Treatability of preservative formulations means actual percent retention of the protein-based preservatives in treated wood blocks, which was listed in Table 1. As shown in Table 1, the measured retentions of Cu and B in wood samples treated with P_1, P_2 and P_3 were very close to the target retention, respectively. The treatability of Cu in pretreated wood samples was 84.6% to 87.9%, while the treatability of B in pretreated wood samples was 89.6% to 95.5%. This fact suggested that the three preservative formulations could effectively penetrate wood blocks since ammonium hydroxide is a good dissociating agent [4,5].

Table 1. Treatability of three formulation feather protein-based wood preservatives.

Formulations	Target retention		Measured retention		Treatability	
	Cu	B	Cu	B	Cu	B
P_1, Cu-B-Pr	12	20	10.16	17.92	84.63	89.60
P_2, Cu-B-Pr-HA	12	20	10.40	18.31	86.69	91.55
P_3, Cu-B-Pr-Go	12	20	10.55	18.69	87.92	93.45

3.2. FT-IR Analysis before/after Decay

To compare the structural changes of wood samples after the preservative treatment, the fingerprint region in the FT-IR spectra of control and treated wood samples are presented in Figure 1. As can be seen from Figure 1, it was found that the spectra at 3350 and 2900 cm^{-1} decreased distinctly, revealing that a relatively high content of the hydroxy and aliphatic acid extractives could interact with preservative ingredients during the treatment process. It was observed that the signals at 1740 cm^{-1} for hemicellulose almost disappeared, suggesting that deacetylation of hemicelluloses occurred during the impregnation stage. The absorption bands at 1590 cm^{-1}, 1505 cm^{-1} attributed to aromatic skeletal vibration breathing with C=O stretching in the lignin fraction signals, were observed to be weakened. Moreover, the peaks at 1460 cm^{-1} and 1370 cm^{-1} corresponding to cellulose, hemicellulose and lignin, are also significantly diminished or decreased sharply, implying that effective interaction occurred between cellulose, hemicellulose and lignin with three groups of preservative.

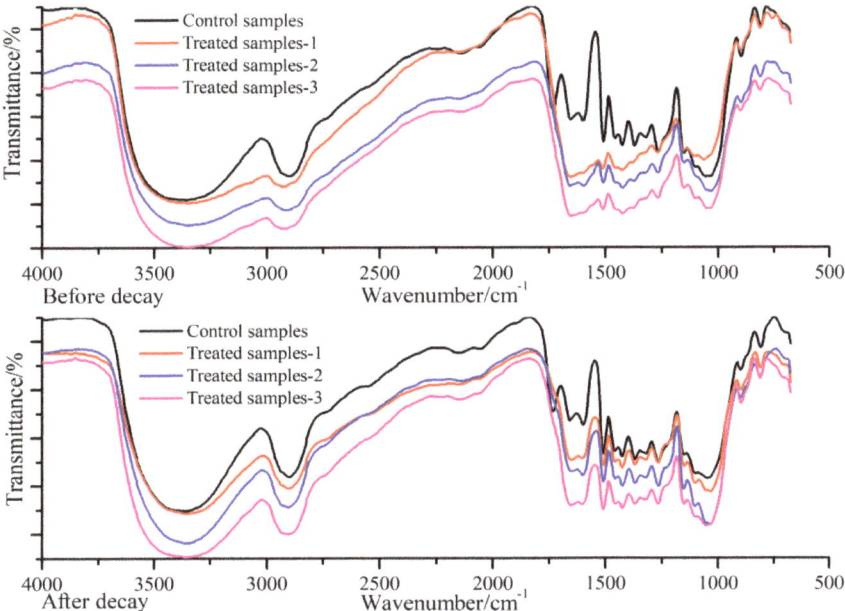

Figure 1. Fourier transform infrared (FTIR) spectrum of wood samples before/after decay.

After a decay resistance test, the absorption peaks at 3350, 2900, 1740, 1160, 1040 cm^{-1} are distinctly weakened in the spectra of control sample, indicating that cellulose and hemicelluloses were partly destroyed during the decay process. By contrast, the main components in the treated samples remained relatively steady after the decay test, suggesting that treated wood samples have been effectively protected. In short, the data presented herein revealed that the protein-based preservative systems were effective formulations and constituted appropriate protections for treated wood blocks.

3.3. Morphology Analysis

The morphology of control and treated samples was investigated by SEM, and the distributions and contents of Cu and B within the wood cell wall were also analyzed by SEM-EDS, as shown in Figure 2. To observe the cross-section morphology by SEM, a small piece was randomly cut from the inside of the wood blocks, after removing at least 5 mm from the edge, completed with the preservative modification treatment and wood drying. The cross-section and radial-section morphologies of the control and treated wood samples are shown in Figure 3. From the magnified images, deposition of Cu-B-protein was not observed inside tracheid in these samples.

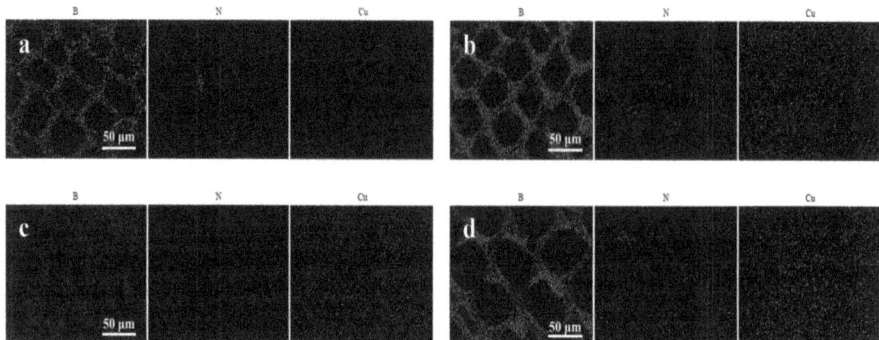

Figure 2. SEM images showing the distribution of B, N and Cu within the wood cell wall. (**a**), control sample, (**b**), sample treated with P_1, (**c**), sample treated with P_2, (**d**), sample treated with P_3.

Cross-section

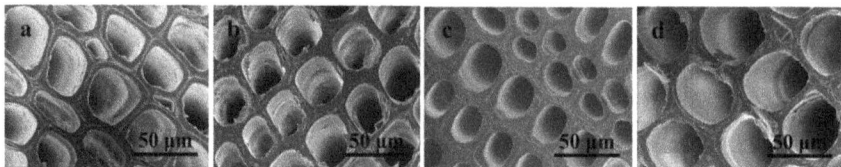

Radial-section

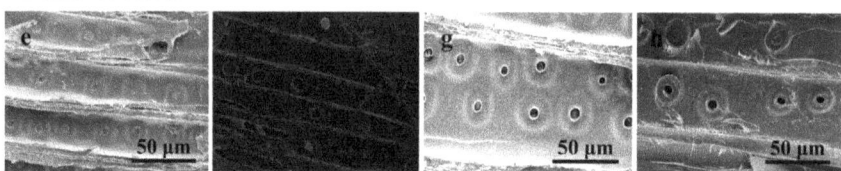

Figure 3. SEM photographs of wood samples. Cross-section: (**a**), control sample, (**b**), sample treated with P_1, (**c**), sample treated with P_2, (**d**), sample treated with P_3; Radial-section: (**e**), control sample, (**f**), sample treated with P_1, (**g**), sample treated with P_2, (**h**), sample treated with P_3.

SEM pictures of control and treated samples with protein-based wood preservatives demonstrated that Cu and B elements have successfully impregnated and penetrated uniformly into the treated wood cell walls. SEM-EDS analysis showed the distributions and contents of active preservative elements within the cell walls before and after decay tests, and the results are shown in Table 2. It was found that elements (Cu and B) distributions in the samples treated by P_2 and P_3 preservatives were higher than those in P_1 treated wood samples. In contrast, lower N contents were observed in P_2 and P_3 treated wood samples. These results showed that the P_2, P_3 preservatives with nano-carriers can chelate more active components (Cu, B) and less protein in preservative formulations, facilitating the permeation of preservatives into the treated wood cell walls. In particular, for the samples treated by P_2 preservative with nano-hydroxyapatite, the protein content in cell walls was the least but Cu and B contents were at high levels. Since protein is also a kind of nutritious matter, preferable wood preservative should chelate less protein and more active components for protecting wood materials from decay. Consequently, it can be concluded that nano-carriers in P_2 preservatives could promote more active ingredients permeating into treated wood cells.

Table 2. Elements distributions with cell wall by SEM-Energy Dispersive Spectrometer (EDS).

	Control Samples	Treated Samples-1	Treated Samples-2	Treated Samples-3
B	1.76	2.23	3.10	3.16
C	43.27	40.00	41.76	39.98
N	0.00	7.37	3.87	6.67
O	52.15	42.28	42.58	40.32
Ca	0.96	1.06	1.03	1.06
Cu	1.86	7.06	7.66	8.81
Total	100.0	100.0	100.0	100.0

After 12 weeks decay test, the entire cell wall can scarcely be found in the control samples (Figure 4), suggesting that serious degradation of cell walls had occurred after the decay process. In contrast, the cell walls were relatively unchanged in the three treated wood samples, further suggesting the effectiveness of protein-based preservatives. The cell wall treated by preservative P_2 is more intact than samples modified by preservative P_1 and P_3. This might be attributed to the high contents of Cu and B but with less protein content of P_2. In general, the morphology results reflected by SEM-EDS and SEM pictures confirmed that the active ingredients in preservatives can effectively penetrate and fixate within the wood blocks, especially in the P_2 preservative.

After 12 weeks decay stage
Cross-section

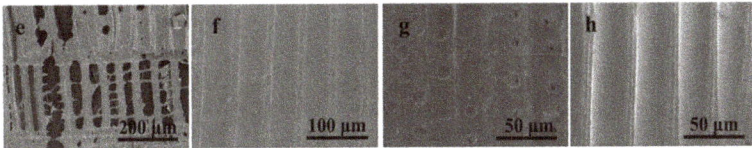

Radial-section

Figure 4. SEM photographs of wood samples after 12 weeks decay stage. Cross-section: (**a**), control sample, (**b**), sample treated with P_1, (**c**), sample treated with P_2, (**d**), sample treated with P_3; Radial-section: (**e**), control sample, (**f**), sample treated with P_1, (**g**), sample treated with P_2, (**h**), sample treated with P_3.

3.4. XRD Analysis

In this study, the crystallinity index (C_rI) of control and treated samples was measured by XRD-6000 instrument (Shimadzu, Japan). The X-ray diffraction patterns of wood samples are presented in Figure 5. The C_rI values were calculated and shown in Table 3. As can be seen from Figure 5, the XRD pattern of wood samples all showed typical cellulose I structure, indicating that the crystal structures of the treated samples were not changed by preservatives during the impregnation processes. The unchanged crystal structures are beneficial for utilization of the modified wood, and contributed to some properties, such as strength of treated wood blocks, remaining unchanged.

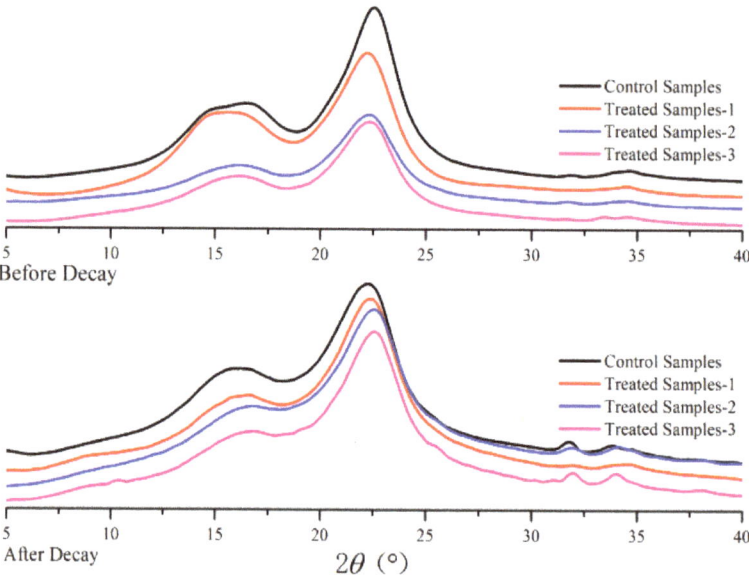

Figure 5. X-ray diffraction (XRD) patterns of wood samples before/after decay.

Table 3. Crystallinities of wood samples before/after decay.

	Degree of Crystallinity	
	Before Decay	After Decay (12 Weeks)
Control samples	60.10	40.33
Treated samples-1	50.69	44.66
Treated samples-2	51.27	46.92
Treated samples-3	50.66	45.49

As can be seen from the Table 3, the C_rI of preservative-treated wood samples all reduced (from 50.66 to 51.27%) distinctly as compared to that (60.1%) of control sample, which might be attributed to the addition of amorphous protein in the preservative. The CrI (50.69% and 50.66%) values of treated wood samples with the preservatives P_1, P_3 were less than that treated with P_2 (51.27%). This might attribute to the function of nano-carrier in P_2 preservative, which promotes more active ingredients chelating less protein and facilitates preservatives into the wood cells.

After 12 weeks of decay tests, the C_rI of three treated wood samples all substantially decreased (44.66%–46.92%), whereas the C_rI of control samples decreased to 40.33%. Fungi can directly attack and degrade the main chemical components of the control wood blocks during the decay process, such as cellulose and hemicelluloses. The CrI of P_2 (46.92%) treated samples was slightly higher than those of P_1, P_3 treated samples (44.66% and 45.49%) after decay, which was attributable to the function of nano-carrier in preservatives. P_2 preservative contained a high content of active ingredients and low protein content and it showed more effective protection for wood materials, demonstrating that P2 is the most effective preservative.

3.5. Thermogravimetric Analysis (TGA)

TGA was used to evaluate the thermal properties of control and modified wood samples. As shown in Figure 6 (before the decay process), the initial pre-carbonization temperature of preservative-treated samples was lower than that of the control samples. This might attribute to the facilitation for char forming derive from Cu and B elements. The final carbon residue of the treated samples (P_1, P_2 and

P_3) was considerably higher than control sample since preservative ingredients can promote the carbonization and retard the thermal decomposition of the wood components.

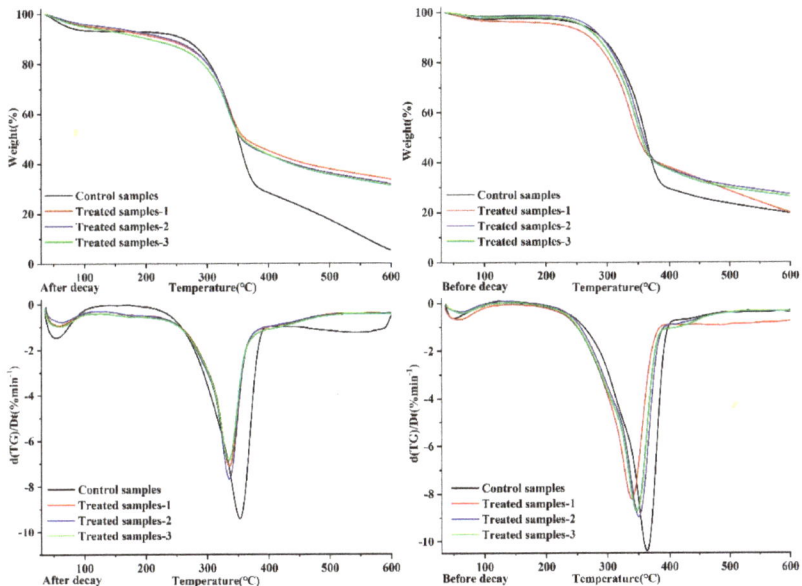

Figure 6. Thermogravimetric (TG)/Differential Thermogravimetric (DTG) spectrum of wood samples before/after decay.

As can be seen from Figure 6 (after the decay process), the content of residual char in the control sample increased as compared to that before decay test. The increased content of "char residues" in the control sample is likely attributed to the high content of lignin in the decayed sample, which is due to the serious degradation of cellulose and hemicelluloses during the decay stage. In contrast, the content of residual char in the treated samples (P_1, P_2 and P_3) after the decay test was close to that before the decay test, demonstrating that preservatives effectively inhibit wood decay, and the loss of treated wood is not obvious after decay treatment. Furthermore, the content of "char residues" in P_2 and P_3 treated wood samples is higher than that of P_1 without nano-carrier. This indicated that wood products pretreated by P_2 and P_3 preservatives have higher thermostability, which can extend the application range of wood products in different conditions.

3.6. Raman Analysis

Raman analysis was performed to reveal the distribution and microscopic changes of the main structural compositions at subcellular level. The morphological and compositional information of control and the treated samples were simultaneously recorded by the CRM, and the intensity of the bands may be used for the calculation of the relative content in the samples [36–38]. Obvious differences between three treated groups and the control sample can be observed in Figure 7 before the decay test. The changes in the contents of carbohydrates and lignin in the wood cell wall implied that the three kinds of preservative (P_1, P_2 and P_3) all can penetrate into the treated wood cell walls.

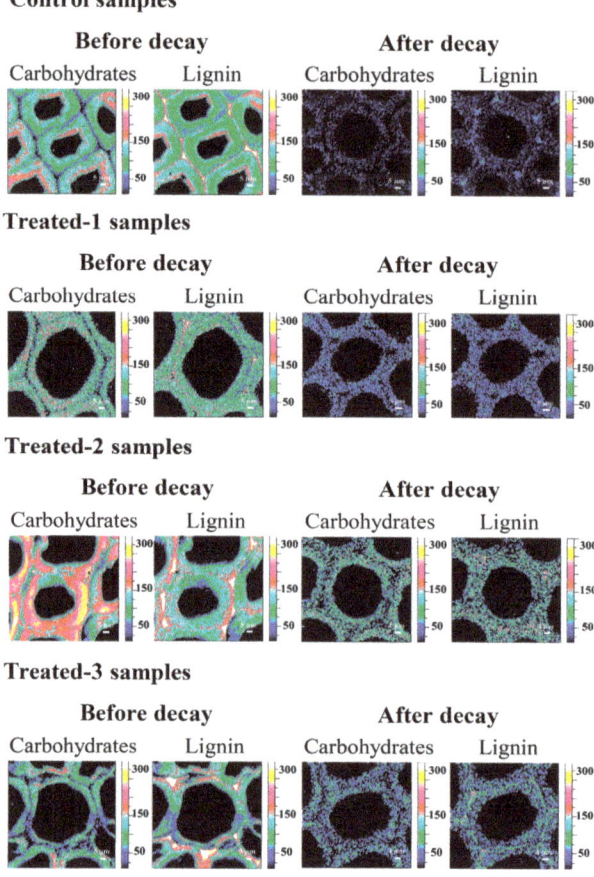

Figure 7. Raman images showing the distribution of carbohydrates and lignin.

As observed in Raman spectra, the control wood cell wall was intact before the decay process, and the concentration of carbohydrates was high in the S_2 layer. After the decay stage, it was found that the distribution of carbohydrates and lignin decreased significantly. However, the cell walls of three treated samples (P_1, P_2, and P_3) were ultimately well preserved after the decay test, although the decrease of carbohydrates and lignin distribution were observed, implying that the preservative can protect the wood from degradation.

It can be seen from Figure 7, after the decay stage, that the concentrations of carbohydrates and lignin in the P_2 treated cell wall are higher than those in P_1 and P_3, suggesting that the P_2 treated cell wall remained relatively intact. This might be attributed to the nano-carrier function in the P_2 preservative, which could promote high contents of effective ingredients chelating within the preservative formulations.

Verification was further performed by Raman spectroscopy, as shown in Figure S1 in Supplementary Materials. In Raman spectra, the average signal intensity in the spectral ranges of 1550–1650 cm^{-1} and 2880–2920 cm^{-1} are respectively applied to assess the carbohydrates and lignin distributions. From Figure S1, an obvious trend can be found, which can be attributed to the distinct decreases of peak intensity generated by the huge reduction in carbohydrates and lignin for the control sample after the decay process. Compared to the control sample, the intensities of the carbohydrates and lignin signals in the treated samples were also reduced, indicating the relative degradation of cell walls treated by three formulation preservatives (P_1, P_2 and P_3). The reduction in peak intensities of carbohydrates in P_2 was the least in the three treated samples, further indicating the better protection effectiveness of P_2 formulation.

In short, Raman analysis demonstrated that the ingredients of preservatives can effectively impregnate into the wood cell wall and adequately protect the treated wood blocks. Furthermore, P_2 preservative formulation is the most optimal one in the three preservative formulations, which is consistent with the aforementioned SEM, XRD analysis.

3.7. Mass Loss Analysis after Decay Test

As can be seen from Table 4 and Figure 8, the mass loss rates of the control sample were significantly much higher than those of the treated samples, indicating that the treated samples (P_1, P_2 and P_3) exhibited strong resistance against decay fungi. It was illustrated that these three formulation preservatives can effectively protect wood blocks. In the treated wood samples, the mass loss rate (9.1%) of sample treated with P_2 formulation was the lowest, suggesting that P_2 exhibited the optimal preservative effect, which might be attributed to the nano-carrier function of nano-hydroxyapatite (HA). It can chelate more content of active ingredients and fix the preservative ingredient (Cu and B) into the cell walls of wood, which is in accordance with the results. In addition, weight percent gain (WPG) of P_1, P_2 and P_3 was 20.5%, 20.3% and 22.0%, respectively. The least mass loss rate and weight percent gain indicated that the P_2 can better protect the wood products from degradation.

Table 4. Mass loss of control and treated samples.

Formulations	WPG/%	Mass Loss/%
P_1, Cu-B-Pr	20.52	12.08
P_2, Cu-B-Pr-HA	20.30	9.10
P_3, Cu-B-Pr-Go	22.02	11.02

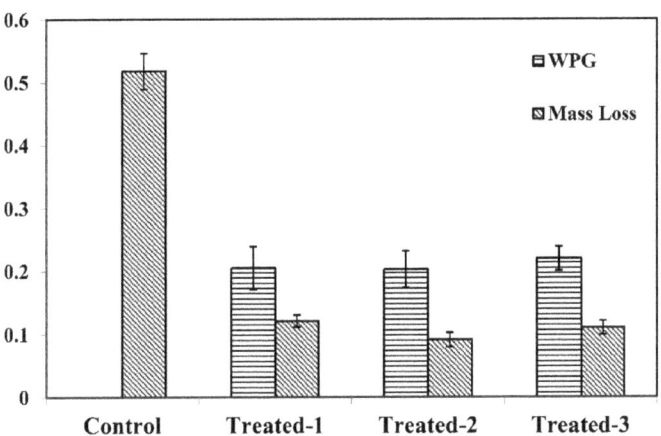

Figure 8. Mass loss and weight percent gain (WPG) control and treated samples.

4. Conclusions

In this study, the results indicated that protein-based preservatives could serve as effective, environmentally friendly and cost-competitive alternatives for traditional wood preservatives. In this formulation system, copper and boron salts are preferably fixed together and exhibit a durable performance on account of the feather protein being introduced as a chelation agent to form insoluble complexes by chelation, instigate chemical reactions with wood components and form a long-term protection mechanism in woodblocks. This enables the feather protein-based preservative to be fairly appropriate for wood construction. Treatability and morphology of the control and treated samples further verified the excellent permeability and feasibility of protein-based preservative formulations. SEM-EDS and Raman analysis of the control and treated samples after decay experiments illustrated the good performance of nano-carriers for the Cu, B penetration and fixation of the protein-based preservative. In particular, the nano-hydroxyapatite preservative formulation could increase the content of Cu and B in preservative at low protein levels. In the future, the protein-based preservatives with nano-carrier (nano-hydroxyapatite) should be further evaluated by field trials to identify their long-term ground-contact applicability.

Supplementary Materials: The following are available online at http://www.mdpi.com/2073-4360/12/1/237/s1. 1. Target retention. 2. Measured retention. 3. Weight percent gain (WPG). Figure S1. Raman spectra of wood samples before and after decay. Table S1. The cost comparison between feather protein-based preservatives with CCA, ACQ.

Author Contributions: Y.X. and J.W. conceived and designed the experiments; Y.X. and C.M. performed the experiments; R.S. and J.W. guided the experiments. Y.X., H.W. and J.W. analyzed the data; R.S., J.W., and S.S., contributed reagents/materials/analysis tools; Y.X. wrote the draft; J.W. revised and proposed amendments to the draft and finalize the manuscript. All these authors had contributed substantially to this work reported. All authors have read and agreed to the published version of the manuscript.

Funding: This research was funded by China Postdoctoral Science Foundation: 2015M581001; National Natural Science Foundation of China: 31860186; National Natural Science Foundation of China: 31360157; National Natural Science Foundation of China: 31872698.

Acknowledgments: The authors are grateful for the financial support by the China Postdoctoral Science Fundation [2015M581001], and the National Natural Science Foundation of China [31860186, 31360157 and 31872698]. We thank Jinzhen Cao's group of Beijing Forestry University for providing *Gloeophyllum trabeum* for allowing us to use as the test fungi in decay trials.

Conflicts of Interest: The authors declare no conflict of interest.

References

1. Zhou, A.; Tam, L.-H.; Yu, Z.; Lau, D. Effect of moisture on the mechanical properties of CFRP-wood composite: An experimental and atomistic investigation. *Compos. Part B Eng.* **2015**, *71*, 63–73. [CrossRef]
2. Zhu, Y.; Zhuang, L.; Goodell, B.; Cao, J.; Mahaney, J. Iron sequestration in brown-rot fungi by oxalate and the production of reactive oxygen species (ROS). *Int. Biodeterior. Biodegrad.* **2016**, *109*, 185–190. [CrossRef]
3. Can, A.; Sivrikaya, H.; Hazer, B. Fungal inhibition and chemical characterization of wood treated with novel polystyrene-soybean oil copolymer containing silver nanoparticles. *Int. Biodeterior. Biodegrad.* **2018**, *133*, 210–215. [CrossRef]
4. Yang, I.; Kuo, M.; Myers, D.J. Soy protein combined with copper and boron compounds for providing effective wood preservation. *J. Am. Oil Chem. Soc.* **2006**, *83*, 239. [CrossRef]
5. Ahn, S.H.; Oh, S.C.; Choi, I.-G.; Kim, H.-Y.; Yang, I. Efficacy of wood preservatives formulated from okara with copper and/or boron salts. *J. Wood Sci.* **2008**, *54*, 495–501. [CrossRef]
6. Yamaguchi, H. Low molecular weight silicic acid-inorganic compound complex as wood preservative. *Wood Sci. Technol.* **2002**, *36*, 399–417. [CrossRef]
7. Humar, M.; Lesar, B. Influence of dipping time on uptake of preservative solution, adsorption, penetration and fixation of copper-ethanolamine based wood preservatives. *Eur. J. Wood Wood Prod.* **2009**, *67*, 265–270. [CrossRef]
8. Kim, H.Y.; Jeong, H.S.; Min, B.C.; Ahn, S.H.; Oh, S.C.; Yoon, Y.H.; Choi, I.G.; Yang, I. Antifungal efficacy of environmentally friendly wood preservatives formulated with enzymatic-hydrolyzed okara, copper, or boron salts. *Environ. Toxicol. Chem.* **2011**, *30*, 1297–1305. [CrossRef]
9. Murguía, M.C.; Machuca, L.M.; Fernandez, M.E. Cationic gemini compounds with antifungal activity and wood preservation potentiality. *J. Ind. Eng. Chem.* **2019**, *72*, 170–177. [CrossRef]
10. Thevenon, M.-F.; Pizzi, A.; Haluk, J.-P. Protein borates as non-toxic, long-term, wide-spectrum, ground-contact wood preservatives. *Holzforsch. Int. J. Biol. Chem. Phys. Technol. Wood* **1998**, *52*, 241–248. [CrossRef]
11. Brocco, V.F.; Paes, J.B.; da Costa, L.G.; Brazolin, S.; Arantes, M.D.C. Potential of teak heartwood extracts as a natural wood preservative. *J. Clean. Prod.* **2017**, *142*, 2093–2099. [CrossRef]

12. Lesar, B.; Kralj, P.; Humar, M. Montan wax improves performance of boron-based wood preservatives. *Int. Biodeterior. Biodegrad.* **2009**, *63*, 306–310. [CrossRef]
13. Ramos, A.; Jorge, F.C.; Botelho, C. Boron fixation in wood: Studies of fixation mechanisms using model compounds and maritime pine. *Holz Roh Werkst.* **2006**, *64*, 445. [CrossRef]
14. Thevenon, M.-F.; Pizzi, A.; Haluk, J.-P. Non-toxic albumin and soja protein borates as ground-contact wood preservatives. *Holz Roh Werkst.* **1997**, *55*, 293–296. [CrossRef]
15. Lebow, S.; Arango, R.; Woodward, B.; Lebow, P.; Ohno, K. Efficacy of alternatives to zinc naphthenate for dip treatment of wood packaging materials. *Int. Biodeterior. Biodegrad.* **2015**, *104*, 371–376. [CrossRef]
16. Toussaint-Dauvergne, E.; Soulounganga, P.; Gerardin, P.; Loubinoux, B. Glycerol/glyoxal: A new boron fixation system for wood preservation and dimensional stabilization. *Holzforschung* **2000**, *54*, 123–126. [CrossRef]
17. Mourant, D.; Yang, D.-Q.; Lu, X.; Riedl, B.; Roy, C. Copper and boron fixation in wood by pyrolytic resins. *Bioresour. Technol.* **2009**, *100*, 1442–1449. [CrossRef]
18. Ahn, S.H.; Oh, S.C.; Choi, I.-G.; Han, G.-S.; Jeong, H.-S.; Kim, K.-W.; Yoon, Y.-H.; Yang, I. Environmentally friendly wood preservatives formulated with enzymatic-hydrolyzed okara, copper and/or boron salts. *J. Hazard. Mater.* **2010**, *178*, 604–611. [CrossRef]
19. Thevenon, M.-F.; Pizzi, A.; Haluk, J.-P. One-step tannin fixation of non-toxic protein borates wood preservatives. *Holz Roh Werkst.* **1998**, *56*, 90. [CrossRef]
20. Thévenon, M.-F.; Pizzi, A.; Haluk, J.; Zaremski, A. Normalised biological tests of protein borates wood preservatives. *Eur. J. Wood Wood Prod.* **1998**, *56*, 162. [CrossRef]
21. Thévenon, M.-F.; Pizzi, A. Polyborate ions' influence on the durability of wood treated with non-toxic protein borate preservatives. *Holz Roh Werkst.* **2003**, *61*, 457–464. [CrossRef]
22. Ratajczak, I.; Mazela, B. The boron fixation to the cellulose, lignin and wood matrix through its reaction with protein. *Holz Roh Werkst.* **2007**, *65*, 231. [CrossRef]
23. Mazela, B.; Domagalski, P.; Mamonova, M.; Ratajczak, I. Protein impact on the capability of the protein-borate preservative penetration and distribution into pine and aspen wood. *Holz Roh Werkst.* **2007**, *65*, 137. [CrossRef]
24. Lykidis, C.; Mantanis, G.; Adamopoulos, S.; Kalafata, K.; Arabatzis, I. Effects of nano-sized zinc oxide and zinc borate impregnation on brown rot resistance of black pine (*Pinus nigra* L.) wood. *Wood Mater. Sci. Eng.* **2013**, *8*, 242–244. [CrossRef]
25. Moradi, F.G.; Hejazi, M.J.; Hamishehkar, H.; Enayati, A.A. Co-encapsulation of imidacloprid and lambda-cyhalothrin using biocompatible nanocarriers: Characterization and application. *Ecotoxicol. Environ. Saf.* **2019**, *175*, 155–163. [CrossRef]
26. Yusoff, S.; Kamari, A.; Aljafree, N. A review of materials used as carrier agents in pesticide formulations. *Int. J. Environ. Sci. Technol.* **2016**, *13*, 2977–2994. [CrossRef]
27. Munir, M.U.; Ihsan, A.; Sarwar, Y.; Bajwa, S.Z.; Bano, K.; Tehseen, B.; Zeb, N.; Hussain, I.; Ansari, M.T.; Saeed, M. Hollow mesoporous hydroxyapatite nanostructures; smart nanocarriers with high drug loading and controlled releasing features. *Int. J. Pharm.* **2018**, *544*, 112–120. [CrossRef]
28. Gholibegloo, E.; Karbasi, A.; Pourhajibagher, M.; Chiniforush, N.; Ramazani, A.; Akbari, T.; Bahador, A.; Khoobi, M. Carnosine-graphene oxide conjugates decorated with hydroxyapatite as promising nanocarrier for ICG loading with enhanced antibacterial effects in photodynamic therapy against Streptococcus mutans. *J. Photochem. Photobiol. B Biol.* **2018**, *181*, 14–22. [CrossRef]
29. Edmunds, C.W. Physico-Chemical Properties and Biodegradability of Genetically Modified *Populus trichocarpa* and *Pinus taeda*. Ph.D. Thesis, North Carolina State University, Raleigh, NC, USA, 2015.
30. Sun, S.-L.; Wen, J.-L.; Ma, M.-G.; Song, X.-L.; Sun, R.-C. Integrated biorefinery based on hydrothermal and alkaline treatments: Investigation of sorghum hemicelluloses. *Carbohydr. Polym.* **2014**, *111*, 663–669. [CrossRef]
31. You, T.-T.; Zhang, L.-M.; Zhou, S.-K.; Xu, F. Structural elucidation of lignin-carbohydrate complex (LCC) preparations and lignin from Arundo donax Linn. *Ind. Crop. Prod.* **2015**, *71*, 65–74. [CrossRef]
32. Chen, J.-H.; Guan, Y.; Wang, K.; Xu, F.; Sun, R.-C. Regulating effect of hemicelluloses on the preparation and properties of composite Lyocell fibers. *Cellulose* **2015**, *22*, 1505–1516. [CrossRef]
33. Chen, J.H.; Guan, Y.; Wang, K.; Xu, F.; Sun, R.C. Regenerated cellulose fibers prepared from wheat straw with different solvents. *Macromol. Mater. Eng.* **2015**, *300*, 793–801. [CrossRef]

34. Yuan, T.-Q.; Zhang, L.-M.; Xu, F.; Sun, R.-C. Enhanced photostability and thermal stability of wood by benzoylation in an ionic liquid system. *Ind. Crop. Prod.* **2013**, *45*, 36–43. [CrossRef]
35. Nguyen, T.T.H.; Li, S.; Li, J.; Liang, T. Micro-distribution and fixation of a rosin-based micronized-copper preservative in poplar wood. *Int. Biodeterior. Biodegrad.* **2013**, *83*, 63–70. [CrossRef]
36. Li, H.-Y.; Sun, S.-N.; Wang, C.-Z.; Sun, R.-C. Structural and dynamic changes of lignin in Eucalyptus cell walls during successive alkaline ethanol treatments. *Ind. Crop. Prod.* **2015**, *74*, 200–208. [CrossRef]
37. Zhang, X.; Ma, J.; Ji, Z.; Yang, G.H.; Zhou, X.; Xu, F. Using confocal Raman microscopy to real-time monitor poplar cell wall swelling and dissolution during ionic liquid pretreatment. *Microsc. Res. Tech.* **2014**, *77*, 609–618. [CrossRef]
38. Zhou, X.; Ma, J.; Ji, Z.; Zhang, X.; Ramaswamy, S.; Xu, F.; Sun, R.C. Dilute acid pretreatment differentially affects the compositional and architectural features of *Pinus bungeana* Zucc. compression and opposite wood tracheid walls. *Ind. Crop. Prod.* **2014**, *62*, 196–203. [CrossRef]

© 2020 by the authors. Licensee MDPI, Basel, Switzerland. This article is an open access article distributed under the terms and conditions of the Creative Commons Attribution (CC BY) license (http://creativecommons.org/licenses/by/4.0/).

Article

A Method to Improve the Characteristics of EPDM Rubber Based Eco-Composites with Electron Beam

Gabriela Craciun [1], Elena Manaila [1,*], Daniel Ighigeanu [1] and Maria Daniela Stelescu [2]

[1] National Institute for Laser, Plasma and Radiation Physics, # 409 Atomistilor St., 077125 Magurele, Romania; gabriela.craciun@inflpr.ro (G.C.); daniel.ighigeanu@inflpr.ro (D.I.)
[2] National R&D Institute for Textile and Leather—Leather and Footwear Research Institute, # 93 Ion Minulescu St., 031215 Bucharest, Romania; dmstelescu@yahoo.com
* Correspondence: elena.manaila@inflpr.ro

Received: 20 December 2019; Accepted: 8 January 2020; Published: 15 January 2020

Abstract: A natural fiber reinforced composite, belonging to the class of eco composites, based on ethylene-propylene-terpolymer rubber (EPDM) and wood wastes were obtained by electron beam irradiation at 75, 150, 300, and 600 kGy in atmospheric conditions and at room temperature using a linear accelerator of 5.5 MeV. The sawdust (S), in amounts of 5 and 15 phr, respectively, was used to act as a natural filler for the improvement of physical and chemical characteristics. The cross-linking effects were evaluated through sol-gel analysis, mechanical tests, and Fourier Transform Infrared FTIR spectroscopy comparatively with the classic method with dibenzoyl peroxide (P) applied on the same types of samples at high temperature. Gel fraction exhibits values over 98% but, in the case of P cross-linking, is necessary to add more sawdust (15 phr) to obtain the same results as in the case of electron beam (EB) cross-linking (5 phr/300 kGy). Even if the EB cross-linking and sawdust addition have a reinforcement effect on EPDM rubber, the medium irradiation dose of 300 kGy looks to be a limit to which or from which the properties of the composite are improved or deteriorated. The absorption behavior of the eco-composites was studied through water uptake tests.

Keywords: EPDM rubber; wood sawdust; electron beam irradiation; dibenzoyl peroxide; cross-liking; physico-chemical characteristics

1. Introduction

Ethylene-propylene-diene monomer (EPDM) rubber is a versatile polymer containing low compound costs. The stability of its saturated backbone structure determines the manifestation of a good resistance at heat and oxidation and also at ozone or weather ageing [1]. Its very good physical and chemical properties make it extremely suitable for obtaining automotive parts, sports goods, packaging materials, etc. [2]. For vulcanizing rubber compounds, many curing systems have been developed: sulfur, peroxides, metal oxides, phenolic resins, quinines. Of these, the first two were the most used for cross-linking of rubber materials until recently [3]. Even if the sulfur vulcanization has been known and applied for over 150 years, the complex chemistry of sulfur vulcanization is still not clearly understood. Both free radical and ionic mechanisms are considered as chemical pathways [3–5]. Both unsaturated and saturated elastomers can be cross-linked by means of organic peroxides but, for the second type, the sulfur curing systems cannot be applied [3]. Currently, the rubber processing using high energy radiations is a method increasingly used for designing new materials based on the modification of polymers [1,6,7]. By the use of gamma rays or electron beams C–C bonds, as in the case of peroxide cross-linking, are formed [1]. Particularly, the electron beam irradiation has many advantages over the mentioned curing systems such as high degrees of cross-linking and extremely strong bonds, which are obtained directly by C–C linkage. The process occurs at room temperature so the degradation generated by temperature is avoided. The curing cycles are shorter than in classical treatments and the productivity is higher, which is very suitable for both thin and thick products (depending on the electrons' energy and the density of the product to be irradiated) and, lastly, a very important aspect of the process is that it does not generate material wastes [8–11]. Due to the radiation ability to initiate chemical reactions at any temperature, under any pressure and in any phase (gas, liquid, or solid) without the use of catalysts, very reactive intermediates are formed [12]. These intermediates can follow several reaction paths, which result in rearrangements and/or the formation of new bonds. Therefore, radiation offers a productive way of forming polymer bridges to bond together very different polymeric and non-polymeric elements of an engineering structure [12]. The fully saturated bonds in the main chain with a lack of quaternary carbon atoms make EPDM rubber suitable for radiation cross-linking that can induce additional cross-linking and/or scission of the polymeric chain [1]. The obtained product characteristics depend on the content and nature of fillers or additives added but also on dose and dose rates applied [1,8,13,14]. The oxidation degradation phenomena that can occur during processing must be taken into account and a method to retard or even to suppress the oxidative degradation is compounded with antioxidants, like Irganox, Tinuvin, or N-phenyl-N'-isopropyl-p-phenylenediamine IPPD [15]. Antioxidants are autoxidation inhibitors, which interfere in the free radical reactions that take place during processing and/or irradiation process. This leads to the incorporation of oxygen into the rubber molecules forming hydroperoxides that feed the chain reaction with new radicals [15–17]. The resistance to thermo-oxidative degradation of irradiated EPDM rubber is greatly improved by the addition of antioxidants [15,18–20]. However, the radiation curing differs from thermal curing, which is carried out at ambient temperature under closely controlled conditions, such as radiation dose, dose rate, penetration depth. This form of curing ultimately results in a more well-defined end product [21].

The EPDM use in so many different applications is due to the capacity to accept large amounts of fillers as silica or carbon black that can significantly improve its properties [1,7,22]. However, the concern regarding the demonstrated adverse effects on occupational health (silicosis, tuberculosis, cancer, autoimmune diseases, etc. [23–25] due to the use of the mentioned reinforcing fillers made as natural fibers are now under attention to replace them [7,26–28]. In addition, the growing global environmental and social concern, and new environmental regulations have forced the search for new composites and green materials, compatible with the environment. For these reasons but also from others related to energy saving, favorable processing properties, dimensional stability, and not least biodegradability potential, the natural fiber reinforced composites were called eco-composites [29] and the wood sawdust was taken under study as a possible active filler substitute [30–32] initially in classical methods of cure consisting of repeated heating cycles in hot presses [8]. The filler amount, particles dimension, and process characteristics as an irradiation dose and dose rate are important for the obtained product properties especially due to the poor interfacial adhesion between the polymeric matrix and hydrophilic lingo-cellulosic fillers observed in classical treatments [8,33,34].

The goal of the paper is to comparatively present the cross-linking effects induced by two different methods of cure, electron beam irradiation, and dibenzoyl peroxide in order to obtain a polymeric eco-composite based on EPDM rubber and wood sawdust. The influence of filler loading and irradiation dose on cross-linking was studied through physical and chemical investigations as sol-gel analysis, mechanical tests, and Fourier Transform Infrared (FTIR) spectroscopy. The absorption behavior of the eco-composites was studied through water uptake tests.

2. Materials and Methods

2.1. Materials and Sample Preparation

The raw materials that were used in the experiments were as follows: (I) Ethylene-propylene-diene terpolymers (EPDM) rubber as eco-composite matrix was of Nordel 4760 type, produced by Dow Chemical Company (Michigan, MI, USA) (mooney viscosity of 70 ML_{1+4} at 120 °C, ethylene content of 70%, 5-ethylidenenorbornene (ENB) content of 4.9 wt %, density of 0.88 g/cm^3 and crystallinity degree of 10%), (II) Polyethylene glycol (PEG) as process aid was of PEG 4000 type supplied by Advance Petrochemicals Ltd. (Ahmedabad, India) (density of 1128 g/cm^3 and melting point in the range of 4–8 °C), (III) Pentaerythritol tetrakis(3-(3,5-di-tert-butyl-4-hydroxyphenyl)propionate) as an antioxidant was of Irganox 1010 type bought from BASF Schweiz (Basel, Switzerland), (IV) Dibenzoyl peroxide as a cross-linking agent was of Perkadox 14-40B type from AkzoNobel Chemicals (Deventer, The Netherlands) (density 1.60 g/cm^3, 3.8% active oxygen content, 40% peroxide content, pH 7), (V) sawdust was of fir wood type obtained from a local sawmill in Romania (Sebes, Romania) (size particles—mash 250–270, single type of wood).

Blends were prepared on an electrically-heated laboratory roller. For preparation of the polymeric composites, the blend constituents were added in the following sequences and amounts: 100 parts of EPDM were rolled until binding for 1–2 min, than 3 phr of PEG 4000, and 1 phr Irganox 1010 were added and embedded for another 3–4 min and finally 5 and 15 phr of wood sawdust were added and mixed for 2–4 min until the homogenization. Blends were removed from the roll in the form of the sheet that is about 2 mm thick. Test specimens were obtained by compression molding at 160 °C and a pressure of 150 MPa using an electrical press for 5 min. Plates were then cooled to room temperature under pressure. Process variables were as follows: temperature between 25–50 ± 5 °C, friction 1:1.1, and total blending time 8–14 min. Plates required for physical and mechanical tests with sizes of 150 × 150 × 2 mm^3 were obtained by pressing in a hydraulic press at 110 ± 5 °C and 150 MPa [8].

Samples vulcanized with dibenzoyl peroxide were prepared in the same way as those for the electron beam, while adding 8 phr of the vulcanizing agent dibenzoyl peroxide Perkadox 14-4B in a hydraulic press at 160 °C for 20 min.

2.2. Experimental Installation and Sample Irradiation

Samples obtained as above and packed in polyethylene film for minimizing the oxidation were irradiated at 75, 150, 300, and 600 kGy in atmospheric conditions and at room temperature of 25 °C using the ALID-7 electron beam accelerator from National Institute for Laser, Plasma and Radiation Physics, Magurele, Romania. The nominal values of the electron beam (EB) parameters were as follows: energy of 5.5 MeV, peak current of 26 mA, output power of 134 W, and 3.75 μs pulse repetition frequency of 50 Hz [8].

The irradiation process performance depends on the rigorous control of the irradiation dose and dose rate [35,36]. In our experiments, the process dose rate was of 3.5 kGy/min. The primary standard graphite calorimeter was used for radiation dosimetry. In order to assure the equality between the entry and the exit irradiation dose of the irradiated samples, but also for an efficient use of the electron beam, the penetration depth was calculated according with the following equation [8,36].

$$E = 2.6 \cdot t \cdot \rho + 0.3 \qquad (1)$$

where E (MeV) is the electron beam energy, t (cm) is the sample thickness, and ρ (g·cm^{-3}) is the sample density (in our case, 1 g·cm^{-3}).

The proper thickness of samples subjected to EB irradiation was calculated as being of 20 mm [8,37].

2.3. Laboratory Tests

Laboratory tests were carried out on EPDM samples with and without sawdust cross-linked by EB irradiation and by dibenzoyl peroxide. The sample codes are as follows: (1) EPDM-EB for samples without sawdust cross-linked by electron beam irradiation, (2) EPDM-EB-S 5 and EPDM-EB-S 15 for samples containing 5 and 15 phr of sawdust cross-linked by EB irradiation, (3) EPDM-P for samples without sawdust cross-linked with dibenzoyl peroxide, (4) EPDM-P-S 5 and EPDM-P-S 15 for samples containing 5 and 15 phr of sawdust cross-linked using dibenzoyl peroxide.

2.3.1. Mechanical Characteristics

The mechanical properties of samples were evaluated using specific and proper equipment and instruments in accordance with the international standards in force as follows: a Schopper tensile tester according to ISO 37/2017 for tensile strength, a hardness tester according to ISO 7619-1/2011 for hardness, and a Schob test instrument according to ISO 4662/2017 for elasticity [8,11,37].

2.3.2. Cross-Linking Evaluation

Sol-gel analysis and cross-link density determination were carried out on EPDM-EB, EPDM-EB-S, EPDM-P, and EPDM-P-S samples as in our previous works [8,11,37]. In order to determine the cross-linked products, gel content (gel fraction) was used as the solvent (toluene) extraction method [8,11,37]. The samples cross-link density (ν) was determined on the basis of equilibrium solvent-swelling measurements in toluene by applying the modified Flory-Rehner equation for tetra functional networks. The Flory-Huggins polymer-solvent interaction term χ_{12} for the EPDM-toluene system was of 0.49 [8,38,39].

2.3.3. Fourier Transform Infrared Spectroscopy (FTIR)

The structure of the EPDM-EB-S and EPDM-P-S composites cross-linked by EB irradiation and dibenzoyl peroxide were analyzed by FTIR measurements using TENSOR 27 spectrophotometer (Bruker, Germany). The absorption spectra were obtained as 30 scans mediation, in the range of 4000–600 cm^{-1}, with a resolution of 4 cm^{-1} [8].

2.3.4. Water Uptake Evaluation

The water absorption in EPDM-EB-S and EPDM-P-S was evaluated as in our previous work [8], by immersion in distilled water until they no longer absorb water, in accordance with ISO 20344/2011.

2.3.5. Rubber-Filler Interaction

The interaction between the EPDM rubber and filler (wood sawdust) was also analyzed as in our previous work [8] using the Kraus theory that helps assess an interfacial interaction in filler-reinforced rubber composites [40–42].

3. Results and Discussion

3.1. Mechanical Characteristics

By EB irradiation, cross-linking and chain scission can occur. The first one appears frequently at lower irradiation doses, up to 150 kGy, while the second one is associated with the breaking of C–C bonds at higher doses [37,43,44]. Due to the aliphatic chain, which has low resistance to ionizing radiation action, the degradation can be predominant as compared to cross-linking in EPDM rubber at higher irradiation doses [37,44,45].

The mechanical properties of the EPDM/EPDM-S composites cross-linked by peroxide (P) and electron beam (EB) are presented comparatively.

As shown in Figure 1, hardness was improved by the addition of sawdust (S) in the case of P cross-linking and by the irradiation dose increasing and S addition, in the case of EB cross-linking. The S addition increased hardness with 8% in the case of P cross-linking and with 13% in the case of EB cross-linking. An irradiation dose up to 300 kGy appear to be sufficient to obtain the reinforcement effect and to the extent of cross-linking in the polymeric material [1,8,46].

In Figure 2, the composites' elasticity behavior is presented. In the cases of both cross-linking methods, the addition of S led to the elasticity decreasing against the samples without S. However, it should be noted that the EB cross-linking method, with or without S, is at least as effective as P cross-linking at a low irradiation dose of 150 kGy and more effective until the irradiation dose of 300 kGy. After that, it can be seen that only the samples containing 15 ppm of S and cross-linked by EB are still more elastic than samples containing 5 and 15 ppm of S and cross-liked by P. As in the case of hardness, low and middle irradiation doses appear to be sufficient to maintain the elasticity decreasing below 6%. If, in the case of P cross-linking is obvious, the presence of S reduces the degree of elasticity and segment mobility of the cured composites. In the case of EB cross-linking, the addition of S does not diminish the strain energy [8,46,47] and does not increase the composite hysteretic behavior [8,47,48].

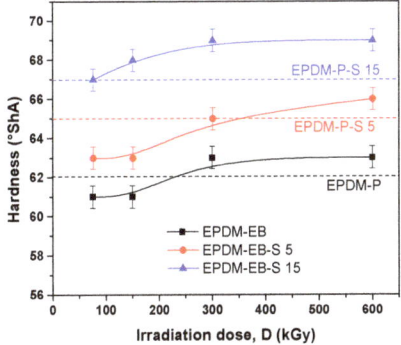

Figure 1. The influence of sawdust amount and the cross-linking method (P/EB) on composite hardness.

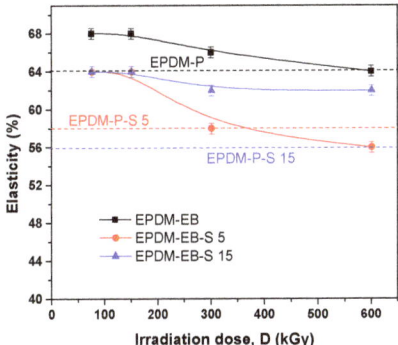

Figure 2. The influence of sawdust amount and cross-linking method (P/EB) on composites elasticity.

The composites' tensile strength behavior is presented in Figure 3. It is easy to observe the superiority of the EB cross-linking method comparatively with the P cross-linking method in the case of EPDM blends, irrespective of the irradiation dose.

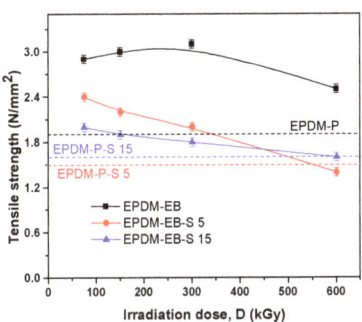

Figure 3. The influence of sawdust amount and cross-linking method (P/EB) on composites' tensile strength.

In the case of P cross-linking, the addition of 5 ppm of S decreased the tensile strength with 20% while the addition of 15 ppm of S decreased the tensile strength with only 10%. In the case of EB cross-linking, at low and medium irradiation doses (up to 150 kGy), the addition of 15 ppm of S maintain the tensile strength over the P cross-linking values, even if comparatively with the values without S addition. The tensile strength decreased between 25% and 35% [1].

The results can conduct to the idea of a not-so-strong interaction between the rubber and filler or an unsatisfactory adhesion of the filler in the polymeric matrix, which is insufficient to constrain the motion of the chains [8,49,50]. It looks like the loading with low S amounts and irradiation with doses up to 300 kGy can be a solution to maintain the composites' properties similar with those obtained in the case of P cross-linking.

As seen in Figure 4, irrespective of the irradiation dose below 300 kGy, the elongations at break of the composites containing S are superior to those measured for samples cross-linked by the P method. Differences between the samples containing the same S amounts and cross-linked by P/EB methods are more than 300%. The elongation at break decreasing with the increase of the irradiation dose and S amount indicates an increase in cross-link density. Furthermore, the addition of S can lead to the appearance of a restriction in the molecular chains' movement [8,51]. Additionally, the striking forces between the filler and the polymer molecules led to the development of a cross-linked structure, which limit the free mobility of the polymer chains. Hence, this increases the resistance to accelerate upon the execution of tension [8,51,52]. The results are comparable with other obtained using different reinforcing fillers [1].

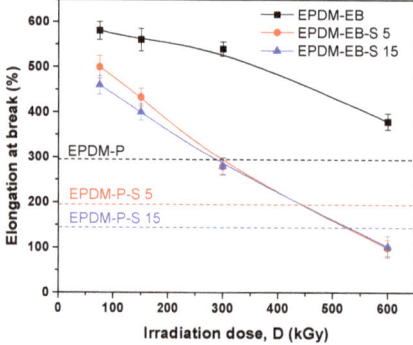

Figure 4. The influence of sawdust amount and cross-linking method (P/EB) on composites' elongation at break.

Additionally, as above, in Figure 5, it can be seen that tearing strength of EPDM-EB and EPDM-EB-S 15 samples up to 300 kGy are over the values obtained in the case of P cross-linking with or without S. More than up to 150 kGy, tearing strength for samples EPDM-EB-S 15 is even over that of EPDM-EB samples. Considerable differences (up to 60%) can be observed between samples containing the same S amounts and cross-linked by P/EB methods.

The results presented above indicate that the EB irradiation and S addition have a reinforcing effect on EPDM rubber and conduct to special and different properties than in P cross-linking. This result can be explained by a different chemical nature of free radicals formed by radiation action that help for the addition to double bonds of unsaturated rubbers unlike those formed in P decomposition [3,53,54]. The relative reactivity or stability of the free radicals generated in P decomposition is related to the hydrogen bond dissociation energy of the parent compound [3,53]. Due to the high values of bond dissociation energy, methyl, phenyl, tert-butoxy, and other alkoxy radicals are highly reactive and are good hydrogen abstractors. Opposite ethyl, tert-butyl, and isopropyl radicals have lower values of bond dissociation energy. Therefore, they are poor hydrogen abstractors [53,55–57]. The reactivity of peroxide radicals depend not only on their structure but also on their size [3].

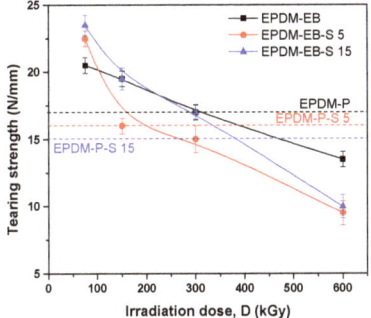

Figure 5. The influence of sawdust amount and cross-linking method (P/EB) on composites tearing strength.

3.2. Gel Fraction and Cross-Link Densities

The cross-linking evaluation was done based on the gel fraction and cross-link density measurements. The results represent the average of five specimens.

As seen in Figure 6, gel fraction exhibit values over 98% (excepting for the EPDM-EB and EPDM-EB-S 15).

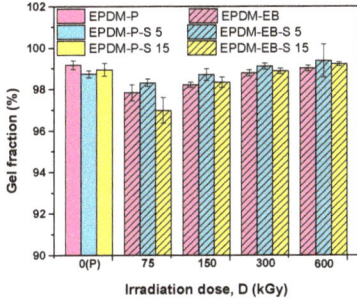

Figure 6. The effect of the cross-linking method (P/EB) and sawdust amount on the gel fraction.

As the irradiation dose increased, the gel fraction also increased. The results obtained at the highest irradiation dose of 300 and 600 kGy are similar with those obtained in the case of peroxide cross-linking. In addition, samples containing 5 phr S cross-linked with EB (EPDM-EB-S 5) exhibit higher gel fractions than those without S (EPDM-EB) and with 15 phr S (EPDM-EB-S 15). Thus, in the case of P cross-linking, it is necessary to add more S (15 phr) to obtain the same results as in the case of EB cross-linking (5 phr S/300 kGy). To obtain improved results, the sample was irradiated over 300 kGy (5 phr S/600 kGy).

The cross-linking process evaluation was also done by calculating cross-link density for P and EB cross-linked samples. The results are presented in Figure 7.

As seen in Figure 7, the cross-link densities of samples cross-linked by EB, with or without S, have increased with the irradiation dose increasing. In addition, for irradiation doses over 150 kGy, the cross-link density grows closely following the increase in filler load [47]. The increase of the network density could be explained by the composite rigidity increasing due to the S presence or by some specific chemical interactions between the S and EPDM matrix. The same behavior was observed in the case of hardness dependence on the S content [47].

Correlating the mechanical properties with the cross-linking evaluation, we can conclude that S acts similar to active fillers in both EPDM-P-S [47] and EPDM-EB-S composites and leads to the improvement of the composite properties [42,47,58–61]. Even in Figure 7, it can be observed that the P cross-linking led to the obtainment of better cross-link densities than EB cross-linking, by correlating the results with those presented in Figures 1–5 up to 300 kGy. The results show that all mechanical properties of EPDM-EB and EPDM-EB-S samples were superior. This can be explained by stiffness and poor elasticity due to the lower mobility of the macromolecular chains at high cross-link densities [42,48,58–61].

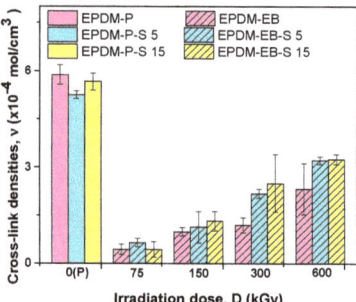

Figure 7. Effect of the cross-linking method (P/EB) and sawdust amount on the gel fraction.

3.3. FTIR Analysis

For a composite structure investigation (identification of different functional groups, presence or absence of specific functional groups), Fourier Transform Infrared (FTIR) spectroscopy was used [62].

Figures 8 and 9 present the infrared spectra obtained in the range of 2000–650 cm^{-1} on EPDM-P/EPDM-P-S and EPDM-EB/EPDM-EB-S composites due to the valence and deformation vibration of the atoms involved in the existing or formed covalent bonds. In addition, details of FTIR spectra in the range of 4000–3000 cm^{-1} on EPDM-EB-S 5 and EPDM-EB-S 15 composites will be presented.

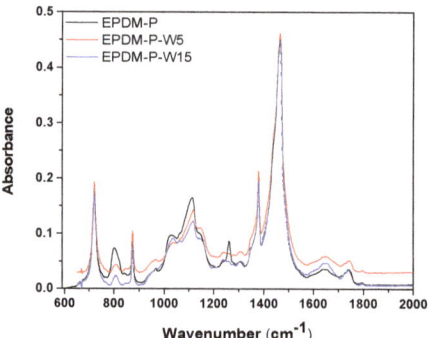

Figure 8. FTIR spectra of EPDM-S composites cross-linked by peroxide (P).

Due to the covalent bonding of the monomer units in EPDM rubber, they can be considered as being separate chemical components. Intra-chain (between adjacent monomer units in the polymer chain) and inter-chain (between monomer units that are not adjacent in the same polymer chain) interactions may be happening [63]. Due to the intra-chain interactions, ethylene and propylene groups are sensitive to the identities of adjacent groups in the polymer chain. On the other hand, polymer crystallinity and morphology are consequences of inter-chain interactions [63].

In the EPDM spectra, three absorption regions were identified as being significant. The first one was dominated by second overtone C–H stretching bands and located between 1100–1350 cm^{-1}. The second one has a higher absorptivity with an order of magnitude, dominated by first-overtone C–H stretching bands, and located between 1570–1850 cm^{-1} and the third one that contains C–H combination bands with much higher absorptivity than those of the bands in regions one and two and located between 1950–2500 cm^{-1} [63].

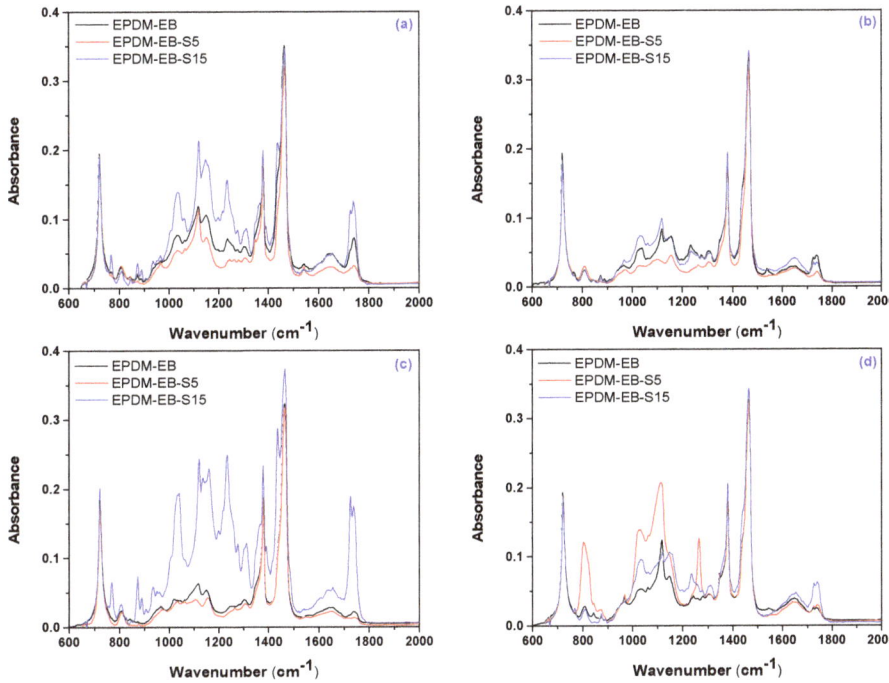

Figure 9. FTIR spectra of EPDM-EB-S composites cross-linked by EB irradiation at 75 kGy (**a**), 150 kGy (**b**), 300 kGy, (**c**) and 600 kGy (**d**) in the range of 650–2000 cm^{-1}.

The FTIR spectra of EPDM-P-S (Figure 8)/EPDM-EB-S (Figures 9 and 10) show the presence of the EPDM specific bands, located in the above specified regions.

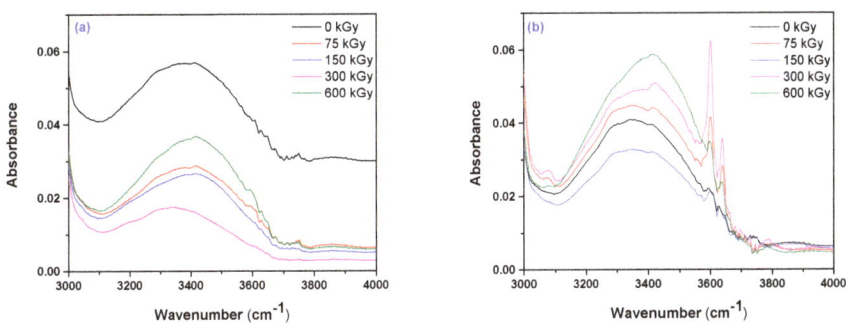

Figure 10. FTIR spectra of EPDM-EB-S 5 (**a**) and EPDM-EB-S 15 (**b**) composites cross-linked by EB irradiation.

Thus, this includes C–H stretching vibration (2918 and 2850 cm^{-1}) [63–67], C=C stretching vibration (1630 cm^{-1}) [68–70], CH$_2$ bending and rocking vibrations (1460 and 720 cm^{-1}) [64–67], and CH$_3$ bending vibration (1376 cm^{-1}) [64–67]. In Figure 8, all fingerprints of dibenzoyl peroxide were found including a weak absorption band in the region 950–800 cm^{-1} due to the O–O stretching vibration, a strong band at 1775 cm^{-1} (saturated aliphatic) due to the C=O stretching vibration, and a band between 1300–1050 cm^{-1} due to the C–O stretching vibrations emphasizing its domination over the O–O bond [71].

The main chemical components of wood sawdust are carbon (60.8%), hydrogen (5.2%), oxygen (33.8%), and nitrogen (0.9%) [72,73]. Cellulose (38%–50%) is the one that gives the wood stiffness [73,74]. Lignin (15–25%) is the cementing agent or resin in wood [73,75] and hemicelluloses (23–32%) are the bonding agent between cellulose and lignin [73]. The primary and secondary hydroxyls, carbonyls, carboxyls, esters, or ethers from cellulose, hemicelluloses, and lignin are examples of active functional groups that can be involved in chemical reactions [76–79].

The wood strength can be significantly reduced by EB irradiation [73] due to a decrease of polymerization degree and crystallinity and to the increase of the hydrolysis rate and yield of cellulose [76–78].

In Figure 8, Figure 9a–c, and Figure 10, the following basic structures of sawdust have been found in the obtained composites: a broad band between 3600–3100 cm^{-1} due to the OH-stretching vibration, which gives important information about the hydrogen bonds [80,81], a strong broad OH stretching (3598–3637 cm^{-1}) that includes inter and intra-molecular hydrogen bond vibrations in cellulose [62], C–H stretching of all hydrocarbon constituents in polysaccharides [62] including methyl and methylene groups (2800–3000 cm^{-1}), and a strong broad superposition with sharp and discrete absorptions in the region from 1000 to 1750 cm^{-1} [82].

Notable differences between EPDM-P, EPDM-P-S, and even between EPDM-P-S 5 and 15 in some regions can be observed in Figure 8. Thus, except the regions 1340–1550 cm^{-1} and 650–780 cm^{-1}, all others absorb EPDM-P-S upper or under the EPDM-P with band intensities varying due to the possible degradation (chain scission/chain link cross-linking) processes and addition of the basic structure from S (especially between 1600–1800 cm^{-1} due to the ring stretching mode strongly associated with the aromatic C–O–CH$_3$ stretching mode, the C=O stretching of conjugated/aromatic ketones, or the aromatic skeletal vibrations [82]) in EPDM-P-S 5 and EPDM-P-S 15 spectra [8,80,81]. Results can be correlated with those presented in Figure 7, where it is observed that the cross-link densities of the EPDM-P-S composites, especially EPDM-P-S 15, are comparable with those of EPDM-P. Thus, the use of other additive or fillers, as S, to replace sulphur (instead of increase the state of cure) are necessary additions to the cross-linking method (the cross-linking using only peroxide is low) [8,82].

By comparing the spectra presented in Figures 9 and 10 with the spectra of cellulose, holocellulose, and lignin [82], the following specific bands have been found: aromatic skeletal vibrations caused by lignin (1510 and 1600 cm^{-1}) and the absorption located at 1730 cm^{-1} caused by holocellulose. This indicates the C=O stretch in non-conjugated ketones, carbonyls, and in ester groups [82,83]. Appearance of the band near 1600 cm^{-1} is a relative pure ring stretching mode associated with the aromatic C–O–CH$_3$ stretching mode [82]. The C=O stretch of conjugated or aromatic ketones absorbs below 1700 cm^{-1} [82] and can be seen as shoulders in the spectra.

Formation of active functional groups able to be involved in chemical reactions and radicals during the EB irradiation is responsible for the bonding process between the EPDM rubber matrix and filler (sawdust) [8,79].

The addition of 5 phr of S (Figure 10a) decreases the 3200–3600 cm^{-1} broad band intensity from 0.058 (for un-irradiated EPDM-S 5) to 0.035 (for EPDM-S 5 irradiated at 600 kGy), 0.029 (for EPDM-EB-S 5 irradiated at 75 kGy), 0.0275 (for EPDM-EB-S 5 irradiated at 150 kGy), and 0.018 (for EPDM-EB-S 5 irradiated at 300 kGy)—shifted band. The addition of 15 phr of S (Figure 10b) places from the beginning the intensity of the broad band at 3200–3600 cm^{-1} corresponding to un-irradiated EPDM-S under all other EPDM-EB-S 15 irradiated, except 150 kGy.

The band assignments in EPDM-P-S and EPDM-EB-S samples from the FTIR spectra presented above are listed in Table 1.

Table 1. Band assignments in EPDM-P-S and EPDM-EB-S samples.

Band Position in the EPDM-P-S/EPDM-EB-S Composites (cm^{-1})	Functional Group
3360–3390, 3598, 3636, 3637 (EPDM-EB-S 15)	O–H stretching vibration (3300–4000 cm^{-1}) from wood sawdust/cellulose [62]
2918	C–H stretching vibration (2800–3000 cm^{-1}) from EPDM [64–67]
2850 (EPDM-P-S, EPDM-EB-S)	C–H stretching vibration (2800–3000 cm^{-1}) from EPDM [64–67] and wood sawdust (polysaccharides/cellulose) [62]
1775 cm^{-1} (EPDM-P-S)	C=O stretching vibration in P [71]
1730–1740 (EPDM-P-S, EPDM-EB-S)	Aromatic skeletal vibrations caused by holocellulose (wood sawdust) [82] C=O stretch in non-conjugated ketones, carbonyls, and ester groups [82,83]
1640	C=C stretching vibration (1630 cm^{-1}) from EPDM [68–70]
1642–1646 (EPDM-EB-S)	Typical bands assigned to cellulose—Vibration of water molecules absorbed in cellulose [62]
1539, 1540 (EPDM-P-S)	Aromatic skeletal vibrations caused by lignin (wood sawdust) [82] C=O stretch in non-conjugated ketones, carbonyls, and in ester groups [82,83]
1460	CH$_2$ bending and rocking vibrations from EPDM [64–67]
1435, 1436 (EPDM-EB-S)	Stretching and bending vibrations of –CH$_2$ bonds, associated with the amount of the crystalline structure of the cellulose [62,84,85]
1376	CH$_3$ bending vibration from EPDM [64–67]
1034 (EPDM-EB-S)	Stretching and bending vibrations of –OH bonds in cellulose [62,84,85]
1300–1050 cm^{-1} (EPDM-P-S)	C–O stretching vibrations emphasizing its domination over the O–O bond in P [71]
950–800 cm^{-1} (EPDM-P-S)	O–O stretching vibration in P [71]
930 (EPDM-EB-S)	Assigned to the amorphous region in cellulose [62,86]
905 (EPDM-EB-S)	C–O bonds in cellulose [40,84,85]
720	CH$_2$ bending and rocking vibrations from EPDM [64–67]

The previously mentioned presence of wood components (cellulose, hemicellulose, and lignin) in EPDM-EB-S samples confirms their origin from the sawdust used for obtaining composites (Table 1). In Figure 10a,b, notable differences in band widths and intensities can be observed due to the cross-linking method and S loading. Thus, in the case of 5 phr S loading, all EB irradiation results are under P curing (Figure 10a). In the case of 15 phr S loading, except for the 150 kGy EB irradiation, all other absorbances were over the P curing absorbance and close to degradation. The results are very well correlated with the mechanical test results (Figures 1–5) where EB irradiation with 150 kGy turned out to be the most effective dose for obtaining better effects than P curing.

3.4. Water Uptake Test Results

The results of water uptake experiments that were carried out on samples with/without sawdust and treated by means of P and EB are presented comparatively in Figure 11. The cross-linking method (EB and P), S loading (5 and 15 phr), and irradiation dose (75, 150, 300, and 600 kGy) are responsible for composite absorption behavior.

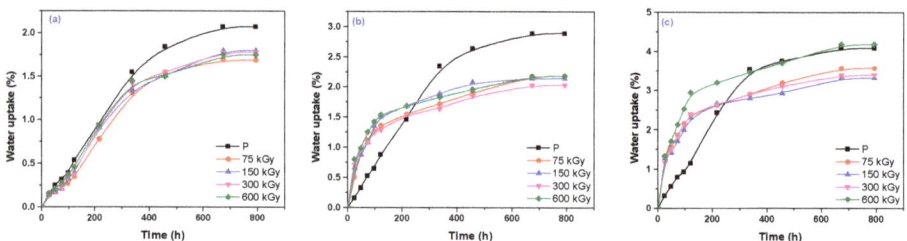

Figure 11. EPDM (**a**), EPDM-S 5 (**b**), and EPDM-S 15 (**c**) water uptake behavior as a function of immersion time.

In Figure 11, it can be observed that water uptake equilibrium was reached earlier by the samples containing S, except in the case of EPDM-S 5 cured with P (Figure 11b). In addition, as the S loading increased, the water uptake increased from up to 2% for EPDM-EB-S 5 (Figure 11b) to up to 4% for EPDM-EB-S 15 (Figure 11c). Samples with or without S and cured by EB irradiation present absorptions lower than those cured by P (Figure 11a,b), except for the sample EPDM-EB-S 15 irradiated at 600 kGy (Figure 11c). The increase of water absorption with the S loading may be explained by both a hydrophilic nature of S and a big interfacial area between the S and the elastomer matrix [8,87]. In rubber composites that contain wood, water is absorbed mainly by the last one because, since rubbers are hydrophobic, their absorbability can be neglected [8,87]. The amount of free –OH groups from cellulose and hemicellulose increases when the S content increases. Their contact with water leads to the formation of hydrogen bonds and, consequently, the weight of the composite increases [8,88]. A reduced amount of –OH groups inside the composite may conduct to a low availability to absorb water [8,88].

There are several studies that have shown that P prevents the aggregation of filler in the rubber matrix forming a network structure and, in the same time, acts as a plasticizer for rubber and as a compatibilizer between the hydrophobic rubber and the hydrophilic filler. These properties are responsible for the improved interaction at the interface between EPDM rubber and sawdust [8,89–91].

3.5. Rubber-Filler Interaction

The Kraus equation was used to analyze the interaction between EPDM rubber matrix and the natural filler, which is sawdust. The results are listed in Table 2 in terms of volume fraction of rubber in the swollen gel (V_{rf}) and degree of restriction of the rubber matrix swelling due to the presence of filler (V_{ro}/V_{rf} ratio) [8,92,93].

Table 2. V_{rf} and V_{ro}/V_{rf} of EPDM-S composites determined in toluene.

Irradiation Dose	Samples	V_{rf}	V_{ro}/V_{rf}
75 kGy	EPDM-EB-S 5	0.1912	0.8609
	EPDM-EB-S 15	0.1516	1.0862
150 kGy	EPDM-EB-S 5	0.2920	0.7787
	EPDM-EB-S 15	0.2964	0.7670
300 kGy	EPDM-EB-S 5	0.2242	1.0760
	EPDM-EB-S 15	0.2358	1.0232
600 kGy	EPDM-EB-S 5	0.3401	0.8922
	EPDM-EB-S 15	0.3202	0.9478

As shown in Table 2, the addition of a higher quantity of sawdust (15 phr) led to a slowly decreasing V_{ro}/V_{rf} ratio, except for the composites obtained at 150 kGy and 300 kGy. According to Kraus theory, reduced values of the V_{ro}/V_{rf} ratio are associated with a good adhesion between rubber and filler and with the appearance of the reinforcement effect [8,92,93].

Even if fillers generally reduce the swelling, high or enhanced values V_{rf} depend on the density of the cross-links and are associated with complex networks that present a lower degree of swelling [8,92,93]. In addition, in Table 2, it can be seen that, for the same irradiation dose of 150 kGy and 300 kGy, the increase of the sawdust amount led to the V_{rf} growing.

The quantitative evaluation of yields of cross-linking and chain scission of the EPDM rubber and EPDM-EB-S composites was done from the plots of $S + \sqrt{S}$ vs. 1/absorbed dose (D) from the Charlesby-Pinner equation for blend compositions (Figure 12) [8,94,95].

$$S + \sqrt{S} = \frac{p_0}{q_0} + \frac{1}{\alpha P_n D} \qquad (2)$$

where S is the sol fraction ($S = 1$-gel fraction), p_0 is the degradation density, average number of main chain scissions per monomer unit and per unit dose, q_0 is the cross-linking density, proportion of monomer units cross-linked per unit dose, P_n is the number averaged degree of polymerization, and D is the radiation dose in Gy.

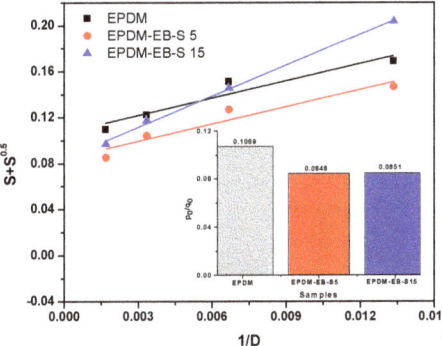

Figure 12. Plot of $S + \sqrt{S}$ vs. 1/absorbed dose (D) for EPDM-EB-S composites.

In Figure 12, it is observed that the EPDM-EB-S 15 sample is the most effective cross-linked by EB irradiation. The cross-linking extent increases linearly with the S content. Low values of p_0/q_0 for high S content are suggestive for the relatively improved radical-radical interactions in the polymer composite [8,94,96,97]. Values under the unit of the p_0/q_0 ratio (0.1069 for EPDM, 0.0848 for EPDM-EB-S 5 and 0.0851 for EPDM-EB-S 15) indicate that the cross-linking prevailed over degradation.

4. Conclusions

Two classes of eco-composites based on ethylene-propylene-terpolymer rubber (EPDM) and wood sawdust (S) were obtained by two different methods: dibenzoyl peroxide cross-linking (P) at a high temperature and EB cross-linking at room temperature of 25 °C using the irradiation dose between 75 kGy and 600 kGy. 5 phr and 15 phr were the amounts of S used as filler. Even if both cross-linking methods insured gel fractions over 98%, the cross-link densities determined on samples cross-linked by the P method were higher when compared to those determined on samples cross-linked by EB. However, the addition of S increases the cross-link density of the composite in the case of the EB cure. Over the irradiation dose of 300 kGy, the composite becomes harder and less elastic, even for 5 phr S. The addition of 15 phr S to the samples irradiated over 300 kGy degrades the tensile properties of the composite. These results correlated with FTIR analysis show that the irradiation dose up to 300 kGy appears to be sufficient to obtain the reinforcement effect in the composite. After that, the process is susceptible to be close to degradation. FTIR analysis showed the cellulose radical formation during the irradiation and also the formation of active functional groups susceptible to chemical reactions that can be associated with the bonding process between the matrix (EPDM rubber) and the filler (S). The water uptake tests have shown very good absorption properties, especially for 5 phr S, irrespective of the irradiation dose.

The sawdust adding was carried out with the purpose of increasing the composite degradation potential without affecting its mechanical properties, as a solution for replacing the materials based exclusively on EPDM.

Author Contributions: Conceptualization and methodology: G.C. and E.M.; investigation: G.C., E.M., D.I. and M.D.S.; data curration: E.M.; writing—review and editing: G.C. and D.I. All authors have read and agreed to the published version of the manuscript.

Funding: This research was funded by the Romanian Ministry of Education and Research through STAR Research, Development and Innovation Program (Space Technology and Advanced Research)—Contract No. 140/2017, Nucleu LAPLAS VI Program—Contract No. 16N/08.02.2019 and Contract No. 6PFE/2018-PERFORM-TEX-PEL funded the research.

Conflicts of Interest: The authors declare no conflict of interest.

References

1. Radi, H.; Mousaa, I.M. Characterization Study of EPDM Rubber Vulcanized by Gamma Radiation in The Presence of Epoxidized Soybean Oil. *Egypt. J. Radiat. Sci. Appl.* **2015**, *28*, 121–134. [CrossRef]
2. Lee, S.H.; Yang, S.W.; Park, E.S.; Hwang, J.Y.; Lee, D.S. High-Performance Adhesives Based on Maleic Anhydride-g-EPDM Rubbers and Polybutene for Laminating Cast Polypropylene Film and Aluminum Foil. *Coatings* **2019**, *9*, 61. [CrossRef]
3. Kruzelak, J.; Sykora, R.; Hudec, I. Vulcanization of Rubber Compounds with Peroxide Curing Systems. *Rubber Chem. Technol.* **2017**, *90*, 60–88. [CrossRef]
4. Akiba, M.; Hashim, A.S. Vulcanization and crosslinking in elastomers. *Prog. Polym. Sci.* **1997**, *22*, 475–521. [CrossRef]
5. Quirk, R.P. Overview of Curing and Cross-linking of Elastomers. *Prog. Rubber Plast. Technol.* **1988**, *4*, 31–45.
6. Samarzija-Jovanovic, S.; Jovanovic, V.; Cincovic, M.M.; Simendic, J.B.; Markovic, G. Comparative study of radiation effect on rubber—Carbon black compounds. *Compos. Part B* **2014**, *62*, 183–190. [CrossRef]
7. Stelescu, M.D.; Airinei, A.; Manaila, E.; Fifere, N.; Craciun, G.; Varganici, C.; Doroftei, F. Exploring the Effect of Electron Beam Irradiation on the Properties of Some EPDM-Flax Fiber Composites. *Polym. Compos.* **2019**, *40*, 315–327. [CrossRef]
8. Manaila, E.; Stelescu, M.D.; Craciun, G.; Ighigeanu, D. Wood Sawdust/Natural Rubber Ecocomposites Cross-Linked by Electron Beam Irradiation. *Materials* **2016**, *9*, 503. [CrossRef]
9. Stelescu, M.D.; Manaila, E.; Craciun, G.; Dumitrascu, M. New green polymeric composites based on hemp and natural rubber processed by electron beam irradiation. *Sci. World J.* **2014**, 684047. [CrossRef]
10. Stelescu, M.D.; Manaila, E.; Craciun, G.; Zuga, N. Crosslinking and grafting ethylene vinyl acetate copolymer with accelerated electrons in the presence of polyfunctional monomers. *Polym. Bull.* **2012**, *68*, 263–285. [CrossRef]
11. Stelescu, M.D.; Manaila, E.; Craciun, G. Vulcanization of ethylene-propyleneterpolymer-based rubber mixtures by radiation processing. *J. Appl. Polym. Sci.* **2013**, *128*, 2325–2336. [CrossRef]
12. Czvikovszky, T. Degradation effects in polymers, in IAEA-TECDOC-1420, Advances in radiation chemistry of polymers. In Proceedings of the Technical Meeting, Notre Dame, IN, USA, 13–17 September 2003; pp. 91–102.
13. Stelescu, M.D.; Manaila, E.; Craciun, G.; Chirila, C. Development and Characterization of Polymer Eco-Composites Based on Natural Rubber Reinforced with Natural Fibers. *Materials* **2017**, *10*, 787. [CrossRef] [PubMed]
14. Craciun, G.; Manaila, E.; Stelescu, M.D. New Elastomeric Materials Based on Natural Rubber Obtained by Electron Beam Irradiation for Food and Pharmaceutical Use. *Materials* **2016**, *9*, 999. [CrossRef] [PubMed]
15. Abdel-Aziz, M.M.; Basfar, A.A. Thermal Stability of Radiation Vulcanized EPDM Rubber. In Proceedings of the 8th Arab International Conference on Polymer Science and Technology, Cairo-Sharm El-Shiekh, Egypt, 27–30 November 2005.
16. Clough, R.L. Isotopic exchange in gamma-irradiated mixtures of $C_{24}H_{50}$ and $C_{24}D_{50}$: Evidence of free radical migration in the solid state. *J. Chem. Phys.* **1987**, *87*, 1588–1595. [CrossRef]

17. Gatos, K.G.; Karger-Kocsis, J. Effects of primary and quaternary amine intercalants on the organoclay dispersion in a sulphur-cured EPDM rubber. *Polymer* **2005**, *46*, 3069–3076. [CrossRef]
18. Charlesby, A. *Atomic Radiation and Polymers*; Pergamon Press: Oxford, UK, 1960.
19. Dole, M. (Ed.) *The Radiation Chemistry of Macromolecules*; Academic Press: New York, NY, USA, 1972; Volume I.
20. Grassia, N.; Scott, G. *Polymer Degradation and Stabilization*; Cambridge University Press: Cambridge, UK, 1985.
21. El-Zayat, M.M.M. Radiation Curing of Rubber/Thermoplastic Composites Containing Different Inorganic Fillers. Master's Thesis, Menoufia University, Shibin Al Kawm, Egypt, 2012.
22. Planes, E.; Chazeau, L.; Vigier, G.; Fournier, J.; Stevenson-Royaud, I. Influence of fillers on mechanical properties of ATH filled EPDM during ageing by gamma irradiation. *Polym. Degrad. Stab.* **2010**, *95*, 1029–1038. [CrossRef]
23. Chaiear, N. Report No. 138 Health and safety in the rubber industry. In *Rapra Review Reports*; Rapra Technology Limited: Shrewsbury, UK, 2001; Volume 12, pp. 21–22. ISBN 1-85957-301-0.
24. Guzmán, E.; Santini, E.; Ferrari, M.; Liggieri, L.; Ravera, F. Interfacial Properties of Mixed DPPC–Hydrophobic Fumed Silica Nanoparticle Layers. *J. Phys. Chem. C* **2015**, *119*, 21024–21034. [CrossRef]
25. Guzmán, E.; Santini, E.; Zabiegaj, D.; Ferrari, M.; Liggieri, L.; Ravera, F. Interaction of Carbon Black Particles and Dipalmitoylphosphatidylcholine at the Water/Air Interface: Thermodynamics and Rheology. *J. Phys. Chem. C* **2015**, *119*, 26937–26947. [CrossRef]
26. Lopez-Manchado, M.A.; Biagiotti, J.; Arroyo, M.; Kenny, J.M. Ternary composites based on PP-EPDM blends reinforced with flax fibers. Part I: Processing and thermal behavior. *Polym. Eng. Sci.* **2003**, *43*, 1018–1030. [CrossRef]
27. Rozik, N.N.; Abd-El.Messieh, S.L.; Yaseen, A.A.; Abd-El-Hafiz, A.S. Dielectric and mechanical properties of natural nanofibers—Reinforced ethylene propylene diene rubber: Carrot foliage and corn gluten. *Polym. Eng. Sci.* **2013**, *53*, 874–881. [CrossRef]
28. La Mantia, F.P.; Morreale, M. Green composites: A brief review. *Compos. Part A* **2011**, *42*, 579–588. [CrossRef]
29. Gaceva, G.B.; Avella, M.; Malinconico, M.; Buzarovska, A.; Grozdanov, A.; Gentile, G.; Errico, M.E. Natural Fiber Eco-Composites. *Polym. Compos.* **2007**, *28*, 98–107. [CrossRef]
30. Cristaldi, G.; Latteri, A.; Recca, G.; Cicala, G. Composites Based on Natural Fibre Fabrics. In *Woven Fabric Engineering*; Dubrovski, P.D., Ed.; IntechOpen: Rijeka, Croatia, 2010; pp. 318–342. Available online: http://www.intechopen.com (accessed on 2 June 2014).
31. Begum, K.; Islam, M.A. Natural fiber as a substitute to synthetic fiber in Polymer Composites: A Review. *Res. J. Eng. Sci.* **2013**, *2*, 46–53.
32. Nora'asheera, M.N. Composites From Polypropylene (PP) Reinforced With Oil Palm Empty Fruit Bunch (OPEFB) Fibre. Bachelor's Thesis, University of Malaysia Pahang, Pahang, Malaysia, 2011.
33. Fakhrul, T.; Mahbub, R.; Islam, M.A. Properties of wood sawdust and wheat Flour Reinforced Polypropylene Composites. *J. Mod. Sci. Technol.* **2013**, *1*, 135–148.
34. Fu, S.Y.; Feng, X.Q.; Lauke, B.; Mai, Y.W. Effects of particle size, particle/matrix interface adhesion and particle loading on mechanical properties of particulate–polymer composites. *Compos. Part B* **2008**, *39*, 933–961. [CrossRef]
35. Fiti, M. *Dozimetria Chimica a Radiatiilor Ionizante (Ionizing Radiation Chemical Dosimetry)*, 1st ed.; Editura Academiei Republicii Socialiste Romania: Bucuresti, Romania, 1973; pp. 24–70.
36. Cleland, M.R. Industrial Applications of Electron Accelerators—Ion Beam Applications. Presented at the CERN Accelerator School/Small Accelerator Course, Zeegse, The Netherlands, 24 May–2 June 2005.
37. Stelescu, M.D.; Airinei, A.; Manaila, E.; Craciun, G.; Fifere, N.; Varganici, C.; Pamfil, D.; Doroftei, F. Effects of Electron Beam Irradiation on the Mechanical, Thermal, and Surface Properties of Some EPDM/Butyl Rubber Composites. *Polymers* **2018**, *10*, 1206. [CrossRef]
38. Anelli, P.; Baccaro, S.; Carenza, M.; Palma, G. Radiation grafting of hydrophilic monomers onto ethylenepropylene rubber. *Radiat. Phys. Chem.* **1995**, *46*, 1031–1035. [CrossRef]
39. Nashar, K.; Gosh, U.; Heinrich, G. Influence of molecular structure of blend components on the performance of thermoplastic vulcanisates prepared in electron induced reactive processing. *Polymer* **2016**, *91*, 204–210. [CrossRef]
40. Kraus, G. Swelling of filler-reinforced vulcanizates. *J. Appl. Polym. Sci.* **1963**, *7*, 861–871. [CrossRef]
41. Mathew, L.; Ulahannan, J.; Joseph, R. Effect of curing temperature, fibre loading and bonding agent on the equilibrium swelling of isora-natural rubber composites. *Compos. Interfaces* **2006**, *13*, 391–401. [CrossRef]

42. Dong, Z.; Liu, M.; Jia, D.; Zhou, Y. Synthesis of natural rubber-*g*-maleic anhydride and its use as a compatibilizer in natural rubber/short nylon fiber composites. *Chin. J. Polym. Sci.* **2013**, *31*, 1127–1138. [CrossRef]
43. Planes, E.; Chazeau, L.; Vigier, G.; Fournier, J. Evolution of EPDM networks aged by gamma irradiation—Consequence on the mechanical properties. *Polymer* **2009**, *50*, 4028–4038. [CrossRef]
44. O'Donell, J.; Whittaker, A.K. The radiation crosslinking and scission of ethylene-propylene copolymer studied by solid-state nuclear magnetic resonance. *Br. Polym. J.* **1985**, *17*, 51–55. [CrossRef]
45. Manaila, E.; Stelescu, M.D.; Craciun, G. Aspects regarding radiation crosslinking of elestomers. In *Advanced Elastomers—Technology, Properties and Applications*; Boczkowska, A., Ed.; InTech: Rijeka, Croatia, 2012; pp. 3–34. ISBN 978-9535107392.
46. Palm, A.; Smith, J.; Driscoll, M.; Smith, L.; Scott Larsen, L. Chemical constituent influence on ionizing radiation treatment of a wood–plastic composite. *Radiat. Phys. Chem.* **2016**, *124*, 164–168. [CrossRef]
47. Stelescu, M.D.; Airinei, A.; Manaila, E.; Craciun, G.; Fifere, N.; Varganici, C. Property correlations for composites based on ethylene propylene diene rubber reinforced with flax fibers. *Polym. Test.* **2017**, *59*, 75–83. [CrossRef]
48. Zhao, J.; Ghebremeskel, G.N. A review of some of the factors affecting fracture and fatigue in SBR and BR vulcanizates. *Rubber Chem. Technol.* **2001**, *74*, 409–427. [CrossRef]
49. Ismail, H.; Shuhelmy, S.; Edgham, M.R. The effect of a silane coupling agent on curing characteristics and mechanical properties of bamboo fibre filled NR composites. *Eur. Polym. J.* **2002**, *38*, 39–45. [CrossRef]
50. Kohls, D.J.; Beauage, G. Rational design of reinforced rubber. *Curr. Opin. Solid State Mater. Sci.* **2002**, *6*, 183–194. [CrossRef]
51. Ahmed, K. Hybrid composites prepared from Industrial waste: Mechanical and swelling behavior. *J. Adv. Res.* **2015**, *6*, 225–232. [CrossRef]
52. Kukle, S.; Gravitis, J.; Putnina, A.; Stikute, A. The effect of steam explosion treatment on technical hemp fibres. In Proceedings of the 8th International Scientific and Practical Conference, Rezekne, Latvia, 20–22 June 2011; Volume 1, pp. 230–237.
53. Naskar, K. Dynamically Vulcanized PP/EPDM Thermoplastics Elastomers: Exploring Novel Routes for Crosslinking with Peroxides. Ph.D. Thesis, University of Twente, Enschede, The Netherlands, 2004.
54. Alvarez Grima, M.M. Novel Co-Agents for Improved Properties in Peroxide Cure of Saturated Elastomers. Ph.D. Thesis, University of Twente, Enschede, The Netherlands, 2007.
55. Dluzneski, P.R. Peroxide Vulcanization of Elastomers. *Rubber Chem. Technol.* **2001**, *74*, 451–492. [CrossRef]
56. Moad, G. The synthesis of polyolefin graft copolymers by reactive extrusion. *Prog. Polym. Sci.* **1999**, *24*, 81–142. [CrossRef]
57. Vroomen, G.L.M.; Visser, G.W.; Gehring, J. Electron Beam curing of EPDM. *Rubber World* **1991**, *205*, 23–26.
58. Dhakal, H.N.; Zhang, Z.Y.; Richardson, M.O.W. Effect of water absorption on the mechanical properties of hemp fibre reinforced unsaturated polyester composites. *Compos. Sci. Technol.* **2007**, *67*, 1674–1683. [CrossRef]
59. Wang, Q.; Wang, F.; Cheng, K. Effect of crosslink density on some properties of electron beam-irradiated styreneebutadiene rubber. *Radiat. Phys. Chem.* **2009**, *78*, 1001–1005. [CrossRef]
60. Majumder, P.S.; Bhowmick, A.K. Structure-property relationship of electronbeam-modified EPDM rubber. *J. Appl. Polym. Sci.* **2000**, *77*, 323–337. [CrossRef]
61. Vijayabaskar, V.; Tikku, V.K.; Bhowmick, K.A. Electron beam modification and crosslinking: Influence of nitrile and carboxyl contents and level of unsaturation on structure and properties of nitrile rubber. *Radiat. Phys. Chem.* **2006**, *75*, 779–792. [CrossRef]
62. Hospodarova, V.; Singovszka, E.; Stevulova, N. Characterization of Cellulosic Fibers by FTIR Spectroscopy for Their Further Implementation to Building Materials. *Am. J. Anal. Chem.* **2018**, *9*, 303–310. [CrossRef]
63. Miller, C.E. Analysis of EPDM Terpolymers by Near-Infrared Spectroscopy and Multivariate Calibration Methods. *Appl. Spectrosc.* **1989**, *43*, 1435–1443. [CrossRef]
64. Gunasekaran, S.; Natarajan, R.K.; Kala, A. FTIR spectra and mechanical strength analysis of some selected rubber derivatives. *Spectrochim. Acta A* **2007**, *68*, 323–330. [CrossRef]
65. Majumder, P.S.; Bhowmick, A.K. Surface and bulk-properties of EPDM rubber modified by electron beam irradiation. *Radiat. Phys. Chem.* **1998**, *53*, 63–78. [CrossRef]

66. Van Gisbergen, J.G.M.; Meijer, H.E.M.; Lemstra, P.J. Structural polymer blends: 2. Processing of polypropylene, EPDM blends: Controlled rheology and morphology fixation via electron beam irradiation. *Polymer* **1989**, *30*, 2153–2157. [CrossRef]
67. Kumutha, K.; Alias, Y. FTIR spectra of plasticized grafted natural rubber–LiCF$_3$SO$_3$ electrolytes. *Spectrochim. Acta A* **2006**, *64*, 442–447. [CrossRef] [PubMed]
68. Barra, G.M.O.; Crespo, J.S.; Bertolino, J.R.; Soldi, V.; Nunes Pires, A.T. Maleic Anhydride Grafting on EPDM: Qualitative and Quantitative Determination. *J. Braz. Chem. Soc.* **1999**, *10*, 31–34. [CrossRef]
69. Vieira, I.; Severgnini, V.L.S.; Mazera, D.J.; Soldi, M.S.; Pinheiro, E.A.; Pires, A.T.N.; Soldi, V. Effects of maleated ethylene propylene diene rubber (EPDM) on the thermal stability of pure polyamides, and polyamide/EPDM and polyamide/poly(ethylene terephthalate) blends: Kinetic parameters and reaction mechanism. *Polym. Degrad. Stab.* **2001**, *74*, 151–157. [CrossRef]
70. Schmidt, V.; Domenech, S.C.; Soldi, M.S.; Pinheiro, E.A.; Soldi, V. Thermal stability of polyaniline/ethylene propylene diene rubber blends prepared by solvent casting. *Polym. Degrad. Stab.* **2004**, *83*, 519–527. [CrossRef]
71. Aarthi, R.; Ramalingam, S.; Periandy, S.; Senthil Kannan, K. Molecular structure-associated pharmacodynamic investigation on benzoyl peroxide using spectroscopic and quantum computational tools. *J. Taibah Univ. Sci.* **2018**, *12*, 104–122. [CrossRef]
72. Horisawa, S.; Sunagawa, M.; Tamai, Y.; Matsuoka, Y.; Miura, T.; Terazawa, M. Biodegradation of nonlignocellulosic substances II: Physical and chemical properties of sawdust before and after use as artificial soil. *J. Wood Sci.* **1999**, *45*, 492–497. [CrossRef]
73. Postek, M.T.; Poster, D.L.; Vladár, A.E.; Driscoll, M.S.; Laverne, J.A.; Tsinas, Z.I.; Al-Sheikhly, M.I. Ionizing Radiation Processing and its Potential in Advancing Biorefining and Nanocellulose Composite Materials Manufacturing. *Radiat. Phys. Chem.* **2017**, *143*, 47–52. [CrossRef]
74. Panshin, A.J.; de Zeeuw, C. *Textbook of Wood Technology: Structure, Identification, Properties, and Uses of the Commercial Woods of the United States and Canada (McGraw-Hill Series in Forest Resources)*, Subsequent ed.; McGraw-Hill College: New York, NY, USA, 1980; pp. 120–125. ISBN1 10-0070484414. ISBN2 13-978-0070484405.
75. Sjöström, E. *Wood Chemistry: Fundamental and Applications*, 2nd ed.; Academic Press, Inc.: London, UK; San Diego, CA, USA, 1993; pp. 1–20. ISBN 9780080925899.
76. Driscoll, M.; Stipanovic, A.J.; Winter, W.T.; Cheng, K.; Manning, M.; Spiese, J.; Galloway, R.; Cleland, M.R. Electron beam irradiation of cellulose. *Radiat. Phys. Chem.* **2009**, *78*, 539–542. [CrossRef]
77. Cheng, K.; Barber, V.A.; Driscoll, M.S.; Winter, W.T.; Stipanovic, A.J. Reducing Woody Biomass Recalcitrance by Electron Beams, Biodelignification and Hot-Water Extraction. *J. Bioprocess. Eng. Biorefin.* **2013**, *2*, 143–152. [CrossRef]
78. Smith, S.; Bergey, N.S.; Salamanca-Cardona, L.; Stipanovic, A.; Driscoll, M. Electron Beam Pretreatment of Switchgrass to Enhance Enzymatic Hydrolysis to Produce Sugars for Biofuels. *Carbohydr. Polym.* **2014**, *100*, 195–201. [CrossRef]
79. Dumitrescu, L.; Matei, A.; Manciulea, I. Composites acrylic copolymers-wood waste. In Proceedings of the 4th International Conference of Advanced Composite Materials Engineering (COMAT 2012), Brasov, Romania, 18–20 October 2012; DERC Publishing House: Tewksbury, MA, USA, 2012; pp. 597–600.
80. Bodirlau, R.; Teaca, C.A. Fourier transform infrared spectroscopy and thermal analysis of lignocellulose fillers treated with organic anhydrides. *Rom. J. Phys.* **2009**, *54*, 93–104.
81. Khan, M.A.; Rahaman, M.S.; Al-Jubayer, A.; Islam, J.M.M. Modification of jute fibers by radiation-induced graft copolymerization and their applications. In *Cellulose-Based Graft Copolymers: Structure and Chemistry*; Thakur, V.K., Ed.; CRC Press: Boca Raton, FL, USA, 2015; pp. 209–235. ISBN 9780429156878.
82. Owen, N.L.; Thomas, D.W. Infrared studies of "hard" and "soft" woods. *Appl. Spectrosc.* **1989**, *43*, 451–455. [CrossRef]
83. Hergert, H.L. Infrared spectra. In *Lignins: Occurrence, Formation, Structure and Reactions*; Sarkanen, K.V., Ludwig, C.H., Eds.; John Wiley & Suns Inc.: New York, NY, USA, 1971; pp. 267–297.
84. Xu, F.; Yu, J.; Tesso, T.; Dowell, F.; Wang, D. Qualitative and Quantitative Analysis of Lignocellulosic Biomass Using Infrared Techniques: A Mini-Review. *Appl. Energy* **2013**, *104*, 801–809. [CrossRef]
85. Fackler, K.; Stevanic, J.S.; Ters, T.; Hinterstoisser, B.; Schwanninger, M.; Salmén, L. FT-IR imaging microscopy to localise and characterise simultaneous and selective white-rot decay within spruce wood cells. *Holzforschung* **2011**, *65*, 411–420. [CrossRef]

86. Poletto, M.; Ornaghi, H.L.; Zattera, A.J. Native Cellulose: Structure, Characterization and Thermal Properties. *Materials* **2014**, *7*, 6105–6119. [CrossRef] [PubMed]
87. Chaudhuri, S.; Chakraborty, R.; Bhattacharya, P. Optimization of biodegradation of natural fiber (Chorchorus capsularis): HDPE composite using response surface methodology. *Iran. Polym. J.* **2013**, *22*, 865–875. [CrossRef]
88. Idrus, M.A.M.M.; Hamdan, S.; Rahman, M.R.; Islam, M.S. Treated Tropical Wood Sawdust-Polypropylene Polymer Composite: Mechanical and Morphological Study. *J. Biomater. Nanobiotechnol.* **2011**, *2*, 435–444. [CrossRef]
89. Qu, P.; Gao, Y.; Wu, G.; Zhang, L. Nanocomposites of poly (lactic acid) reinforced with cellulose nanofibrils. *BioResources* **2010**, *5*, 1811–1823.
90. Luo, S.; Cao, J.; Wang, X. Investigation of the interfacial compatibility of PEG and thermally modified wood flour/polypropylene composites using the stress relaxation approach. *BioResources* **2013**, *8*, 2064–2073. [CrossRef]
91. Luo, S.; Cao, J.; Wang, X. Properties of PEG/thermally modified wood flour/polypropylene (PP) composites. *For. Stud. China* **2012**, *14*, 307–314. [CrossRef]
92. Desmet, G. Functionalization of Cotton-Cellulose Via High Energy Irradiation Initiated Grafting and Cyclodextrin Immobilization. Master's Thesis, Faculty of Engineering University of Ghent, Ghent, Belgium, 2010.
93. Nallasamy, P.; Mohan, S. Vibrational Spectra of CIS-1,4-Polyisoprene. *Arab. J. Sci. Eng.* **2004**, *29*, 17–26.
94. Dubey, K.A.; Bhardwaj, Y.K.; Chaudhari, C.V.; Kumar, V.; Goel, N.K.; Sabharwal, S. Radiation processed ethylene vinyl acetate-multiple walled carbon nanotube nano-composites: Effect of MWNT addition on the gel content and crosslinking density. *Express Polym. Lett.* **2009**, *3*, 492–500. [CrossRef]
95. Charlesby, A.; Pinner, S.H. Analysis of the solubility behaviour of irradiated polyethylene and other polymers. *Proc. R. Soc. A* **1959**, *249*, 367–386. [CrossRef]
96. Sharifa, J.; Yunus, W.M.Z.W.; Dahlan, K.Z.H.M.; Ahmad, M.H. Preparation and properties of radiation crosslinked natural rubber/clay nanocomposites. *Polym. Test.* **2005**, *24*, 211–217. [CrossRef]
97. Zagorski, Z.P. EB-crosslinking of elastomers, how does it compare with radiation crosslinking of other polymers? *Radiat. Phys. Chem.* **2004**, *71*, 261–265. [CrossRef]

© 2020 by the authors. Licensee MDPI, Basel, Switzerland. This article is an open access article distributed under the terms and conditions of the Creative Commons Attribution (CC BY) license (http://creativecommons.org/licenses/by/4.0/).

Review

Wood Composites and Their Polymer Binders

Antonio Pizzi [1,*], Antonios N. Papadopoulos [2] and Franco Policardi [3]

1. LERMAB-ENSTIB, University of Lorraine, 88000 Epinal, France
2. Department of Forestry and Natural Environment, International Hellenic University, 66100 Drama, Greece; antpap@for.ihu.gr
3. Faculty of Electrical Engineering, University of Ljubljana, Tržaška cesta 25, SI-1000 Ljubljana, Slovenia; franc.policardi@fe.uni-lj.si
* Correspondence: antonio.pizzi@univ-lorraine.fr

Received: 4 May 2020; Accepted: 12 May 2020; Published: 13 May 2020

Abstract: This review presents first, rather succinctly, what are the important points to look out for when preparing good wood composites, the main types of wood composites manufactured industrially, and the mainly oil-derived wood composite adhesives and binders that dominate and have been dominating this industry. Also briefly described are the most characteristic biosourced, renewable-derived adhesives that are actively researched as substitutes. For all these adhesives, synthetic and biosourced, the reviews expose the considerable progresses which have occurred relatively recently, with a host of new approaches and ideas having been proposed and tested, some even implemented, but with even many more already appearing on the horizon.

Keywords: wood composites; wood composite binders; synthetic wood adhesives; biosourced wood adhesives; environment-friendly; new approaches

1. Introduction

Wood composites is a growing field of products that are increasingly present for a variety of applications, with an undiminished upward trend now for very many decades. One must first define what is strictly intended as a wood composite as there are a host of different products that, while they could be classified as wood composites, are really on the margin of what is defined as such in the jargon of the wood profession. In general, two main and distinct groups of wood composites exist, namely, strictly speaking, just wood panel composites and the rest, this latter being glulam, fingerjoints, etc., with parallam and scrimber really belonging in the margin of both classes. In general, also, a wood composite is a composite in which the wood is in a markedly dominant proportion. This would exclude wood–plastic composites where the proportion of the plastic is equal, or almost equal, to the proportion of the wood present, and they are excluded from this review.

It must be clearly pointed out that one cannot speak about wood composites without speaking in depth of the polymer binders and adhesives used to hold them together. The history of wood composites themselves is inextricably intertwined with the history and the development of the polymer binders that hold them together and their manufacture. In fact, not only has there been continuous development of new or improved binders that has allowed the development of wood composites but, as presented later in this review, it is the continual renewal, new discovery, and upgrading of such binders that has allowed and allows progress in wood composites. It is for this reason that this review is divided into three parts. The first describes the main wood composites existing today, the second the synthetic adhesives that have dominated and still dominate this field, and the third the variety of new adhesives, either synthetic or biosourced, or renewables, which is the real area of intellectual ferment and exploration today. This is defined with the clear understanding that for any adhesives or composites where in-depth reviews already exist, the reader will be referred to these. This approach is adopted as the field is really very vast, as befits a wood binders industry that represents in excess of 65%, by volume, of all the adhesives produced in the world, all applications included.

It must be also pointed out that the technique to prepare a good wood composite is the mix of two very different technologies, namely the use of a good adhesive, hence the capacity of formulating, engineering, and preparing one, and the technology of manufacture of the composite, with this being particularly true in the case of wood panels. Thus, to obtain a good/acceptable wood composite is a 50/50 balance between these two technologies. This must be clearly kept in mind because one can prepare really awful wood panel composites even when using an excellent adhesive if the composite assembly technology is faulty and, conversely, one can still prepare an acceptably good composite by playing with its manufacturing technology, even if the adhesive is rather mediocre, if not outright poor. Thus, the mastering of the two technologies is essential. The manufacturing techniques of wood composites are mechanical and will not be treated in this review, which is polymers-oriented; nonetheless, the most essential examples will be presented in brief. In-depth, detailed reviews of the manufacturing technology of pressing and on how to master this to obtain a good wood panel composite already exist, and the readers are referred to these extensive specialized reviews [1].

2. Wood, Wood Plasticization and Wood Panels

Wood is a natural composite made of approximately 60–65% carbohydrate fibers (approx. two-thirds cellulose and one-third hemicelluloses), 25–30% of a random polyphenolic branched polymer, lignin, functioning as a fiber binder, and 10% of residues, extractives, or cellular waste infiltrates (oleoresins, tannins, starches, some inorganic salts, etc.) coating its porous cellular surfaces. Wood panel manufacturing is based on the densification of a particle mat and its consolidation in a hot press. At a molecular level, wood cells are deformed above their elastic domain and plastic deformation or damage occurs [1]. Both mechanisms result in an irreversible deformation. As wood viscoelastic properties depend on internal mat conditions, when the environment is such that the viscous component becomes very important, flow takes place, and this leads to a true plastic deformation. If conditions do not allow or inhibit molecular movement, damage will appear. Thus, mechanisms that occur in wood densification and are regarded as a manifestation of plasticity would, under different circumstances, be considered as damage. Referring to wood composites, most of the final board properties depend on how wood cells buckle [2].

Among the many factors influencing the performance of the composite, the relative moisture content of the panel's surfaces and core is the most important as one can easily play on it to improve a composite by its pressing technology, even if the adhesive is not a great one. This concept leads to the importance of the density profile across the thickness of a wood composite panel. The density profile as a function of panel thickness is an important measure and a forecasting tool of the likely characteristics of a panel. At parity of overall panel density, the shape of the density profile determines which are the characteristics that the panel will have. The different panel density profiles are shown in Figure 1, in an exaggerated manner, for example purposes.

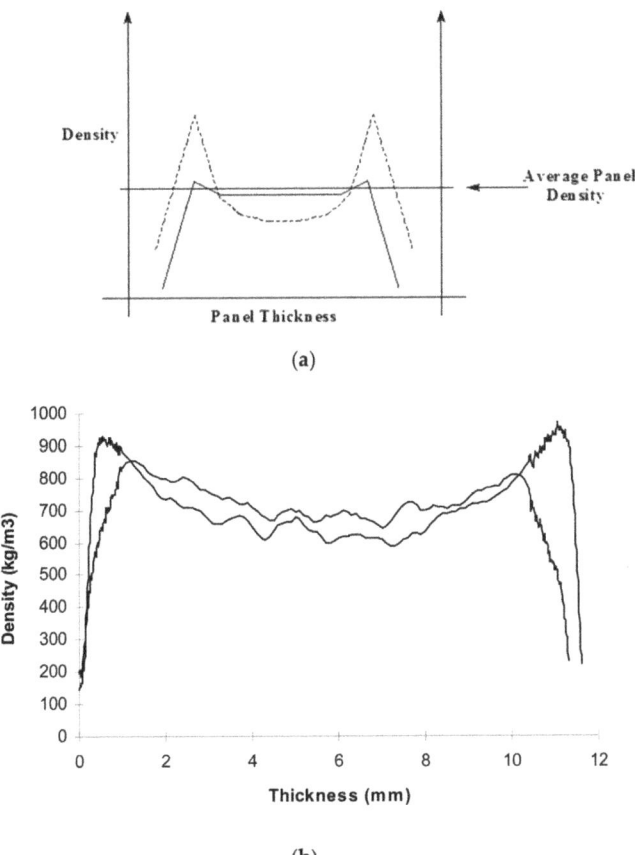

Figure 1. (a) Schematic, exaggerated representation of two different panel density profiles along the panel thickness leading to very different board properties; (b) an example of two real, different density profiles of two different wood particleboards [1].

First of all, the density of the surfaces is practically always higher than the density of the core of the board, if the board is well made. The low density of the outermost surfaces of the raw panel is always eliminated in industrial panels just by sanding, so that the surfaces can present the highest density possible, hence, as hard as possible. Panels which have relatively higher surface densities and lower core densities at parity of total panel density have better bending strength and better screw-holding capability. Conversely, panels which have relatively lower surface densities in relation to the core density have better internal bond strength (tensile strength perpendicular to the panel plane) and better durability. Thus, what type of strength characteristics and, hence, which density profile it should be manufactured with often depends on the application which is intended for the board. Many parameters influence the compression ratio in the press and, hence, the density profile of the panel. In the introduction of this chapter, it has been stated that one can still make good panels when using a relatively poor adhesive. This can be explained on the basis of the density profile. Internal bond strength depends almost exclusively on how good the adhesive is, and on how high is the density of the panel layer of lower density. When the adhesive is poor, it is then sufficient to use pressing conditions of the panel that increase compression of the core to bring it to a higher density value, and lowers the density of the surfaces without changing the panel overall density. A flatter profile is then obtained, such as the flatter profile in Figure 1, where the weakest layer has a higher density, hence giving a much higher internal bond strength, this being the real measure of how good the panel and bonding are. This can be achieved quite readily by just changing, for example, the relative percentage moisture content (MC%) of surfaces and core before pressing. Thus, changing the surface to core MC% of 14%:10%, which is quite common, to just 13%:11% or 12%:10% will considerably densify the board core and upgrade the internal bond strength of the board [1]. In this respect, other parameters also have strong influence on this effect, such as the rate of press closure, maximum pressure, and others [1]. In industrial practice, however, a multipurpose board satisfying national standards is often produced unless the client has pre-specified the type of characteristics they want.

3. Types of Wood Composites

Many types of wood panels, for a variety of applications, are manufactured today. Their definitions, in brief, are as follows, but the reader is referred to far more detailed and in-depth reviews on the technology of manufacturing of these products [1,3,4].

Particleboard: A flat hot-pressed wood composite panel composed of randomly oriented wood chips bonded by hot-pressing by using thermosetting adhesive resins, mainly urea–formaldehyde (UF), melamine–urea–formaldehyde (MUF), phenolic resins (PF and TF), and isocyanates (pMDI). The board is generally composed of three distinct layers, the surface layers being composed of finer wood chips than the coarser core layers. Some processes yield a continuously chip-size-graded board along the surface/core/surface thickness. The panel generally has a density of 650–700 kg/m^3 and the average amount of resin solids in the board core section is of between 6% and 12% on dry wood (although lower and much higher percentages are also sometimes used). Panels of the same type but composed of wood chips of greater length and greater width but similar thickness are called also flakeboard and waferboard.

Oriented Strand Board (OSB): A flat hot-pressed three-layers wood composite panel composed of oriented wood wafers bonded by hot-pressing by using thermosetting adhesive resins. The very thin wafers (length and width are very much bigger than in particleboard and of the order of 100 mm × 20 mm, respectively) are oriented in the same direction within the same layer and at 90° of each other in adjacent layers yielding a particularly strong panel very suitable for structural applications. It is the modern competitor of plywood but at a much lower price. The lower surface area of the wafers, in relation to other types of panel, yields panels that need to be bonded with only 4–5% adhesive solids on dry wood. OSB is, today, the main substitute panel for the rather more expensive plywood, but presenting the same advantages. It is a panel for structural use.

Medium Density Fiberboard (MDF): a flat hot-pressed composite panel composed of wood fibers obtained by thermomechanical wood pulping and traditionally bonded with an adhesive to a density of around 750–800 kg/m^3. MDFs of much lower densities are also known. It is a panel mainly bonded with urea–formaldehyde resins, and used for furniture and interior use. Its production has experienced a considerable growth.

Hardboard (high density fiberboard): A flat-pressed wood composite panel composed of randomly oriented wood fibers obtained by thermomechanical wood pulping and traditionally bonded without any adhesive by hot-pressing simply by the very high density (900–1100 kg/m^3) and the high-temperature-induced flow of the lignin component of the fibers. Panels containing a small amount of adhesives (2–3% adhesive solids on dry fiber), generally PF resins, are often produced today to upgrade the properties of the panel.

Plywood: A flat hot-pressed multilayer wood panel composed of oriented wood veneers bonded by hot-pressing by using thermosetting adhesive resins. The veneer wood grains are oriented at 90° of each other in adjacent layers, yielding a particularly strong panel. As a consequence, this is the panel with the best strength/weight ratio but is rather expensive in relation to the equally strong OSB.

Laminated Veneer Lumber (LVL): A flat-pressed multilayer wood panel similar to plywood composed of oriented wood veneers but differently from plywood oriented all in the same direction in all the layers and bonded by hot-pressing by using thermosetting adhesive resins.

Laminated beams (glulam), parallam, scrimber, and fingerjoints: A flat-pressed multilayer wood beam with thick wood planks constituting the layers, used for structural exterior applications and bonded with PRF (phenol–resorcinol–formaldehyde) cold-setting resins, or MUF cold-setting resins, or even with certain types of polyurethanes (PURs), especially single-component PURs. The individual wood planks are bonded to the necessary length to compose the beam by fingerjoints bonded with one of the same three adhesives above.

Parallam and scrimber are similar products. Parallam is a beam made by a continuous manufacturing process composed of bigger size wood needles (very elongated wood particles) reassembled with a structural exterior grade adhesive, the favorite adhesive when heat curing isocyanates (pMDI) and PRFs when cold-curing. Scrimber was instead conceived by the CSIRO in Australia with the idea of "scrimming", thus breaking down the structure of the wood by crushing it only as far as necessary and producing bundles of interconnected but still aligned strands to allow it to be formed by coating them with an adhesive into a desired end-product, rather than destroying the natural alignment of wood fibers and realigning them as more conventional processes do.

A recent review of the type of panels used in European industry is also available, and the reader is referred to this [4].

4. Current Wood Composite Adhesives

The most current big volume wood composite adhesives are briefly listed as follows. Synthetic adhesives do still dominate this market. What is already on the market will be described later for biobased adhesives and for totally new approaches in synthetic adhesives. The reader is however referred to far more in-depth reviews for each of them.

4.1. Urea–formaldehyde Adhesives

Urea–formaldehyde (UF) adhesives are, by far, the adhesive dominating the wood panel composites market for the preparation of interior grade composites used for furniture and a wide variety of other applications. An approximate volume of 11 million tons/year of these adhesives for wood are used worldwide. They are obtained by reaction of urea and formaldehyde according to the simplified Scheme 1:

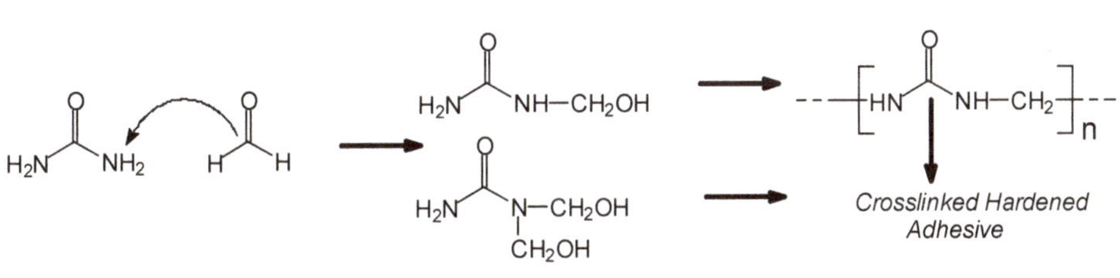

Scheme 1. Basis reaction of urea and formaldehyde to prepare UF resin.

Their technology has considerably progressed during the last few decades under the pressure of ever more stringent formaldehyde emission regulations. Notwithstanding their drawbacks, such as lack of resistance to exterior weather conditions and emission of formaldehyde, they are however very difficult to substitute due to their relatively low cost, their excellent adhesive performance, and their ease of handling. The reader is referred to more specific, in depth and detailed reviews on their chemistry and conditions of applications [5–8].

4.2. Melamine-Formaldehyde and Melamine–Urea–Formaldehyde Adhesives

Melamine is an expensive chemical, thus, today, these resins are traded as melamine–urea–formaldehyde (MUF) resins of equal performance as the older pure melamine–formaldehyde (MF) resins. They are of two types: (i) mUF, thus UF resins with just between 2–5% melamine that are nothing else than upgraded interior grade UF resins, this being also one method of decreasing a UF's formaldehyde emission.; (ii) exterior- and semi-exterior-grade MUF adhesives containing between 30% to 40% of melamine. Their schematic representation can be shown in Scheme 2.

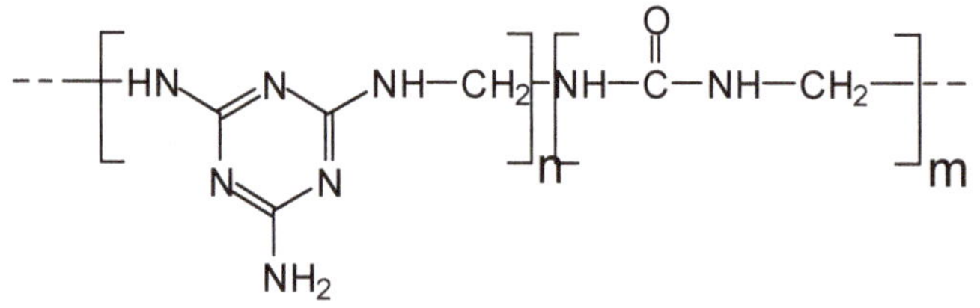

Scheme 2. Schematic representation of a MUF resin.

Traditionally, a few decades ago, these resins were considered as semi-exterior-grade, but the progress in their technology has been so considerable that they can compete well in performance with the more classical exterior-grade phenol–formaldehyde (PF) adhesives. Here, too, the reader is referred to more in-depth reviews [7–9].

4.3. Phenol–Formaldehyde Adhesives

Phenol–formaldehyde (PF) adhesives are, by volume, the second most important wood composite adhesive, with up to 3 million tons/year being used worldwide. Their schematic structure can be represented as in Scheme 3

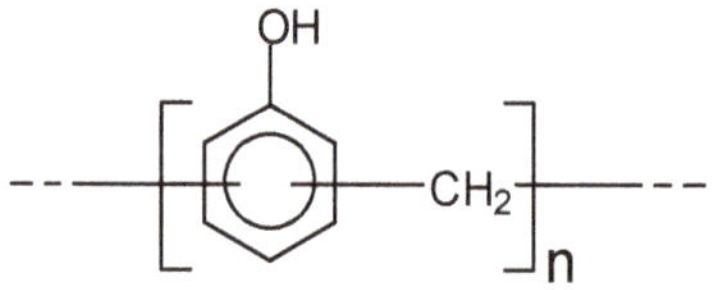

Scheme 3. Schematic representation of a PF resin.

They are fully exterior-grade adhesives to bind truly weather-resistant wood composites and extensively used for particleboard, OSB, and marine plywood. They dominate where panels are used for housing, but have traditionally had the defect of having slower curing at higher temperature than for MUF adhesives. The progress has been considerable also in these resins, with PF adhesives now pressing as fast as MUF resins. Here, too, the reader is referred to more in-depth reviews [8,10,11].

4.4. Phenol-Resorcinol-Formaldehyde (PRF) Adhesives

Differently from the first three classes of wood composite adhesives presented above, PRF adhesives are cold-setting adhesives. They can be schematically represented in Scheme 4

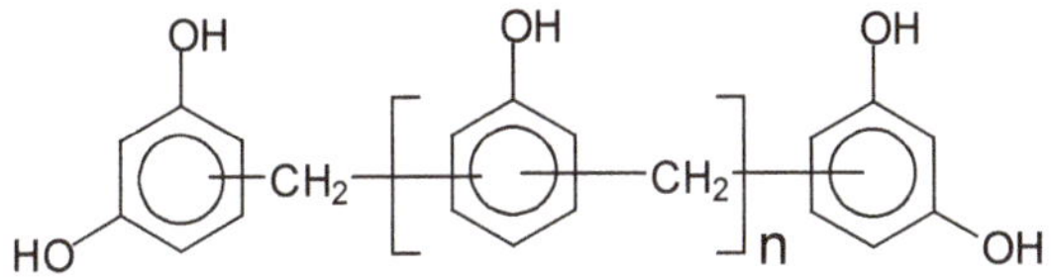

Scheme 4. Schematic representation of a PRF resin.

They are expensive, due to the high cost of resorcinol that gives them their characteristic to cure at ambient temperature. They are thus used for glulam, finger-jointing, and similar products, and for ambient temperature-made LVL. They are binders for fully exterior-grade, weather-resistant composites. Due to their high cost and the particular products they are used to bind, their volume is relatively low, around 30 thousand tons/year, but they are high-value resins. Here, too, the reader is referred to more in-depth reviews [8,10,12].

4.5. Polymeric Isocyanates

The concern about the formaldehyde emission vapor levels from UF adhesives has brought isocyanate adhesives to the fore, where formaldehyde emission does not occur as no formaldehyde is added. For wood composites only polymeric 4,4′-diphenyl methane diisocyanate (pMDI) is the isocyanate used, which is a liquid of 100% solids. It can be schematically represented as (Scheme 5)

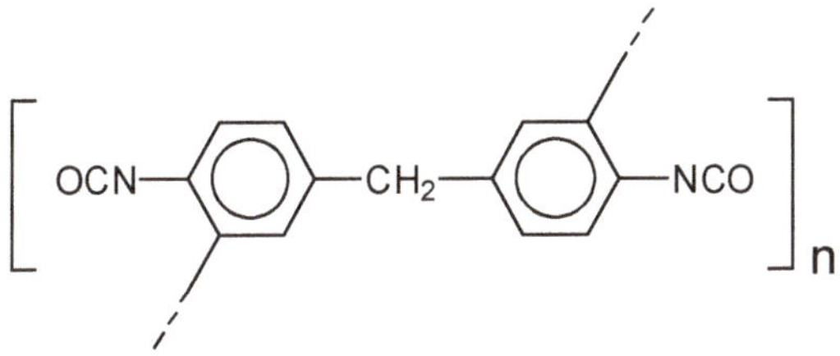

Scheme 5. Schematic representation of pMDI.

pMDI is an excellent adhesive and can be used in markedly smaller proportions than UF, MUF, and PF adhesives to bind wood composites. It is, though, relatively more expensive than the other three major formaldehyde-based adhesives, this somewhat counterbalancing the advantage of it being needed in lower proportions. Even this adhesive has more recently been under some pressure from stringent environmental regulations due to its vapor and relative toxicity. A use where it has a particular advantage is in mixing, in smaller proportions, with any of the three formaldehyde-based adhesives to upgrade their performance. Here, too, the reader is referred to more in-depth reviews [5,8,13].

4.6. One-Component Polyurethanes (PURs)

One-component PURs are the main competitor of PRF adhesives for the same types of applications. Their structure can be schematically represented as (Scheme 6)

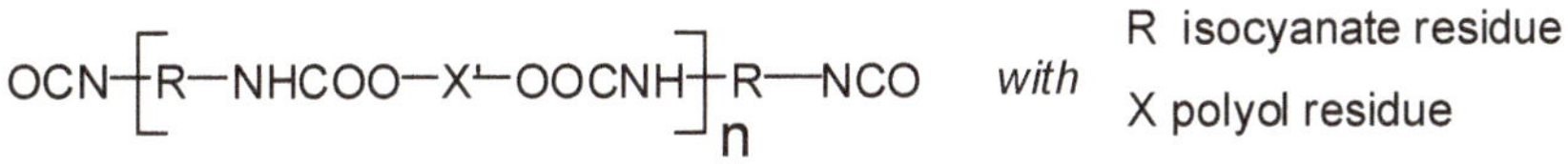

OCN—[R—NHCOO—X′—OOCNH—R—NCO]$_n$ with   R isocyanate residue
X polyol residue

Scheme 6. Schematic representation of monocomponent PUR.

Their advantage is their considerably better ease of handling as no mixing with hardeners or catalysts are needed. Their main drawback is in the structure itself, as they are prone to yielding bonded joints subject to both creep and temperature-dependent creep. However, some of them have been formulated to partially overcome this drawback. They are relatively expensive, but this is not a drawback as their competitor, the PRF adhesive, is also expensive. Here, too, the reader is referred to more in-depth reviews [14].

5. New, Biobased, Renewable and Synthetic Wood Polymeric Adhesives

5.1. Biobased Wood Composite Adhesives

There are several biobased adhesives based on renewable natural materials that are at the forefront of new developments. Some of these are already industrial, sometime for many years, such as tannin adhesives and some soy adhesives, while others are on the way of industrialization, and many others are, as yet, at the experimental stage.

First of all, it is necessary to define what is meant by biobased wood adhesives or adhesives from renewable, natural, non-oil-derived raw materials. This is necessary because in its broadest meaning, the term might be considered to include urea–formaldehyde resins, urea being a non-oil-derived raw material. This, of course, is not the case. The term "biobased adhesive" has come to be used in a very well specified and narrow sense to only include those materials of natural, non-mineral origin which can be used as such or after small modifications to reproduce the behavior and performance of synthetic resins. Thus, only a limited number of materials can be currently included, at a stretch, in the narrowest sense of this definition. These are tannins, lignin, carbohydrates, unsaturated oils, proteins and protein hydrolysates, dissolved wood, and wood welding by self-adhesion.

Regarding tannin adhesives, tannins extracted from bark or wood of trees are traditionally used for leather, and these adhesives have been in industrial use since the early 1970s in a few countries of the southern hemisphere. There are a number of detailed reviews on the use of tannins for wood adhesives. The reader is referred to these detailed studies [8,15]. However, here, existing technologies and industrial use of tannin wood adhesives are presented.

As extensive studies already exist, regarding this application of tannin, only a few of the main achievements of tannin-based adhesives for wood products will be highlighted: (1) The development, optimization, and industrialization of non-fortified but chemically modified thermosetting tannins for particleboard, other particle products, and plywood [16–18]; (2) The technology for rapidly pressing tannin adhesives for particleboard, which is also industrial [19]; (3) The development and industrialization of tannin–urea–formaldehyde adhesives for plywood and, in particular, as impregnators for corrugated board starch binders [20,21]; (4) The development and industrialization of cold-setting tannin–resorcinol–formaldehyde adhesives for glulam and finger-jointing [22]; (5) The large-scale development and industrialization of quick-setting "honeymoon" separate application cold-setting adhesives for tannin-bonded glulam and fingerjoints [23–25]; (6) The development and industrialization of zinc salts to accelerate the hardening of non-fortified tannin adhesives for plywood [5,26–28]; (7) Successful formulation, development, and industrialization of pine bark tannin adhesives for particleboards and for glulam and finger-jointing in Chile [18,29]; (8) The development of isocyanate/tannin copolymers as difficult-to-bond hardwood adhesives and for plywood and other applications [30,31]; (9) The development of very low formaldehyde tannin adhesives for particleboards and other wood panels; (10) The development of the use of hardeners other than formaldehyde for thermosetting tannin adhesives [5,32–34]; (11) The discovery and development of self-condensation of tannin for adhesives [35–42].

All industrialized technologies today are based on paraformaldehyde or hexamethylene tetramine (hexamine) [42–45]. The latter is much more user- and environmentally friendly.

As regards new approaches in tannin-based wood adhesives, a number of experimental improvements have been studied, dictated by the new environment in which wood adhesives must operate. First of all, the relative scarcity of tannins produced in the world, compared to the tonnage of synthetic adhesives used in the panel industry, has led to a great deal of research on the extension of the tannin resource in order to have larger tonnage. As the potential material for tannin extraction shows that millions of tons of this material can be extracted each year worldwide, some companies have started to build additional extraction plants. This movement is still relatively small, but is ongoing. The second approach, to extend the tannin with another abundant and natural material, has led to the preparation of adhesives based on in situ copolymers of tannins and lignin [46] or copolymers of tannin and protein or soy flour [47], the use of tannin–furfuryl alcohol adhesive formulations, furfuryl alcohol being also a biobased material [48].

The second new constraint is the demand of most companies to eliminate formaldehyde emissions from tannin adhesive. This quest has taken two approaches: (i) total elimination of formaldehyde by substituting it with aldehydes that are less or nontoxic and nonvolatile [33,49], such as glyoxal, glutaraldehyde, or vanillin, the latter giving a fully biobased tannin adhesive and, more recently also, fully biobased carbohydrate extracts from very diffuse African trees that yield, on hot-pressing, both hydroxymethyl furfural and hydroxymethyl furfuryl alcohol as hardeners [34]; (2) the use of non-aldehyde hardeners such as tris(hydroxymethyl)nitromethane [50] and tris(hydroxymethyl)aminomethane [51] or even by combination with furfuryl alcohol, the latter functioning both as a hardener and a contributor to a tannin/furan copolymer [48,52]; (3) the use of hexamine with the formation of $-CH_2-NH-CH_2-$ bridges between the tannin molecules, where the secondary amine is capable of absorbing any emission of formaldehyde from the heating of the wood itself or any other emission of formaldehyde to produce truly zero-formaldehyde emission panels [42–45]; (4) lastly, the hardening of the tannins by self-condensation without the addition of a hardener, self-condensation catalyzed by the wood substrate itself in the case of fast-acting procyanidin tannins, such as pine bark tannins, and for slower tannins, by addition of silica or silicate or other accelerators [8,35–41] allowing the preparation of wood particleboard of indoor quality.

A very new approach, giving very encouraging results, was obtained from the research on sodium periodate-specific oxidation of carbohydrates pursued in the ambit of the research on soy flour adhesives (see later) [53,54]. In this approach, periodate oxidation of the monomeric and oligomeric carbohydrates present in tannin extract and of glucose added to it caused cleavage of the monomeric and polymeric carbohydrates forming a variety of nontoxic and nonvolatile monomeric and oligomeric aldehydes. These reacted with the tannin polyphenolic part of the tannin extract to give good bonding results for particleboards [55].

Lignin adhesives: Extensive reviews on a number of proposed technologies of formulation and application do exist, and the reader is referred to these in earnest [56–65]. Lignin is the most abundant natural polymer after cellulose. Its polyphenolic nature has always generated interest for preparing wood adhesives. None of the many adhesive systems based on pure lignin resins, hence, without synthetic resin addition, have succeeded commercially at an industrial level. Some were tried industrially, but for one reason or another, too long a pressing time, high corrosiveness for the equipment etc., they did not meet with commercial success. Still notable among these is the Nimz system based on the networking of lignin in presence of hydrogen peroxide [8,56,66] and the Shen system based on the self-coagulation and crosslinking of lignin by a strong mineral acid in the presence of some aluminum salt catalysts [5,8,56]. Of interest in the MDF field is also the system of adding laccase enzyme-activated lignin to the fibers or activating the lignin, in situ, in the fibers, also by enzyme treatment with the addition of 1% polymeric isocyanate [66,67]. The more traditional approaches are the use of methylolated lignin or lignosulphonates in PF resins [68] or even the preparation of copolymerized lignin–phenol–formaldehyde resins for plywood. In this respect, adhesion of plywood by the first of these with limited formylated lignin addition (up to 20%) has been used extensively in the past, in Canada, for plywood [69,70]. More recently, an LPF resin according to the second approach has been commercialized for plywood containing up to 50% lignin [71], and improvements on this by using ionic liquid-treated lignin to prepare LPF and LP-glyoxal resins have also been reported with encouraging results [72–74]. A further new approach for lignin wood adhesives uses pre-methylolated lignin with small amounts of a synthetic PF resin and of isocyanates (PMDIs), the lignin proportions being up to 65% of total solids [75,76]. Some new and rather promising technologies on totally biolignin adhesives have, however, also been developed recently. These are (a) adhesives for particleboard, thin hardboard, and other agglomerate wood panels based on a mix of tannin/hexamine with pre-glyoxalated lignin [46], and (b) similar formulations for high resin content, high-performance agricultural fiber composites [77–83].

The newest approach was obtained from the research on sodium periodate specific oxidation of carbohydrates pursued in the ambit of the research on soy flour adhesives (see later) [53,54]. In this approach, periodate-specific oxidation of pre-demethylated lignin by cleavage of some bonds of its aliphatic chains at the level of its beta-O-4 bonds and vicinal C–OH groups caused cleavage of the lignin structure, yielding a variety of nontoxic and nonvolatile aldehyde groups in the lignin structure itself. These reacted with the lignin aromatic nuclei to give a lignin adhesive, yielding good bonding results for plywood [84].

Protein Adhesives: First of all, it must be noted that excellent wood composite UF resins containing up to 15% of proteins have already been commercially available for quite a long time now, around a decade. These are the AsWood resins for wood panel products. However, this field is dominated by the abundant research work on soy protein and soy flour adhesives. A considerable number of different approaches have been tried with encouraging results. First technologies based on the pre-reaction of soy protein hydrolysate pre-reacted with formaldehyde, and this being mixed with a PF resin and with isocyanate (pMDI) gave encouraging results [85,86]. The same with pre-glyoxalated soy and/or wheat gluten with PF and pMDI [87,88]. An even more interesting, non-traditional system is based on the pre-reaction of the soy protein hydrolysate with maleic anhydride to form an adduct that is then reacted in the panel with polyethylene imine [89,90]. This system also works well, as one company has started using it industrially in the United States, but suffers from the drawback of being expensive. Equally, work on protein–glutaraldehyde adhesives has been published [91].

Three totally new experimental approaches to yield totally biobased soy flour and soy protein wood adhesives have very recently come to the fore. The first is based on the reaction of a hydrolyzable tannin (tannic acid), or a condensed tannin with soy protein isolate and/or soy flour. This gave rather encouraging results for plywood, particleboard, and hardboard [47,92–94]. The second unusual approach is by the formation of soy protein isolate-based polyamides formed by the reaction of soy protein isolate with maleic anhydride and hexamethylene diamine [95] as a totally new resin for such an application. The third approach to obtain totally biobased adhesives is even more revolutionary. It is based on the specific oxidation of carbohydrates by sodium periodate. The bonding results using soy flour gave good bonding results for plywood [53]. The periodate treatment was found to cause cleavage of monomeric and polymeric carbohydrates in soy flour, forming a variety of nontoxic and nonvolatile monomeric and oligomeric aldehydes. These reacted with the protein part of the soy flour to give good bonding results [54]. Even more interesting was that periodate oxidation formed aldehyde groups on the soy protein itself, this also contributing to good bonding results [53,54].

Carbohydrate Adhesives: Carbohydrates can be used as wood panel adhesives in three main ways: (i) as modifiers of existing PF and UF adhesives, with considerable literature on this traditional approach already existing [96–102], (ii) by forming degradation compounds which then can be used as adhesives building blocks, and (iii) directly as wood adhesives. The second route leads to furanic resins. Furanic resins, notwithstanding that their basic building blocks, furfuraldehyde and furfuryl alcohol, are derived from acid treatment of the carbohydrates in waste vegetable material, are considered today as purely synthetic resins [103]. This opinion needs to change, as they are real natural-derived bio-resins, and extensively used in applications other than wood composite binders. Appropriate reviews dedicated just to them do exist [103]. However, both compounds are relatively expensive and very dark-colored, and furanic resins have made their industrial mark in fields where their high cost is not a disadvantage. They can be used very successfully for panel adhesives, but the relatively higher toxicity of furfuryl alcohol before it is reacted needs to be addressed if these resins are to be considered for wood composites. In this context, furanic–aldehyde resins for wood composites starting from furfuryl alcohol and nontoxic, nonvolatile aldehydes have also been reported [104].

Several research groups [105,106] have recently described the use of liquefied products from cellulosic materials, literally liquefied wood, which showed good wood adhesive properties. Lignocellulosic and cellulosic materials were liquefied in the presence of sulfuric acid under normal pressure using either phenol or ethylene glycol.

The oxidation by periodate ion, resulting in a 1,2-glycol scission, is one of the most widely used reactions in carbohydrate chemistry. The mild reaction and the aqueous solvent conditions for periodate oxidation are particularly apt for use with water-soluble carbohydrates. The development and wide application of the reaction are due to its high degree of selectivity [53,54,107–109]. While the periodate specific oxidation reaction and outcome are better known for carbohydrate monomers [54,107] and dimers [54] (Figure 2) such as glucose and sucrose, equivalent reactions are also known for higher carbohydrate oligomers up to cellulose itself [53,54,108,109] (Figure 2). In the case of cellulose and long carbohydrate oligomers alone, the reactions that have been shown to occur are the condensation of the aldehydes formed with other carbohydrate chains to yield crosslinking, leading to solid panels in the case of cellulose (Figure 2) [109].

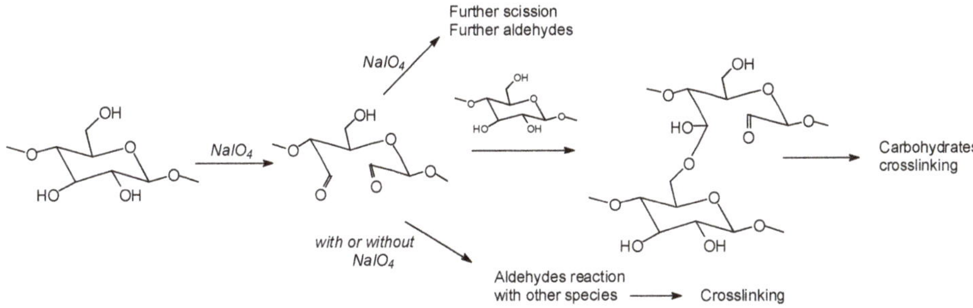

Figure 2. Specific oxidation of carbohydrates by sodium periodate leading to nontoxic and nonvolatile biosourced aldehydes and to different crosslinked products according to the conditions used [54].

By increasing the level of oxidation with further periodate in the presence of an aldehyde-reactive species such as a soy protein [53,54], a flavonoid tannin [55], or other reactive species, including lignin, crosslinking can also occur, leading to feasible wood adhesives [53–55,84] already described above (Figure 2).

A recent approach to wood composite binders is based on glucose- and sucrose-based non-isocyanate polyurethanes (NIPUs). The finding that glucose and sucrose reacted with dimethyl carbonate (or other cyclic dicarbonates) and a diamine to form NIPU resins [110] (Figure 3) has led to the application of these resins as wood panel composite adhesives [111] with very encouraging results. An equally new approach is the use of the reaction of glucose with maleic anhydride and a diamine to rather form polyamide binders, rather than NIPU, that also yielded wood particleboard binders showing good results [95].

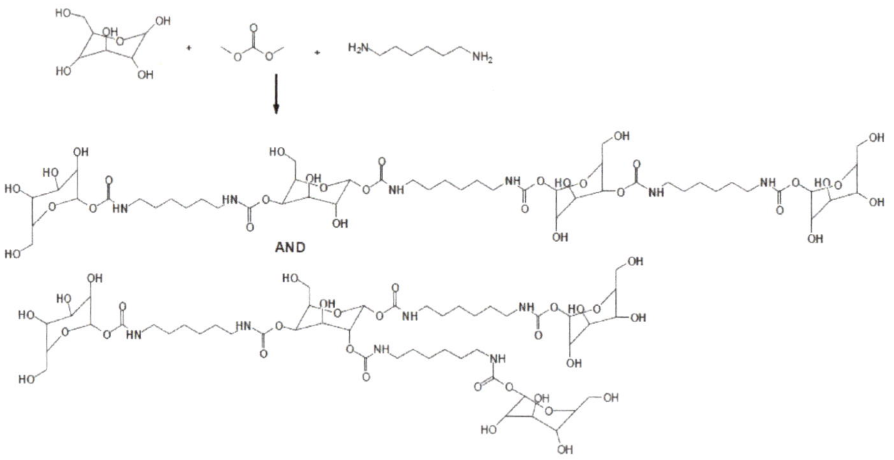

Figure 3. Schematic representation of the formation of non-isocyanate polyurethane (NIPU) oligomers from monosaccharides reacted with dimethyl carbonate and hexamethylene diamine. The two tetramers shown have been identified and are among the multitude of other NIPU oligomers identified [110].

Equally exciting and revolutionary has been the development of the use of citric acid as a wood composite binder. This is articulated in a number of different approaches, all involving carbohydrates in one manner or another. The concept of citric acid binders for wood composites was first advanced by Japanese researchers yielding very encouraging bonding results [112,113]. The use of citric acid was then extended later to application to wood welding with increased waterproofing of the wood weld line [114] and to bonding of flat veneers such as plywood and LVL [115]. This latter work showed, by chemical analysis, that citric acid was able to function as a binder of the carbohydrates in wood by reacting with them as well as to react with lignin, across two joint veneers [115], with species as in Figure 4 being identified, and was a viable plywood binder. Again, sugar, such as in sucrose + citric acid, was demonstrated as well to be a very viable adhesive for wood composite panels [116–119] in which not only furfural and hydroxymethyl furfural were formed by the action of the acid on the sugar but also crosslinking of the sugar and the holocellulose carbohydrates by the citric acid was observed (Figure 5), corroborating the same crosslinking with wood carbohydrates, glucose, and with lignin by the action of citric acid directly on wood, as found by Del Menezzi et al. (Figure 4) [115].

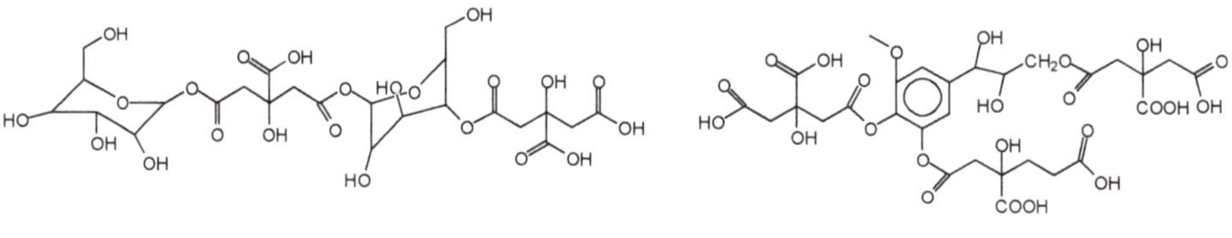

Figure 4. Example of citric acid bridges linking glucoses or holocellulose chains, and actually identified reaction products of citric acid with lignin. The compounds shown have been identified among several others of the same type, showing that both holocellulose and lignin are involved in wood bonding and citric acid bridge-forming by citric acid alone [116].

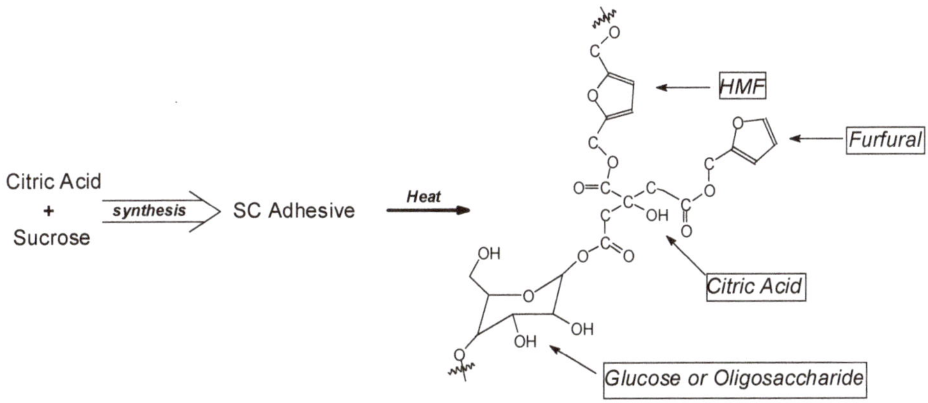

Figure 5. Proposed bridges and possible structures formed by the reaction citric acid with sucrose and with the furfural and hydroxymethyl furfural generated from sucrose by the citric acid treatment in sugar + citric acid adhesives [118].

5.2. Other Bioadhesives for Wood Composites

There are other biomaterials that have been proposed and used for the preparation of bioadhesives. None of the resins presented in this section have reached, at least up to now, industrial trials, but they are interesting and unusual concepts that are worthwhile to pursue and report on.

First, cashew nutshell liquid, mainly composed of cardanol but containing also other compounds. Its dual nature, a resorcinol aromatic ring with an unsaturated fatty acid chain, makes a potential natural raw material for the synthesis of water-resistant resins and polymers. The resorcinol group and/or the double bonds in the chain can be directly used to form hardened networks. Alternatively, more suitable functional groups, such as aldehyde groups and others, can be generated by ozonolysis on the alkenyl chain. The first reaction step yields a cardanol hydroperoxide as major product, that following reduction by glucose or by zinc/acetic acid, yields a high proportion of cardanol-derived aldehyde groups. These crosslink with the aromatic groups of cardanol itself and, thus, a self-condensation of the system, yielding hardened networks [120]. A good review on cardanol thermoset polymers other than for wood composites does exist [121].

Equally, adhesives based on vegetable oils have also been proposed. Encouraging techniques involving unsaturated oils for wood and wood fiber adhesives have been reported [122]. Wheat straw particleboards were made using UF and acrylated epoxidized soy oil (AESO) resins with two resin content levels: 8% and 13%. The physical and mechanical properties of these boards showed that AESO-bonded particleboards have higher physical and mechanical properties than UF-bonded boards, especially regarding internal bonding and thickness swelling [122]. Bioresins based on soybean and other oils have also been mentioned in the literature [123]. These liquid resins were obtained from plant and animal triglycerides by suitably functionalizing the triglyceride with chemical groups (e.g., epoxy, carboxyl, hydroxyl, vinyl, amine, etc.) that render it polymerizable. The reference claims that composites were made using natural fibers such as hemp, straw, flax, and wood in fiber, particle, and flake form, but no results were given. In an older work, an epoxidized oil resin was evaluated as a wood adhesive in composite panels, and it could be tightly controlled through the appropriate selection of triglycerides and polycarboxylic anhydrides [124]. The literature on this resin [123] claims that crosslinking can be varied through the addition of specialized catalysts and several samples were prepared at a range of temperatures (120–180°C) that exhibited high water tolerance, even at elevated temperatures. Their main drawbacks appear to be a slow hot press time and their relatively high cost.

Fungal mycelium bonding is another very new and unusual concept that has been tested and presented for bonding wood panel composites. These new mycelium-based biocomposites (MBBs) were obtained from local agricultural (hemp shives) and forestry (wood chips) byproducts which were bonded together with natural growth of fungal mycelium [125]. As a result, hemp mycocomposites (HMCs) and wood mycocomposites (WMCs) were manufactured and their mechanical, water absorption, and biodegradation properties determined. Compression strength was better for WMCs by about 60% compared to that of HMCs. Water absorption and swelling were relatively high, and these composites were extensively biodegradable, an advantage for certain applications and a disadvantage for others.

Finally, in this category, one can count wood friction welding without adhesives. Rapid friction of two wood surfaces, one against the other, partially melts and mobilizes the intercellular wood material, mainly lignin and xylan hemicelluloses, to form a high-density and high-strength composite interphase formed by intertwined wood fibers and molten material [126–129]. A basic review on this subject exists [130]. The system works well both as linear welding or rotational dowel welding. Several systems have been recently developed to improve its water resistance [114,131,132]. Linear wood friction welding is an impressive but rather limited technique, as it requires expensive machinery and it can weld only pieces of rather limited length, around one meter. Rotational dowel welding, instead, is very flexible, needing only inexpensive but good quality hand or fixed drills, and presents no limitations of application. It has been and is already being used in small workshops for some biofurniture, assembly of small wood pieces, and room refitting [133–142]. Some temperature-welded plywood has also been prepared, but its hot press time is far too long to be of industrial significance [143,144].

There are numerous other natural biomaterials, oligomers, and polymers that have been used as binders in fields other than wood composites, but that would be worthwhile to test for wood composites as well. Among these, one can mention humins, isosorbide, itaconic acid, urushiol, and others [145–149].

6. New Approaches to Synthetic Adhesives

It must be clearly pointed out that while the environmental concern that predominates today has led to a considerable increase in research and publication on biobased adhesives for wood composites, the research on synthetic adhesives also has made huge progress with a host of new ideas and approaches been thought out and presented. Not all can be presented here, but a few of the more interesting examples are selected. There are two distinct trends in this context: (i) the ongoing improvement, also considerable, of traditional synthetic adhesives, especially in UF resins with, moreover, very valuable new ideas being presented [150–156] and MUF resins (see later), and (ii) the presentation of partially non-traditional or even totally non-traditional synthetic adhesives, some also with biobased content.

For the first type of approach, the first tendency has been to prepare engineered UF resins of progressively lower molar ratio, at levels much lower than 1:1 [157], which has become rather common in industry today. This was due to the attempt to minimize formaldehyde emissions from wood panels bonded with UF resins. One of the drawbacks of the much lower than 1:1 molar ratio has been identified in the increase in the tendency of the UF resin to form increasingly present crystalline domains upon hardening as a result of hydrogen bonds between linear molecules [153,158–160]. At higher molar ratios, the hardened resin is amorphous, affording better adhesion and better bonding performance. The overly high crystallinity drawback was very recently solved, and solved well [153] by blocking the formation of hydrogen bonds using transition metal ion-modified bentonite nanoclay through in situ intercalation and, thus, converting the crystalline domains of the UF resins to amorphous polymers. Addition of 5% nanoclay to the UF yielded in excess of 50% better adhesion, and almost 50% lower formaldehyde emission, thus resulting in a marked improvement in performance with a low level of crystallinity.

In the same trend, the potential introduction of a very acid pH condensation step in the preparation of UF resins, inducing the formation of occasionally considerable amounts of uron (a cyclic intramolecular urea methylene ether) in the UF resins of lower formaldehyde emission has attracted some research interest [149]. This initial work indicated that introduction of such an acid step can lead to UF resins of improved bonding strength, but also of higher post-cure formaldehyde emission. The results also indicated that minimization of the formation of urons yielded better UF resins when the strongly acid condensation step is introduced in the reaction. Subsequent work on the very acid step showed favorite uron formation at pH values higher than 6, and lower than 4 at which the equilibrium urons/dimethylol ureas are shifted in favor of the cyclic uron species [151]. If the pH is slowly shifted from one pH range to the other, the equilibrium shifts in favor of the formation of a majority of methylol urea. A rapid change in pH does not cause this to any great extent. UF resins with high uron proportions showed this structure linked by methylene bridges to urea and other urons, and also as methylol urons, the reactivity of the methylol group of this latter being much lower than that of methylol urea. Thermomechanical analysis (TMA) tests and tests on wood particleboard prepared with uron-rich resins to which urea was added at the end of the reaction yielded bonds of good strength. Equally, mixing a uron-rich resin with a low F/U molar ratio UF resin yielded resins of greater strength than a simple UF of corresponding molar ratio, indicating that UF resins of lower formaldehyde emission with still acceptable strength could be prepared with these resins [151]. Subsequent research on this aspect compared five resins produced in different manners: four via the traditional alkaline–acid process and one using the strongly acid step process [152]. The differences between the syntheses were mainly based on different formaldehyde/urea molar ratios during the synthesis, temperatures, and the number of urea additions. The resins differed in some characteristics, namely percentage of unreacted oligomers, chemical composition, viscosity, and reactivity [152]. The internal bonds of the particleboards bonded with them was similar for all the resins prepared with the alkaline–acid process, but were better than those bonded with the strong acid step; nonetheless, formaldehyde emission appeared to be independent of the type of synthesis used.

Very reactive iminoamino methylene base intermediates ($CH_2=NH-CH_{2+}$) obtained by the decomposition of hexamethylenetetramine (hexamine) [43–45] stabilized by the presence of strong anions such as SO_4^{-2} and HSO_4^-, or hexamine sulfate were shown to markedly improve the water and weather resistance of hardened melamine–urea–formaldehyde (MUF) resins used as wood adhesives, and of the wet internal bond strength performance of wood composite boards bonded with them [45,161,162]. The effect was shown to be induced by very small amounts, between 1 and 5 wt % of this material on resin solid content. This strong effect allowed the use of MUF resins of much lower melamine content and also provided good performance of the bonded joints. Because the main effect was also present at the smaller proportion of hexamine as hexamine sulfate, it was not due at all to any increase in the molar ratio of the resin as a consequence of hexamine sulfate addition [45,162]. The effect of hexamine sulfate was closely linked to the strong buffering action it has on MUF resins. Its role is mainly to induce regularity of the reaction and the stability of conditions during resin networking, due to the buffer [161,162]. Shifting of the polycondensation/degradation equilibrium to the left appeared to be the determining factor. This was a consequence of maintaining a higher, constant pH during curing due to the buffer action. The resins are faster-curing than when catalyzed by ammonium sulfate. The effect is valid within the narrow buffering range of pH used for resin hardening. Polycondensation is far too slow to occur at a much higher pH, and degradation is, instead, more predominant at much lower pH. The network formed is then more crosslinked and less tainted by degradation when curing occurs within the correct pH range [45,161,162]. The result is a much better performance of the wood board after water attack. The effects induced by hexamine sulfate are of longer duration than those of other potential buffers. This is due to the hexamine sulfate heat stability under standard hot curing conditions of the resin. Alternate systems were found and shown to have a comparable effect. This approach has already been used to a limited extent in some industrial MUF adhesives.

Of interest is the upgrading of UF, MUF, and PF resins by addition of PMDI (polymeric 4,4′-diphenyl methane diisocyanate) directly in the water-based formaldehyde-based resin glue mix before application to wood [30,31]. This was shown to lead to crosslinking by both formation of the traditional methylene bridges and simultaneously coupled with them by the formation of urethane bridges [30,31]. This is now a currently used industrial adhesive system, especially for UF and MUF resins, and proved to be to be the only system at the time to prepare polyurethane bridges in water. It was based on the stability of PMDI in water for about 5 h before being water-deactivated. Variations of the theme, such as the use of water emulsifiable blocked PMDI [163–165] and, more recently, of microencapsulated pMDI [166] have also been tried. They both worked and were attempted to see if an improvement on the original, simpler, system could be achieved. The bonding performance for both is the same as the simpler system, but the stability of the isocyanate in water is much improved affording, perhaps, the preparation of a stable premix of better shelf- and pot-life.

Very recently, the catalytic influence of TiO_2 in accelerating both synthesis and cure of MF resins has been found [154]. The effect is noticeable and could be due to a variety of causes. There are several possible explanations for this effect. Thus, firstly, Ti is known to form coordination complexes with carbonyl groups (as an aldehyde to pass the C=O to a C–OH, with the intermediate CH_2–O–TiO_2, thus an effect of acceleration due to the Ti charge being stronger than that of H^+. This effect is well known in PF resins, for example. In this case, one should assume that the Ti in TiO_2 behaves as a divalent ion. With this approach, TiO_2 might even complex more than one molecule of formaldehyde at the same time, possibly rapidly increasing the number of aldehyde molecules added onto the same or to different NH_2 groups, also explaining the acceleration of the rate of reaction. The second possibility is the McMurry reaction, where catalysis by Ti joins two carbonyl groups by elimination of the two oxygens, thus from $H_2C=O + O=CH_2$ passing to $CH_2=CH_2$. What would happen, then, is not the formation of just $CH_2=CH_2$ but the reaction of the NH_2 group of melamine with the C of two formaldehyde molecules and the formation of a –NH–CH_2=CH_2–NH– bridge between two molecules of melamine and, thus, two formaldehyde at the time bridging instantly, again explaining the catalysis of the reaction.

The third possibility is that if TiO_2 forms a stable complex, then the initial attack of the aldehyde on the melamine will be fast, but after that, all will remain blocked [26–28]. It has been proven that this is not the case as the reaction accelerates. It means that if complexes are formed, the rate of exchange in solution is very rapid [167] as the complex is not stable (and the TiO_2 complexes cannot be very stable, otherwise they could not function as catalysts in a number of organic chemistry reactions) and, thus, only the effect of the initial fast attack remains, hence, the acceleration in rate [26–28].

Lastly a further explanation might well be valid, thus that once blocked the reaction of the aldehyde with the melamine the bridges forming in the resin are instead through the Ti itself if the complexes formed are stable. First, as –NH–CH_2O–Ti–OCH_2–NH– with the other two valences of the Ti have been linked in the same way to other NH–CH_2–O groups, thus forming a tridimensional knot, this would easily explain the strong acceleration. It is very possible that several, if not all, of these mechanisms are at work, although to a different extent.

Having discussed interesting advances in wood composite aminoplastic adhesives, one has to say that progress has been equally impressive in PF adhesives in the last two decades. Just two cases are reported here, although there are others that are also of interest.

The first regards the curing and reaction rate acceleration mechanism of the resins for wood composites induced by certain esters (also called the "alpha-set" in jargon). The concept of ester acceleration was initially promoted by the Borden company in the early 1950s, but applied in a different manner to foundry core binders, not to wood composites [168]. The approach was rather modified to be adapted to wood composite binders with propylene carbonate and triacetin (glycerol triacetate) being, finally, the two esters retained for wood composite binders [169–173]. Controversy followed, with some research groups claiming that nothing was occurring [174,175], and others claiming that it was the same type of acceleration being caused by inorganic carbonates, such as sodium and potassium carbonates [176], which was disproved [172], while other groups instead also observed some structural resin modifications [173,177–179], with the controversy finally being resolved by isolating and identifying the structural modifications caused to the resin by the esters [180].

The mechanism was found to involve the phenate ion of the resin to apparently yield a carbonyl or carboxyl group attached to the aromatic ring. Either directly or by subsequent rapid rearrangement after the initial attack on the ortho site, these C=O groups were found on sites different from the ortho site. The appearance determined from NMR shift calculation indicated preferential positioning or repositioning to the para site and, surprisingly, to the meta sites of the phenolic ring. The shifts of these C=O groups correspond to those of an anhydride and to no other intermediate structures previously thought of. Anhydride-like bridges were clearly shown by MALDI-TOF mass spectrometry to contribute to oligomer structures in which linkages between phenol rings were mixed methylene bridges and anhydride bridges. These structures appeared to be temporary, possibly due to the instability of the anhydride bridges; hence, they were in small proportions at any given moment of the reaction. ^{13}C NMR and MALDI-TOF analysis clearly indicated that these structures were, at some moment, an integral part of the structure of the liquid resin and that they existed parallel to the methylene bridges pertaining to a normal PF resin structure [180]. While the complexity and unusualness of the mechanism is interesting, the main importance of ester acceleration is to be able to use PF resins with simpler and shorter manufacture time and of much faster hot press time of the wood composite, a real industrial advantage.

Low-condensation phenol–formaldehyde (PF) resins co-reacted under alkaline conditions with up to 42% molar urea on phenol during resin preparation yielded PUF resins capable of faster hardening times than equivalent pure PF resins prepared under identical conditions and presented better performance than the latter [181,182]. The water resistance of the prepared PUF resins seemed comparable to pure PF resins when used as adhesives for wood composite panels. Part of the urea is copolymerized to yield the alkaline PUF resin, but unreacted urea was still present in the resin, especially at the higher levels of urea addition. Increase of the initial formaldehyde to phenol molar ratio considerably decreased the proportion of unreacted urea and increased the proportion of PUF resin. The copolymerized urea functions as a prebranching molecule in the forming, hardening resin network. PUF resins are capable of further noticeable curing acceleration by addition of ester accelerators; namely, glycerol triacetate (triacetin), to reach gel times as fast as those characteristic of catalyzed aminoplastic resins, but with wet strength values of the wood composites bonded with them, characteristic of exterior PF resins. Synergy between the relative amounts of copolymerized urea and ester accelerator occurs at the lower levels of the two additives. However, this synergy decreases at the higher percentages of urea and triacetin. The relative performance of PUF adhesives was checked by preparation of wood particleboard, and the capability of the accelerated PUF resins to achieve press times as fast as those of aminoplastic (UF and others) resins has been confirmed [181,182]. This system has been in industrial utilization, to a moderate extent, for more than a decade.

For the non-traditional synthetic adhesives, the (at first shocking) concept is that urea–formaldehyde can also be classed as a bioadhesive. Urea is bioderived from nitrogen in the air, but while formaldehyde exists in nature, its industrial production is not "bio" at all. Thus, there is interest in eliminating formaldehyde through replacement with something less or nontoxic and, especially, nonvolatile (to eliminate formaldehyde emission). The first attempts to solve this problem led to the preparation of urea–glyoxal (UG) resins for wood composite adhesives [183,184]. Urea–glyoxal resins are already known and used in the textile industry, but the formulations needed to be extensively changed for wood composites. The problem encountered was that UG resins are much slower curing in hot-pressed than UF resins, their energy of activation for curing being markedly higher. The first approach to solve this drawback was to prepare and use melamine–glyoxal (MG) resins, melamine being much more reactive with aldehydes than urea. This yielded an improvement to the point that these MG resins could at least be used for paper impregnation, the resin being in direct contact with the hot platen of the press and, thus, at higher temperature [185]. However, while faster, they were still too slow as adhesives for bonding wood panel composites. The breakthrough to solve these very limiting drawbacks came with the introduction of ionic liquids (ILs) as hardeners of UG resins. These markedly decreased the energy of activation of curing of UG adhesives and the resultant IL UG adhesives were used to prepare wood particleboards at pressing times and with results comparable with UF resins [186]. The approach then progressed to IL MG and, finally, to IL melamine–glyoxal–glutaraldehyde (IL MGG′) adhesives that gave a very acceptable performance as adhesives for wood composites [187].

Equally noticeable in this context is the renewed trend and renewed interest in resins based on urea furfural and urea–furfuryl alcohol having become, again, a basis of study, notwithstanding that these are old technologies used for other applications [188,189]. New, however, is the interest in hydroxymethyl furfural, with urea–hydroxymethyl furfural [190] and phenol–hydroxymethyl furfural [191] as well as tannin–hydroxymethyl furfural [192]. While these are acceptable for phenol and natural polyphenols such as tannins, that are anyhow dark adhesives, the natural developing dark colors of the furanic materials constitute a serious commercial drawback for urea resins that are traditionally transparent or white. Resins based on furfuryl alcohol + an aldehyde have also been tested with encouraging results for plywood bonding, these being furfuryl–alcohol formaldehyde, furfuryl alcohol–glyoxal, and furfuryl alcohol–glutaraldehyde [104].

Even more different approaches to mainly synthetic thermoset adhesives have been tried and presented. For example, glucose or sucrose reacted with glycerol triacetate and a diamine yielded a variety of mixed oligomers (Figure 6) and a very acceptable wood adhesive for plywood. The system was found by pure chance, but it worked, and the oligomers and polymers formed were identified and characterized [193].

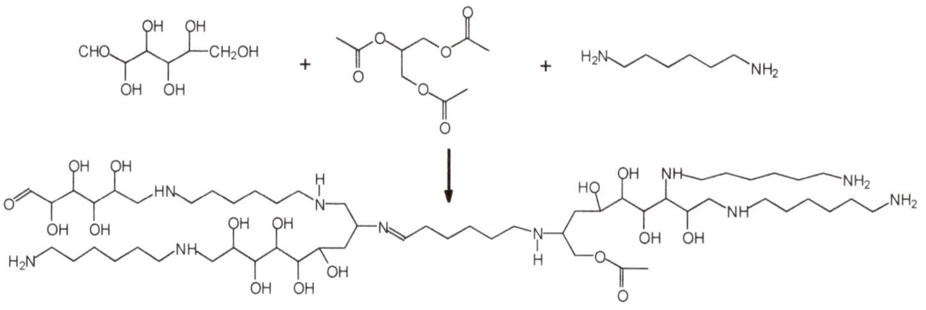

Figure 6. Example of a mixed higher molecular weight oligomer leading to adhesive crosslinking formed, among others, in the new adhesive system derived by the reaction of glucose or sucrose with triacetin and hexamethylene diamine [194].

7. Thermoplastics as Binders for Wood Composites

Very promising directions for the bonding of plywood and LVL based on environmentally friendly or even recycled products, is the field where thermoplastics (polyethylene, polypropylene, poly(vinyl chloride), and their copolymers) are used as the wood composite binders. Such thermoplastic polymers, as a substitute for conventional UF and PF thermosetting adhesives, can be used for veneer bonding in various forms, such as textile fiber waste (i.e., polyurethane, polyamide-6), recycled plastic shopping bags, or film. This area constitutes an area of major research effort especially in recovering and reusing waste polymeric materials. One example of this is the use of waste polyurethanes for the binding of particleboard and plywood [195,196]. There are excellent reviews on this wide field and the reader is referred to these for a more in-depth look of this area [197–200].

Furthermore, dry adhesive film, in particular, is now used for plywood as it is simpler to apply than wet adhesives; all of the untidy and unpleasant mixing and spreading operations in wet gluing are thus wholly removed from the factory floor by the use of such dry adhesive films. The dry adhesive film contains, in each square meter of surface, precisely the same quantity of adhesive, equal quality, uniform composition, yielding exactly the same bond strength and the same standard thickness [194].

8. Conclusions

The field of wood composites is a huge field of research of vast economic importance, fast moving both in the conception of newer composites but even more in the conception and development of newer and sometimes even revolutionary types of binders for them. While the wood composite industry, mainly for reason of supply, is still dominated by traditional oil-derived adhesives, both in these fields as well as in the strongly upcoming field of biobased adhesives, there has been almost incredible progress as well as developments dictated by the intellectual ferment induced by a number of outside constraints. These are the stricter government regulations to reduce and even eliminate formaldehyde and other materials that are to some extent toxic, consumer awareness and the consequent drive of industry to favor more environment-friendly materials and, finally, the drive of industry to decrease or even eliminate their dependence on petrochemicals, due to the real or imagined future decrease of oil reserves with its consequent increase in the price of raw materials for purely traditionally manufactured wood binders. As it stands, what presented in this review is only a brief overview of what has happened and is happening in this field, with possibly lesser or more difficult to implement approaches not being mentioned or not mentioned enough. Progress in this fascinating field of primary economic importance has been accelerating, and the number of new ideas, approaches, and new proposed binder systems is continuously increasing, providing a glimpse of an exciting and interesting research future.

Author Contributions: Conceptualization and methodology A.P., A.N.P. and F.P.; writing—original draft preparation, A.P.; writing—review and editing, A.P., A.N.P. and F.P. All authors have read and agreed to the published version of the manuscript.

Funding: This research received no external funding.

Conflicts of Interest: The authors declare no conflict of interest.

References

1. Pizzi, A. Wood and Fiber Panels Technology. In *Lignocellulosic Fibers and Wood Handbook: Renewable Materials for to-Day's Environment*; Belgacem, M.N., A.Pizzi, A., Eds.; Scrivener-Wiley: Beverley, MA, USA, 2017; Chapter 15; pp. 385–406.
2. Geimer, R.L.; Mahoney, R.J.; Loehnertz, S.P.; Meyer, R.W. *Influence of Processing Induced Damage on the Strength of Flakes and Flakeboards*; Research Paper FPL 463; USDA Forest Products Laboratory: Madison, WI, USA, 1985.
3. Maloney, T.M. *Modern Particleboard & Dry-Process Fiberboard Manufacturing*; Backbeat Books; Rowman & Littlefield: Lanham, MD, USA, 1993.
4. Mantanis, G.; Athanassiadou, E.; Barbu, M.; Wijnendaele, K. Adhesive systems used in the European particleboard, MDF and OSB industries. *Wood Mater. Sci. Eng.* **2018**, *13*, 104–116. [CrossRef]
5. Pizzi, A. *Wood Adhesives Chemistry and Technology*; Marcel Dekker: New York, NY, USA, 1983.
6. Pizzi, A. Urea-formaldehyde adhesives. In *Handbook of Adhesive Technology*, 2nd ed.; Pizzi, A., Mittal, K.L., Eds.; Marcel Dekker: New York, NY, USA, 2003; Chapter 31; pp. 635–652.
7. Pizzi, A. Urea and melamine aminoresin adhesives. In *Handbook of Adhesive Technology*, 3rd ed.; Pizzi, A., Mittal, K.L., Eds.; Taylor and Francis: New York, NY, USA, 2017; Chapter 10; pp. 283–320.
8. Pizzi, A. *Advanced Wood Adhesives Technology*; Marcel Dekker: New York, NY, USA, 1994.
9. Pizzi, A. Melamine-formaldehyde adhesives. In *Handbook of Adhesive Technology*, 2nd ed.; Pizzi, A., Mittal, K.L., Eds.; Marcel Dekker: New York, NY, USA, 2003; Chapter 32; pp. 653–680.
10. Pizzi, A. Phenolic resin adhesives. In *Handbook of Adhesive Technology*, 3rd ed.; Pizzi, A., Mittal, K.L., Eds.; Taylor and Francis: New York, NY, USA, 2017; Chapter 8; pp. 223–262.
11. Pizzi, A. Phenolic resin adhesives. In *Handbook of Adhesive Technology*, 2nd ed.; Pizzi, A., Mittal, K.L., Eds.; Marcel Dekker: New York, NY, USA, 2003; Chapter 26; pp. 541–572.
12. Pizzi, A. Resorcinol adhesives. In *Handbook of Adhesive Technology*, 2nd ed.; Pizzi, A., Mittal, K.L., Eds.; Marcel Dekker: New York, NY, USA, 2003; Chapter 29; pp. 599–614.

13. Frazier, C.E. Isocyanate wood binders. In *Handbook of Adhesive Technology*, 2nd ed.; Pizzi, A., Mittal, K.L., Eds.; Marcel Dekker: New York, NY, USA, 2003; Chapter 33; pp. 681–694.
14. Lay, D.G.; Cranley, P.; Pizzi, A. Polyurethane adhesives. In *Handbook of Adhesive Technology*, 3rd ed.; Pizzi, A., Mittal, K.L., Eds.; Taylor and Francis: New York, NY, USA, 2017; Chapter 11; pp. 321–348.
15. Pizzi, A. Tannin-based wood adhesives. In *Wood Adhesives Chemistry and Technology*; Pizzi, A., Ed.; Marcel Dekker: New York, NY, USA, 1983; Volume 1, pp. 178–246.
16. Pizzi, A. Wattle-based adhesives for exterior grade particleboard. *For. Prod. J.* **1978**, *28*, 42–47.
17. Pizzi, A.; Scharfetter, H. The chemistry and development of tannin-based wood adhesives for exterior plywood. *J. Appl. Polym. Sci.* **1978**, *22*, 1745–1761. [CrossRef]
18. Valenzuela, J.; von Leyser, E.; Pizzi, A.; Westermeyer, C.; Gorrini, B. Industrial production of pine tannin-bonded particleboard and MDF. *Eur. J. Wood Prod.* **2012**, *70*, 735–740. [CrossRef]
19. Pizzi, A. Glue blenders effect on particleboard using wattle tannin adhesives. *Holzforsch. Holzverwert.* **1979**, *31*, 85–86.
20. Pizzi, A. Hot-setting tannin-urea-formaldehyde exterior wood adhesives. *Adhes. Age* **1977**, *20*, 27–32.
21. Custers, P.A.J.L.; Rushbrook, R.; Pizzi, A.; Knauff, C.J. Industrial applications of wattle-tannin/urea-formaldehyde fortified starch adhesives for damp-proof corrugated cardboard. *Holzforsch. Holzverwert.* **1979**, *31*, 131–132.
22. Pizzi, A.; Roux, D.G. The chemistry and development of tannin-based weather- and boil-proof cold-setting and fast-setting adhesives for wood. *J. Appl. Polym. Sci.* **1978**, *22*, 1945–1954. [CrossRef]
23. Pizzi, A.; Rossouw, D.D.T.; Knuffel, W.; Singmin, M. "Honeymoon" phenolic and tannin-based fast setting adhesive systems for exterior grade fingerjoints. *Holzforsch. Holzverwert.* **1980**, *32*, 140–151.
24. Pizzi, A.; Cameron, F.A. Fast-set adhesives for glulam. *For. Prod. J.* **1984**, *34*, 61–65.
25. Mansouri, H.R.; Pizzi, A.; Fredon, E. Honeymoon fast-set adhesives for glulam/fingerjoints of higher natural materials content. *Eur. J. Wood Prod.* **2009**, *67*, 207–210. [CrossRef]
26. Pizzi, A. Phenolic resins by reactions of coordinated metal ligands. *J. Polym. Sci. Polym. Lett.* **1979**, *17*, 489–492. [CrossRef]
27. Pizzi, A. Phenol and tannin-based adhesive resins by reactions of coordinated metal ligands, Part 1: Phenolic chelates. *J. Appl. Polym. Sci.* **1979**, *24*, 1247–1255. [CrossRef]
28. Pizzi, A. Phenol and tannin-based adhesive resins by reactions of coordinated metal ligands, Part II: Tannin adhesives preparation, characteristics and application. *J. Appl. Polym. Sci.* **1979**, *24*, 1257–1268. [CrossRef]
29. von Leyser, E.; Pizzi, A. The formulation and commercialization of glulam pine tannin adhesives in Chile. *Holz Roh-und Werkst.* **1990**, *48*, 25–29. [CrossRef]
30. Pizzi, A.; Walton, T. Non-emulsifiable, water-based diisocyanate adhesives for exterior plywood, Part 1: Novel reaction mechanisms and their chemical evidence. *Holzforschung* **1992**, *46*, 541–547. [CrossRef]
31. Pizzi, A.; Valenzuela, J.; Westermeyer, C. Non-emulsifiables, water-based, diisocyanate adhesives for exterior plywood, Part 2: Industrial application. *Holzforschung* **1993**, *47*, 69–72. [CrossRef]
32. Böhm, R.; Hauptmann, M.; Pizzi, A.; Friederich, C.; Laborie, M.-P. The chemical, kinetic and mechanical characterization of Tannin-based adhesives with different crosslinking systems. *Int. J. Adhes. Adhes.* **2016**, *68*, 1–8. [CrossRef]
33. Santiago-Medina, F.J.; Foyer, G.; Pizzi, A.; Calliol, S.; Delmotte, L. lignin-derived non-toxic aldehydes for ecofriendly tannin adhesives for wood panels. *Int. J. Adhes. Adhes.* **2016**, *70*, 239–248. [CrossRef]
34. Ndiwe, B.; Pizzi, A.; Tibi, B.; Danwe, R.; Konai, N.; Amirou, S. African tree bark exudate extracts as biohardeners of fully biosourced thermoset tannin adhesives for wood panels. *Ind. Crops Prod.* **2019**, *132*, 253–268. [CrossRef]
35. Pizzi, A.; Meikleham, N.; Dombo, B.; Roll, W. Autocondensation-based, zero-emission, tannin adhesives for particleboard. *Holz Roh-und Werkst.* **1995**, *53*, 201–204. [CrossRef]
36. Meikleham, N.; Pizzi, A.; Stephanou, A. Induced accelerated autocondensation of polyflavonoid tannins for phenolic polycondensates, Part 1: ^{13}C NMR, ^{29}Si NMR, X-ray and polarimetry studies and mechanism. *J. Appl. Polym. Sci.* **1994**, *54*, 1827–1845. [CrossRef]
37. Pizzi, A.; Meikleham, N.; Stephanou, N. Induced accelerated autocondensation of polyflavonoid tannins for phenolic polycondensates—Part II: Cellulose effect and application. *J. Appl. Polym. Sci.* **1995**, *55*, 929–933. [CrossRef]

38. Garcia, R.; Pizzi, A.; Merlin, A. Ionic polycondensation effects on the radical autocondensation of polyflavonoid tannins-An ESR study. *J. Appl. Polym. Sci.* **1997**, *65*, 2623–2632. [CrossRef]
39. Garcia, R.; Pizzi, A. Polycondensation and autocondensation networks in polyflavonoid tannins, Part 1: Final networks. *J. Appl. Polym. Sci.* **1998**, *70*, 1083–1091. [CrossRef]
40. Garcia, R.; Pizzi, A. Polycondensation and autocondensation networks in polyflavonoid tannins, Part 2: Polycondensation vs. autocondensation. *J. Appl. Polym. Sci.* **1998**, *70*, 1093–1110. [CrossRef]
41. Garcia, R.; Pizzi, A. Cross-linked and entanglement networks in thermomechanical analysis of polycondensation resins. *J. Appl. Polym. Sci.* **1998**, *70*, 1111–1116. [CrossRef]
42. Pichelin, F.; Nakatani, M.; Pizzi, A.; Wieland, S.; Despres, A.; Rigolet, S. Structural beams from thick wood panels bonded industrially with formaldehyde free tannin adhesives. *For. Prod. J.* **2006**, *56*, 31–36.
43. Kamoun, C.; Pizzi, A. Mechanism of hexamine as a non-aldehyde polycondensation hardener, Part 1: Hexamine decomposition and reactive intermediates. *Holzforsch. Holzverwert.* **2000**, *52*, 16–19.
44. Kamoun, C.; Pizzi, A. Mechanism of hexamine as a non-aldehyde polycondensation hardener, Part 2: Recomposition of intermediate reactive compound. *Holzforsch. Holzverwert.* **2000**, *52*, 66–67.
45. Kamoun, C.; Pizzi, A.; Zanetti, M. Upgrading of MUF resins by buffering additives—Part 1: Hexamine sulphate effect and its limits. *J. Appl. Polym. Sci.* **2003**, *90*, 203–214. [CrossRef]
46. Navarrete, P.; Mansouri, H.R.; Pizzi, A.; Tapin-Lingua, S.; Benjelloun-Mlayah, B.; Rigolet, S. Synthetic-resin-free wood panel adhesives from low molecular mass lignin and tannin. *J. Adhes. Sci. Technol.* **2010**, *24*, 1597–1610. [CrossRef]
47. Ghahri, S.; Pizzi, A.; Mohebby, B.; Mirshoktaie, A.; Mansouri, H.R. Soy-based, tannin-modified plywood adhesives. *J. Adhes.* **2018**, *94*, 218–237. [CrossRef]
48. Abdullah, U.H.B.; Pizzi, A. Tannin-Furfuryl alcohol wood panel adhesives without formaldehyde. *Eur. J. Wood Prod.* **2013**, *71*, 131–132. [CrossRef]
49. Ballerini, A.; Despres, A.; Pizzi, A. Non-toxic, zero-emission tannin-glyoxal adhesives for wood panel. *Holz Roh-und Werkst.* **2005**, *63*, 477–478. [CrossRef]
50. Trosa, A.; Pizzi, A. A no-aldehyde emission hardener for tannin-based wood adhesives. *Holz Roh-und Werkst.* **2001**, *59*, 266–271. [CrossRef]
51. Grigsby, W.J.; McIntosh, C.D.; Warnes, J.M.; Suckling, I.D.; Anderson, C.R. Adhesives. U.S. Patent 7,319,115 B2, 2008.
52. Trosa, A.; Pizzi, A. Industrial hardboard and other panels binder from tannin/furfuryl alcohol in absence of formaldehyde. *Holz Roh-und Werkst.* **1998**, *56*, 213–214. [CrossRef]
53. Frihart, C.R.; Lorenz, L. Specific oxidants improve the wood bonding strength of soy and other plant flours. *J. Polym. Sci. A Polym. Chem.* **2019**, *57*, 1017–1023. [CrossRef]
54. Frihart, C.R.; Pizzi, A.; Xi, X.; Lorenz, L. Reactions of Soy flour and Soy protein by non-volatile aldehydes generation by specific oxidation. *Polymers* **2019**, *11*, 1478. [CrossRef]
55. Xi, X.; Pizzi, A.; Frihart, C.R.; Lorenz, L.; Gerardin, C. Tannin plywood adhesives by non-volatile aldehydes generation from specific oxidation of mono- and disaccharides. *Int. J. Adhes. Adhes.* **2020**, *98*, 102499. [CrossRef]
56. Nimz, H.H. Lignin-based adhesives. In *Wood Adhesives Chemistry and Technology*; Pizzi, A., Ed.; Marcel Dekker: New York, NY, USA, 1983; Chapter 5; Volume 1, pp. 247–288.
57. Blanchet, P.; Cloutier, A.; Riedl, B. Particleboard made from hammer milled black spruce bark residues. *Wood Sci. Technol.* **2000**, *34*, 11–19. [CrossRef]
58. Lopez-Suevos, F.; Riedl, B. Effects of Pinus pinaster bark extracts content on the cure properties of tannin-modified adhesives and on bonding of exterior grade MDF. *J. Adhes. Sci. Technol.* **2003**, *17*, 1507–1522. [CrossRef]
59. Kim, S.; Kim, H.-J. Curing behaviour and viscoelastic properties of pine and wattle tannin-based adhesives studied by dynamic mechanical thermal analysis and FT-IR-ATR spectroscopy. *J. Adhes. Sci. Technol.* **2003**, *17*, 1369–1384. [CrossRef]
60. Calvé, L.R. Fast cure and pre-cure resistant cross-linked phenol-formaldehyde adhesives and methods of making same. Canada Patent 2042476, 1999.
61. Shimatani, K.; Sono, Y.; Sasaya, T. Preparation of moderate-temperature setting adhesives from softwood kraft lignin. Part 2. Effect of some factors on strength properties and characteristics of lignin-based adhesives. *Holzforschung* **1994**, *48*, 337–342. [CrossRef]

62. Gardner, D.; Sellers, T., Jr. Formulation of a lignin-based plywood adhesive from steam-exploded mixed hardwood lignin. *For. Prod. J.* **1986**, *36*, 61–67.
63. Newman, W.H.; Glasser, W.G. Engineering plastics from lignin-XII. Synthesis and performance of lignin adhesives with isocyanate and melamine. *Holzforschung* **1985**, *39*, 345–353. [CrossRef]
64. Azarov, V.I.; Koverniskii, N.N.; Zaitseva, G.V. Izvestjia Vysshikh Uchnykh Zavedenii. *Lesnai Zhurna* **1985**, *5*, 81–83.
65. Viikari, L.; Hase, A.; Quintus-Leina, P.; Kirsi, K.; Tuominen, S.; Gaedda, L. Lignin Based Adhesives and a Process for the Preparation Thereof. European Patent EP 0953029 A1, 1999.
66. Kharazipour, A.; Haars, A.; Shekholeslami, M.; Hüttermann, A. Enzymgebundene holzwerkstoffe auf der basis von lignin und phenoloxidasen. *Adhäsion* **1991**, *35*, 30–36.
67. Kharazipour, A.; Mai, C.; Hüttermann, A. Polyphenols for compounded materials. *Polym. Degrad. Stabil.* **1998**, *59*, 237–243. [CrossRef]
68. Antov, P.; Savov, V.; Mantanis, G.I.; Neykov, N. Medium-density fibreboards bonded with phenol-formaldehyde resin and calcium lignosulfonate as an eco-friendly additive. *Wood Mat. Sci. Eng.* **2020**. [CrossRef]
69. Calvé, L.R. Phenolic-lignin wood adhesives for plywood. In Proceedings of the 19th IUFRO World Congress, Montreal, QC, Canada, 6–10 August 1990.
70. Calvé, L.R. Fast Cure and Pre-Cure Resistant Cross-Linked Phenol-Formaldehyde Adhesives and Methods of Making Same. US Patent 5,173,527, 1992.
71. Valkonen, S. Lignin-based binders: An industrial reality, latest developments. In Proceedings of the International Conference on Wood Adhesives, Atlanta, GA, USA, 25–27 October 2017.
72. Younesi-Kordkheili, H.; Pizzi, A. Some Properties of Particleboard Panels Bonded with Phenol- Lignin- Glyoxal Resin. *J. Adhes.* **2019**. [CrossRef]
73. Younesi-Kordkheili, H.; Pizzi, A. Properties of plywood panels bonded with ionic liquid-modified lignin–phenol–formaldehyde resin. *J. Adhes.* **2018**, *94*, 143–154. [CrossRef]
74. Younesi-Kordkheili, H.; Pizzi, A.; Hornabakhsh-Raouf, A.; Nemati, F. The Effect of Modified Soda Bagasse Lignin by Ionic Liquid on Properties of Urea-Formaldehyde Resin as Wood Adhesive. *J. Adhes.* **2017**, *93*, 914–925. [CrossRef]
75. El Mansouri, N.E.; Pizzi, A.; Salvado, J. Lignin-based polycondensation resins for wood adhesives. *J. Appl. Polym. Sci.* **2007**, *103*, 1690–1699. [CrossRef]
76. El Mansouri, N.E.; Pizzi, A.; Salvado, J. Lignin-based wood panel adhesives without formaldehyde. *Holz Roh-und Werkst.* **2007**, *65*, 65–70. [CrossRef]
77. Pizzi, A.; Kueny, R.; Lecoanet, F.; Massetau, B.; Carpentier, D.; Krebs, A.; Loiseau, F.; Molina, S.; Ragoubi, M. High resin content natural matrix-natural fibre biocomposites. *Ind. Crops Prod.* **2009**, *30*, 235–240. [CrossRef]
78. Mansouri, H.R.; Navarrete, P.; Pizzi, A.; Tapin-Lingua, S.; Benjelloun-Mlayah, B.; Rigolet, S. Synthetic-resin-free wood panel adhesives from low molecular mass lignin and tannin. *Eur. J. Wood Prod.* **2011**, *69*, 221–229. [CrossRef]
79. Navarrete, P.; Mansouri, H.R.; Pizzi, A.; Tapin-Lingua, S.; Benjelloun-Mlayah, B.; Pasch, H.; Rode, K.; Delmotte, L.; Rigolet, S. Low formaldehyde emitting biobased wood adhesives manufactured from mixtures of tannin and glyoxalated lignin. *J. Adh. Sci. Technol.* **2012**, *26*, 1667–1684. [CrossRef]
80. Bertaud, F.; Tapin-Lingua, S.; Pizzi, A.; Navarrete, P.; Petit-Conil, M. Development of green adhesives for fibreboards manufacturing, using tannins and lignin from pulp mill residues. *Cellul. Chem. Technol.* **2012**, *46*, 449–455.
81. Sauget, A.; Nicollin, A.; Pizzi, A. Fabrication and mechanical analysis of mimosa tannin and commercial flax fibers biocomposites. *J. Adhes. Sci. Technol.* **2013**, *27*, 2204–2218. [CrossRef]
82. Zhu, J.; Abhyanker, H.; Njuguna, J.; Perreux, D.; Thiebaud, F.; Chapelle, D.; Pizzi, A.; Sauget, A.; Nicollin, A. Tannin-based flax fibre-reinforced bio-composites for structural application in Superlight Electric Vehicles. *Ind. Crops Prod.* **2013**, *50*, 68–76. [CrossRef]
83. Nicollin, A.; Li, X.; Girods, P.; Pizzi, A.; Rogaume, Y. Fast pressing composite using tannin-furfuryl alcohol resin and vegetal fibers reinforcement. *J. Renew. Mater.* **2013**, *1*, 311–316. [CrossRef]
84. Chen, X.; Xi, X.; Pizzi, A.; Fredon, E.; Du, G.; Gerardin, C. Oxidized Demethylated Lignin as a Bio-Based Adhesive for Wood Bonding. *J. Adhes.* **2020**. [CrossRef]

85. Wescott, J.M.; Frihart, C.R.; Lorenz, L. Durable soy-based adhesives. In *Proceedings Wood Adhesives 2005*; Forest Products Society: Madison, WI, USA, 2006.
86. Lorenz, L.; Frihart, C.R.; Wescott, J.M. Analysis of soy flour/phenol-formaldehyde adhesives for bonding wood. In *Wood Adhesives 2005: November 2–4, 2005... San Diego, California*; Forest Products Society: Madison, WI, USA, 2006.
87. Amaral-Labat, G.A.; Pizzi, A.; Goncalves, A.R.; Celzard, A.; Rigolet, S. Environment-friendly soy flour-based resins without formaldehyde. *J. Appl. Polym. Sci.* **2008**, *108*, 624–632. [CrossRef]
88. Lei, H.; Pizzi, A.; Navarrete, P.; Rigolet, S.; Redl, A.; Wagner, A. Gluten protein adhesives for wood panels. *J. Adhesion Sci. Technol.* **2010**, *24*, 1583–1596. [CrossRef]
89. Liu, Y.; Li, K. Chemical modification of soy protein for wood adhesives. *Macromol. Rapid Comm.* **2002**, *23*, 739–742. [CrossRef]
90. Liu, K.; Li, K. Development and characterization of adhesives from soy protein for bonding wood. *Int. J. Adhes. Adhes.* **2007**, *27*, 59–67. [CrossRef]
91. Xi, X.; Pizzi, A.; Gerardin, C.; Liao, J.; Amirou, S.; Abdalla, S. Glutaradehyde-wheat protein-based adhesives for plywood. *J. Adhes.* **2019**. [CrossRef]
92. Lagel, M.C.; Pizzi, A.; Redl, A. Phenol-wheat protein-formaldehyde adhesives for wood-based panels. *ProLigno* **2014**, *10*, 3–17.
93. Ghahri, S.; Pizzi, A.; Mohebby, B.; Mirshoktaie, A.; Mansouri, H.R. Improving Water Resistance of Soy-Based Adhesive by Vegetable Tannin. *J. Polym. Environ.* **2018**, *26*, 1881–1890. [CrossRef]
94. Ghahri, S.; Pizzi, A. Improving Soy-Based Adhesives for Wood Particleboard by Tannins Addition. *Wood Sci. Technol.* **2018**, *52*, 261–279. [CrossRef]
95. Xi, X.; Pizzi, A.; Gerardin, C.; Chen, X.; Amirou, S. Soy Protein Isolate based Polyamides as Wood Adhesives. *Wood Sci. Technol.* **2020**, *54*, 89–102. [CrossRef]
96. Conner, A.H.; River, B.H.; Lorenz, L.F. Carbohydrates-modified PF resins for wood panels. *J. Wood Chem. Technol.* **1986**, *6*, 591–596. [CrossRef]
97. Conner, A.H.; Lorenz, L.F.; River, B.H. Carbohydrate-modified PF resins formulated at neutral conditions. *ACS Symp. Ser.* **1989**, *385*, 355–369.
98. Moubarik, A.; Pizzi, A.; Charrier, F.; Allal, A.; Badia, M.-A.; Mansouri, H.R.; Charrier, B. Mechanical characterization of industrial particleboard panels glued with cornstarch- mimosa tannin- urea formaldehyde resins. *J. Adhes. Sci. Technol.* **2013**, *27*, 423–429. [CrossRef]
99. Moubarik, A.; Mansouri, H.R.; Pizzi, A.; Charrier, F.; Allal, A.; Charrier, B. Corn flour-mimosa tannin-based adhesives without formaldehyde for interior particleboard production. *Wood Sci. Technol.* **2013**, *47*, 1–9. [CrossRef]
100. Moubarik, A.; Allal, A.; Pizzi, A.; Charrier, F.; Charrier, B. Characterization of a formaldehyde-free cornstarch-tannin wood adhesives for interior plywood. *Eur. J. Wood Prod.* **2010**, *68*, 427–433. [CrossRef]
101. Moubarik, A.; Charrier, B.; Allal, A.; Charrier, F.; Pizzi, A. Development and optimisation of a new formaldehyde-free cornstarch and tannin adhesive. *Eur. J. Wood Prod.* **2010**, *68*, 167–177. [CrossRef]
102. Moubarik, A.; Pizzi, A.; Allal, A.; Charrier, F.; Charrier, B. Cornstarch and tannin in phenol-formaldehyde resins for plywood production. *Ind. Crops Prod.* **2009**, *30*, 188–193. [CrossRef]
103. Belgacem, M.N.; Gandini, A. Furan-based adhesives. In *Handbook of Adhesive Technology*, 2nd ed.; Pizzi, A., Mittal, K.L., Eds.; Marcel Dekker: New York, NY, USA, 2003; Chapter 30; pp. 615–634.
104. Xi, X.; Wu, Z.; Pizzi, A.; Gerardin, C.; Lei, H.; Du, G. Furfuryl alcohol-aldehyde plywood adhesive resins. *J. Adhes.* **2018**. [CrossRef]
105. Alma, M.H.; Yoshioka, M.; Yao, Y.; Shiraishi, N. Preparation of sulfuric acid-catalyzed phenolated wood resin. *Wood Sci. Technol.* **1998**, *32*, 297–308. [CrossRef]
106. Alma, M.H.; Yoshioka, M.; Yao, Y.; Shiraishi, N. The preparation and flow properties of HC1 catalyzed phenolated wood and its blends with commercial novolak resin. *Holzforschung* **1996**, *50*, 85–90. [CrossRef]
107. Bobbitt, J.M. Periodate oxidation of carbohydrates. *Adv. Carbohydr. Chem.* **1956**, *11*, 1–41.
108. Guigo, N.; Mazeau, K.; Putaux, J.L.; Heux, L. Surface modification of cellulose microfibrils by periodate oxidation and subsequent reductive amination with benzylamine: A topochemical study. *Cellulose* **2014**, *21*, 4119–4133. [CrossRef]
109. Codou, A.; Guigo, N.; Heux, L.; Sbirrazzuoli, N. Partial periodate oxidation and thermal cross-linking for the processing of thermoset all-cellulose composites. *Compos. Sci. Technol.* **2015**, *117*, 54–61. [CrossRef]

110. Xi, X.; Pizzi, A.; Delmotte, L. Isocyanate-free Polyurethane Coatings and Adhesives from Mono- and Di-Saccharides. *Polymers* **2018**, *10*, 402. [CrossRef]
111. Xi, X.; Wu, Z.; Pizzi, A.; Gerardin, C.; Lei, H.; Zhang, B.; Du, G. Non-Isocyanate Polyurethane Adhesive from sucrose used for particleboard. *Wood Sci. Technol.* **2019**, *53*, 393–405. [CrossRef]
112. Umemura, K.; Ueda, T.; Munawar, S.; Kawai, S. Application of citric acid as natural adhesive for wood. *J. Appl. Polym. Sci.* **2012**, *123*, 1991–1996. [CrossRef]
113. Umemura, K.; Kawai, S. Development of Wood-Based Materials Bonded with Citric Acid. *For. Prod. J.* **2015**, *65*, 38–42. [CrossRef]
114. Amirou, S.; Pizzi, A.; Delmotte, L. Citric acid as waterproofing additive in butt joints linear wood welding. *Eur. J. Wood Prod.* **2017**, *75*, 651–654. [CrossRef]
115. Del Menezzi, C.; Amirou, S.; Pizzi, A.; Xi, X.; Delmotte, L. Reactions with Wood Carbohydrates and Lignin of Citric Acid as a Bond Promoter of Wood Veneer Panels. *Polymers* **2018**, *10*, 833. [CrossRef] [PubMed]
116. Umemura, K.; Sugihara, O.; Kawai, S. Investigation of a new natural adhesive composed of citric acid and sucrose for particleboard. *J. Wood Sci.* **2013**, *59*, 203–208. [CrossRef]
117. Widyorini, R.; Nugraha, P.; Rahman, M.; Prayitno, T. Bonding ability of a new adhesive composed of citric acid-sucrose for particleboard. *BioResources* **2016**, *11*, 4526–4535. [CrossRef]
118. Zhao, Z.; Sakai, S.; Chen, Z.; Zhu, N.; Huang, C.; Sun, S.; Zhang, M.; Umemura, K.; Yong, X. Further Exploration of Sucrose–Citric Acid Adhesive: Investigation of Optimal Hot-Pressing Conditions for Plywood and Curing Behavior. *Polymers* **2019**, *11*, 1996. [CrossRef]
119. Sun, S.; Zhao, Z.; Umemura, K. Further Exploration of Sucrose-Citric Acid Adhesive: Synthesis and Application on Plywood. *Polymers* **2019**, *11*, 1875. [CrossRef]
120. Tomkinson, J. Adhesives based on natural resources. In *Wood Adhesion and Glued Products: Wood Adhesives*; Dunky, M., Pizzi, A., Van Leemput, M., Eds.; European Commission, Directorate General for Research: Brussels, Belgium, 2002; pp. 46–65.
121. Jial, P.; Song, F.; Li, Q.; Xia, H.; Shu, X.; Zhou, Y. Recent Development of Cardanol Based Polymer Materials: A Review. *J. Renew. Mat.* **2019**, *7*, 601–619.
122. Tasooji, M.; Tabarsa, T.; Khazaeian, A.; Wool, R.P. Acrylated epoxidised soy oil as an alternative to urea-formaldehyde in making wheat sraw particleboard. *J. Adhes. Sci. Technol.* **2010**, *24*, 1717–1727. [CrossRef]
123. Wool, R.P. *Proceedings of the second European Panel Products Symposium*; Academic Press: Bangor, UK, 1998.
124. Miller, R.; Shonfeld, U. *Company Literature*; Esbacher Weg 15; Preform Raumgliederungssysteme GmBH: Feuchtwangen, Germany, 2002.
125. Zimele, Z.; Irbe, I.; Grinins, J.; Bikovens, O.; Verovkins, A.; Bajare, D. Novel Mycelium-based Biocomposites (MBB) as Building Materials. *J. Renew. Mat.* **2020**, in press.
126. Gfeller, B.; Zanetti, M.; Properzi, M.; Pizzi, A.; Pichelin, F.; Lehmann, M.; Delmotte, L. Wood bonding by vibrational welding. *J. Adhes. Sci. Technol.* **2003**, *17*, 1425–1590.
127. Pizzi, A.; Leban, J.-M.; Kanazawa, F.; Properzi, M.; Pichelin, F. Wood dowels bonding by high speed rotation welding. *J. Adhes. Sci. Technol.* **2004**, *18*, 1263–1278. [CrossRef]
128. Kanazawa, F.; Pizzi, A.; Properzi, M.; Delmotte, L.; Pichelin, F. Influence parameters in wood dowels welding by high speed rotation. *J. Adhes. Sci. Technol.* **2005**, *19*, 1025–1038. [CrossRef]
129. Mansouri, M.; Leban, J.-M.; Pizzi, A. End-grain butt joints obtained by friction welding of high density eucalyptus wood. *Wood Sci. Technol.* **2010**, *44*, 399–406. [CrossRef]
130. Leban, J.M.; Pizzi, A.; Properzi, M.; Pichelin, F.; Gelhaye, P.; Rose, C. Wood Welding. A challenging alternative to conventional wood gluing. *Scand. J. For. Res.* **2005**, *20*, 534–538. [CrossRef]
131. Pizzi, A.; Mansouri, H.R.; Leban, J.M.; Delmotte, L.; Omrani, P.; Pichelin, F. Enhancing the exterior performance of wood linear and rotational welding. *J. Adhes. Sci. Technol.* **2011**, *25*, 2717–2730. [CrossRef]
132. Pizzi, A.; Zhou, X.; Navarrete, P.; Segovia, C.; Mansouri, H.R.; Placentia Pena, M.I.; Pichelin, F. Enhancing water resistance of welded dowel wood joints by acetylated lignin. *J. Adhes. Sci. Technol.* **2013**, *27*, 252–262. [CrossRef]
133. Bocquet, J.-F.; Pizzi, A.; Resch, L. Full-scale (Industrial) wood floor assembly and structures by welded through dowels. *Holz Roh-und Werkst.* **2007**, *65*, 149–156. [CrossRef]

134. Bocquet, J.F.; Pizzi, A.; Despres, A.; Mansouri, H.R.; Resch, L.; Michel, D.; Letort, F. Wood joints and laminated wood beams assembled by mechanically welded wood dowels. *J. Adhes. Sci. Technol.* **2007**, *21*, 301–317. [CrossRef]
135. Segovia, C.; Pizzi, A. Performance of dowel-welded wood furniture linear joints. *J. Adhes. Sci. Technol.* **2009**, *23*, 1293–1301. [CrossRef]
136. Segovia, C.; Pizzi, A. Performance of dowel-welded T-joints for wood furniture. *J. Adhes. Sci. Technol.* **2009**, *23*, 2073–2084. [CrossRef]
137. Oudjene, M.; Khalifa, M.; Segovia, C.; Pizzi, A. Application of numerical modelling to dowel-welded wood joints. *J. Adhes. Sci. Technol.* **2010**, *24*, 359–370. [CrossRef]
138. Segovia, C.; Renaud, A.; Pizzi, A. Performance of dowel-welded L-joints for wood furniture. *J. Adhes. Sci. Technol.* **2010**, *25*, 1829–1837. [CrossRef]
139. Renaud, A. Minimalist Z-chair assembly by rotational dowel welding. *Eur. J. Wood Prod.* **2009**, *67*, 111–112. [CrossRef]
140. O'Loising, C.; Oudjene, M.; Shotton, E.; Pizzi, A.; Fanning, P. Mechanical behaviour and 3D stress analysis of multilayered wooden beams made with welded-through wood dowels. *Compos. Struct.* **2012**, *94*, 313–321.
141. O'Loising, C.; Oudjene, M.; Ait-Adler, H.; Fanning, P.; Pizzi, A.; Shotton, E.; Meghlat, E.-M. Experimental study of timber-to-timber composite beam using welded-through wood dowels. *Constr. Build. Mater.* **2012**, *36*, 245–250.
142. Segovia, C.; Zhou, X.; Pizzi, A. Wood blockboards for construction by wood welding with pre-oiled dowels. *J. Adhes. Sci. Technol.* **2013**, *27*, 577–585. [CrossRef]
143. Mansouri, M.; Leban, J.-M.; Pizzi, A. High density panels by wood veneers welding without any adhesives. *J. Adhes. Sci. Technol.* **2010**, *24*, 1529–1534. [CrossRef]
144. Cristescu, C.; Karlsson, O. Changes in content of furfurals and phenols in self-bonded laminated boards. *BioResources* **2013**, *8*, 4056–4071. [CrossRef]
145. Je, H.; Won, J. Natural urushiol as a novel under-water adhesive. *Chem. Eng. J.* **2020**, in press.
146. Bobade, S.K.; Paluvai, N.R.; Mohanty, S.; Nayaka, S.K. Bio-Based Thermosetting Resins for Future Generation: A Review. *Polym. -Plast. Technol. Eng.* **2016**, *55*, 1863–1896. [CrossRef]
147. Sangregorio, A.; Guigo, N.; van der Waal, J.C.; Sbirrazzuoli, N. All 'green' composites comprising flax fibres and humins resins. *Compos. Sci. Technol.* **2019**, *171*, 70–77. [CrossRef]
148. Sangregorio, A.; Muralidhara, A.; Guigo, N.; Marlair, G.; Angelici, C.; Thygesen, G.; de Jong, E.; Sbirrazzuoli, N. Humins based resin for wood modification and property improvement. *Green Chem.* **2020**. [CrossRef]
149. Hammami, N.; Jarroux, N.; Robitzer, M.; Majdoub, M.; Habas, J.-P. Optimized Synthesis According to One-Step Process of a Biobased Thermoplastic Polyacetal Derived from Isosorbide. *Polymers* **2016**, *8*, 294. [CrossRef] [PubMed]
150. Gu, J.; Higuchi, M.; Morita, M.; Hse, C.-Y. Synthetic conditions and chemical structure of urea-formaldehyde resins. I: Properties of the resins synthesized by three different procedures. *Mokuzai Gakkaishi* **1995**, *41*, 1115–1121.
151. Soulard, C.; Kamoun, C.; Pizzi, A. Uron and Uron-Urea-formaldehyde resins. *J. Appl. Polym. Sci.* **1999**, *72*, 277–289. [CrossRef]
152. Gonçalves, C.; Pereira, J.; Almeida, M.; Carvalho, L.H. Impact of alkaline–acid and strongly acid process on the synthesis of urea–formaldehyde resins and derived composites: A comparison study. *Eur. J. Wood Prod.* **2019**, *77*, 1177–1187. [CrossRef]
153. Lubis, M.A.R.; Park, B.-D. Enhancing the Performance of Low Molar Ratio Urea–Formaldehyde Resin Adhesives via in-situ Modification with Intercalated Nanoclay. *J. Adhes.* **2020**, in press. [CrossRef]
154. Hatam, A.; The role of TiO_2 on the catalysis of reaction and cure of urea-formaldehyde resins. Personal communication, 2020.
155. Dorieh, A.; Mahmoodi, N.; Mamaghani, M.; Pizzi, A.; Zeydi, M.M.; Moslemi, A. New insight into the use of latent catalysts for the synthesis of urea formaldehyde adhesives and the mechanical properties of medium density fiberboards bonded with them. *Eur. Polym. J.* **2019**, *112*, 195–205. [CrossRef]
156. Dorieh, A.; Mahmoodi, N.; Mamaghani, M.; Pizzi, A.; Zeydi, M.M. Comparison of the properties of urea-formaldehyde resins by the use of formalin or urea formaldehyde condensates. *J. Adhes. Sci. Technol.* **2018**, *32*, 2537–2551. [CrossRef]

157. Pizzi, A.; Lipschitz, L.; Valenzuela, J. Theory and practice of the preparation of low formaldehyde emission UF adhesives for particleboard. *Holzforschung* **1994**, *48*, 254–261. [CrossRef]
158. Dunker, A.K.; Johns, W.E.; Rammon, R.; Farmer, B.; Johns, S.Y. Slightly Bizarre Protein Chemistry: Urea-Formaldehyde Resin from a Biochemical Perspective. *J. Adhes.* **1986**, *19*, 153–176. [CrossRef]
159. Levendis, D.; Pizzi, A.; Ferg, E.E. The correlation of strength and formaldehyde emission with the crystalline/amorphous structure of UF resins. *Holzforschung* **1992**, *45*, 260–267. [CrossRef]
160. Ferg, E.E.; Pizzi, A.; Levendis, D. A ^{13}C NMR analysis method for urea-formaldehyde resin strength and formaldehyde emission. *J. Appl. Polym. Sci.* **1993**, *50*, 907–915. [CrossRef]
161. Zanetti, M.; Pizzi, A.; Kamoun, C. Upgrading of MUF particleboard adhesives and decrease of melamine content by buffer and additives. *Holz Roh-und Werkst.* **2003**, *61*, 55–65. [CrossRef]
162. Zanetti, M.; Pizzi, A. Upgrading of MUF resins by buffering additives – Part 2: Hexamine sulphate mechanisms and alternate buffers. *J. Appl. Polym. Sci.* **2003**, *90*, 215–226. [CrossRef]
163. Despres, A.; Pizzi, A.; Delmotte, L. ^{13}C NMR investigation of the reaction in water of UF resins with blocked emulsified isocyanates. *J. Appl. Polym. Sci.* **2006**, *99*, 589–596. [CrossRef]
164. Wieland, S.; Pizzi, A.; Hill, S.; Grigsby, W.; Pichelin, F. The reaction in water of UF resins with isocyanates at short curing times: A ^{13}C NMR investigation. *J. Appl. Polym. Sci.* **2006**, *100*, 1624–1632. [CrossRef]
165. Wieland, S.; Pizzi, A.; Grigsby, W.; Warnes, J.; Pichelin, F. Microcrystallinity and colloidal peculiarities of UF/isocyanates hybrid resins. *J. Appl. Polym. Sci.* **2007**, *104*, 2633–2636. [CrossRef]
166. Lubis, M.A.R.; Park, B.D.; Lee, S.M. Microencapsulation of polymeric isocyanate for the modification of urea-formaldehyde resins. *Int. J. Adhes. Adhes.* **2020**, in press. [CrossRef]
167. Martin, A.E.; Calvin, M. *Chemistry of Metal Chelate Compounds*; Prentice-Hall: New York, NY, USA, 1952.
168. Lemon, P.H.R.B. An improved sand binder for steel castings. *Int. J. Mater. Prod. Technol.* **1990**, *5*, 25–35.
169. Pizzi, A.; Stephanou, A. On the chemistry, behaviour and cure acceleration of phenol-formaldehyde resins under very alkaline conditions. *J. Appl. Polym. Sci.* **1993**, *49*, 2157–2160. [CrossRef]
170. Stephanou, A.; Pizzi, A. Rapid curing lignins-based exterior wood adhesives, Part 2: Acceleration mechanisms and application to panel products. *Holzforschung* **1993**, *47*, 501–506. [CrossRef]
171. Pizzi, A.; Stephanou, A. Phenol-formaldehyde wood adhesives under very alkaline conditions, Part 2: Acceleration mechanism and applied results. *Holzforschung* **1994**, *48*, 150–156. [CrossRef]
172. Pizzi, A.; Stephanou, A. Completion of alkaline cure acceleration of phenol-formaldehyde resins: Acceleration by organic anhydrides. *J. Appl. Polym. Sci.* **1994**, *51*, 1351–1352. [CrossRef]
173. Pizzi, A.; Garcia, R.; Wang, S. On the networking mechanisms of additives accelerated PF polycondensates. *J. Appl. Polym. Sci.* **1997**, *66*, 255–266. [CrossRef]
174. Higuchi, M.; Tohmura, S.; Sakata, I. Acceleration of the cure of phenolic resin adhesives 5. Catalytic actions of carbonates and formamide. *Mokuzai Gakkaishi* **1994**, *40*, 604–611.
175. Conner, A.H.; Lorenz, L.F.; Hirth, K.C. Accelerated cure of phenol–formaldehyde resins: Studies with model compounds. *J. Appl. Polym. Sci.* **2002**, *86*, 3256–3263. [CrossRef]
176. Kamo, N.; Okamura, H.; Higuchi, M.; Morita, M. Condensation reactions of phenolic resins V: Cure-acceleration effects of propylene carbonate. *J. Wood Sci.* **2004**, *50*, 236–241. [CrossRef]
177. Park, B.-D.; Riedl, B.; Hsu, E.W.; Shields, J.A. Differential scanning calorimetry of phenol–formaldehyde resins cure-accelerated by carbonates. *Polymer* **1999**, *40*, 1689–1699. [CrossRef]
178. Park, B.-D.; Riedl, B. ^{13}C-NMR study on cure-accelerated phenol–formaldehyde resins with carbonates. *J. Appl. Polym. Sci.* **2000**, *77*, 841–851. [CrossRef]
179. Park, B.-D.; Riedl, B.; Hsu, E.W.; Shields, J.A. Application of cure-accelerated phenol-formaldehyde (PF) adhesives for three-layer medium density fiberboard (MDF) manufacture. *Wood Sci. Technol.* **2001**, *35*, 311–323. [CrossRef]
180. Lei, H.; Pizzi, A.; Despres, A.; Pasch, H.; Du, G. Esters acceleration mechanisms in phenol-formaldehyde resin adhesives. *J. Appl. Polym. Sci.* **2006**, *100*, 3075–3093. [CrossRef]
181. Zhao, C.; Pizzi, A.; Garnier, S. Fast advancement and hardening acceleration of low condensation alkaline PF resins by esters and copolymerized urea. *J. Appl. Polym. Sci.* **1999**, *74*, 359–378. [CrossRef]
182. Zhao, C.; Pizzi, A.; Kuhn, A.; Garnier, S. Fast advancement and hardening acceleration of low condensation alkaline PF resins by esters and copolymerized urea. Part 2: Esters during resin reaction and effect of guanidine salts. *J. Appl. Polym. Sci.* **2000**, *77*, 249–259. [CrossRef]

183. Deng, S.; Du, G.; Li, X.; Zhang, J.; Pizzi, A. Performance and reaction mechanism of zero formaldehyde-emission urea-glyoxal (UG) resin. *J. Taiwan Inst. Chem. Eng.* **2014**, *45*, 2029–2038. [CrossRef]
184. Deng, S.; Pizzi, A.; Du, G.; Zhang, J.; Zhang, J. Synthesis, structure, and characterization of Glyoxal-Urea-Formaldehyde cocondensed resins. *J. Appl. Polym. Sci.* **2014**, *131*, 41009–41019. [CrossRef]
185. Deng, S.; Pizzi, A.; Du, G.; Lagel, M.C.; Delmotte, L.; Abdalla, S. Synthesis, structure characterization and application of melamine-glyoxal adhesive resins. *Eur. J. Wood Prod.* **2018**, *76*, 283–296. [CrossRef]
186. Younesi-Kordkheili, H.; Pizzi, A. Acid Ionic Liquids as a New Hardener in Urea-Glyoxal Adhesive Resins. *Polymers* **2016**, *8*, 57. [CrossRef]
187. Xi, X.; Pizzi, A.; Amirou, S. Melamine-Glyoxal-Glutaraldehyde Wood Panel Adhesives without Formaldehyde. *Polymers* **2018**, *10*, 22. [CrossRef]
188. Novotny, E.E.; Johnson, W.W. Urea-Furfural Resins and Process of Making Them. U.S. Patent 1,827,824, 1931.
189. Stierli, R.F.; Newton, N.Y. Urea-Formaldehyde-Furfuryl Alcohol Resins. U.S. Patent 2,487,394, 1949.
190. Lagel, M.C.; Berner Fachhochschule, Biel, Switzerland. personal communication, 2016.
191. Xu, C.; Zhang, Y.; Yuan, Z. Formaldehyde-Free Phenolic Resins, Downstream Products, Their Synthesis and Use. U.S. patent 10,266,633, 2019.
192. Santiago-Medina, F.J.; Pizzi, A.; Abdalla, S. Hydroxymethylfurfural hardening of pine tannin wood adhesives. *J. Renew. Mat.* **2018**, *5*, 435–447. [CrossRef]
193. Xi, X.; Pizzi, A. No-Aldehydes Glucose/Sucrose-Triacetin-Diamine Wood Adhesives for Particleboard. *J. Renew. Mat.* **2020**, in press.
194. Bekhta, P.; Sedliačik, J. Environmentally-Friendly High-Density Polyethylene-Bonded Plywood Panels. *Polymers* **2019**, *11*, 1166. [CrossRef]
195. Berthevas, P.; Santoro, G.; Wevers, R.; Gruenbauer, H.; Pizzi, A. Recycled polyurethane foam powder can be used in conjunction with PMDI in particleboards to obtain the required properties while reducing costs. In Proceedings of the Ninth European Panel Products Symposium, Llandudno, UK, 5–7 October 2005; pp. 40–47.
196. Mansouri, H.R.; Pizzi, A. Recycled polyurethane micronized powders as active extenders of UF and PF wood panel adhesives. *Holz Roh-und Werkst.* **2007**, *65*, 293–299. [CrossRef]
197. Ashori, A. Wood–plastic composites as promising green-composites for automotive industries: A review. *Bioresour. Technol.* **2008**, *99*, 4661–4667. [CrossRef]
198. Clemons, C. Wood plastic composites in the United States. The interfacing of two industries. *For. Prod. J.* **2002**, *52*, 10–18.
199. Klyosov, A.A. *Wood Plastic Composites*; Wiley Interscience: New York, NY, USA, 2007; pp. 1–726.
200. Schwarzkopf, M.J.; Burnard, M.D. Wood-Plastic Composites—Performance and Environmental Impacts. In *Environmental Impacts of Traditional and Innovative Forest-based Bioproducts*; Kutnar, A., Muthu, S., Eds.; Springer: Singapore, 2016.

© 2020 by the authors. Licensee MDPI, Basel, Switzerland. This article is an open access article distributed under the terms and conditions of the Creative Commons Attribution (CC BY) license (http://creativecommons.org/licenses/by/4.0/).

MDPI
St. Alban-Anlage 66
4052 Basel
Switzerland
Tel. +41 61 683 77 34
Fax +41 61 302 89 18
www.mdpi.com

Polymers Editorial Office
E-mail: polymers@mdpi.com
www.mdpi.com/journal/polymers

www.ingramcontent.com/pod-product-compliance
Lightning Source LLC
LaVergne TN
LVHW070403100526
838202LV00014B/1378